Commercial
Test Prep

Study and Prepare for the Commercial Airplane, Helicopter, Gyroplane, Glider, Balloon, Airship and Military Competency FAA Knowledge Tests

- Effective June 19, 2000
- All FAA Commercial Questions included
- Organized by subject
- Answers, Explanations, References and additional Study Material included for each chapter
- Includes the official FAA Computerized Testing Supplement
- Plus... helpful tips and instructions for the FAA Knowledge Test

Aviation Supplies & Academics, Inc.
Newcastle, Washington

Commercial Pilot Test Prep
2001 Edition

Aviation Supplies & Academics, Inc.
7005 132nd Place SE
Newcastle, Washington 98059-3153
(425) 235-1500
www.asa2fly.com

© 2000 ASA, Inc.

FAA Questions herein are from United States government sources and contain current information as of: June 19, 2000.

None of the material in this publication supersedes any documents, procedures or regulations issued by the Federal Aviation Administration.

ASA assumes no responsibility for any errors or omissions. Neither is any liability assumed for damages resulting from the use of the information contained herein.

Important: This Test Prep should be sold with and used in conjunction with *Computerized Testing Supplement for Commercial Pilot* (FAA-CT-8080-1B).

ASA reprints the FAA test figures and legends contained within this government document, and it is also sold separately and available from aviation retailers nationwide. Order #ASA-CT-8080-1B.

ASA-TP-C-01
ISBN 1-56027-392-5

Printed in the United States of America

01 00   5 4 3 2 1

About ASA: Aviation Supplies & Academics, Inc. (ASA) is an industry leader in the development and sale of aviation supplies and publications for pilots, flight instructors, flight engineers, and aviation maintenance technicians. We manufacture and publish more than 200 products for the aviation industry. Aviators are invited to call 1-800-ASA-2-FLY for a free copy of our catalog. Visit ASA on the web: **www.asa2fly.com**

About the Contributors

Charles L. Robertson
Associate Professor, UND Aerospace
University of North Dakota

Charles Robertson as flight instructor, associate professor and manager of training at UND Aerospace, contributes a vital and substantial combination of pilot and educator to ASA's reviewing team. After graduating with education degrees from Florida State University in 1967, and Ball State University in 1975, he began his USAF career as Chief of avionics branch, 58th Military Airlift Squadron, and went on to flight instruction, training for aircraft systems, and airport managing, while gaining many thousands of hours flying international passenger and cargo, aerial refueling and airlift missions. As Division Chief in 1988, Robertson directed the Strategic Air Command's "Alpha Alert Force," coordinating daily flight training operations. He holds the CFI Airplane Land, Multi-Engine, Single-Engine and Instrument, the ATP Airplane Land and Multi-Engine, Commercial Pilot, Advanced and Instrument Ground Instructor licenses.

Jackie Spanitz
Director of Curriculum Development
Aviation Supplies & Academics, Inc.

Jackie Spanitz earned a bachelor of science degree with Western Michigan University (WMU), in Aviation Technology and Operations—Pilot option. In her masters program at Embry-Riddle Aeronautical University, she earned a degree in Aeronautical Science, specializing in Management. As Director of Curriculum Development for ASA, Jackie oversees new and existing product development, ranging from textbooks and flight computers to flight simulation software products, and integration of these products into new and existing curricula. She provides technical support, research for product development, and project management. Jackie holds the CFI Airplane, Land, Single-Engine and Instrument, Commercial Airplane, Land, Single-Engine, Multi-Engine and Instrument, and Advanced and Instrument Ground Instructor certificates; she is the author of *Guide to the Biennial Flight Review, Private Pilot Syllabus, Instrument Rating Syllabus*, and *Commercial Pilot Syllabus*, and technical editor for ASA's Test Prep series.

Cliff Seretan

Cliff Seretan began flying in 1979 to find fulfillment beyond a successful management career in state government. Over the next several years, he added on certificates and ratings while gaining experience through flying coast-to-coast in light aircraft; then, his flight instructor certificate enabled him to more economically pursue two of his passions—flying and teaching. Cliff taught primary and advanced flight students in the Northeast in a variety of aircraft. In the last few years, he has had the opportunity to diversify his business with aviation management as well as analysis and development of aviation computer products and flight simulator programs. Cliff holds a Commercial Certificate with an Instrument Rating for Single and Multi-Engine Land Airplanes and CFI for Airplane Land and Instrument. With undergraduate and graduate degrees in the Arts, Sciences and Management from New York University, Connecticut College and the State University of New York at Stony Brook, he brings to ASA a blend of aviation and business skills that combine a unique perspective with in-depth subject knowledge.

Contents

Instructions

Preface vii
Update Information viii
Description of the Tests ix
   Military Competency Exam x
   Process for Taking a Knowledge Test x
      Computer Testing Designees xii
   Use of Test Aids and Materials xiii
   Cheating or Other Unauthorized Conduct xiv
   Validity of Airman Test Reports xiv
      Retesting Procedures xiv
Eligibility Requirements for the
   Commercial Pilot Certificate xv
Certificate and Logbook Endorsement xvi
Test-Taking Tips xvii
Suggested Materials for the
   Commercial Pilot Certificate xix
ASA Test Prep Layout xx

Chapter 1 Basic Aerodynamics

Aerodynamic Terms 1–3
Axes of Rotation and the Four Forces
     Acting in Flight 1–5
   Lift 1–6
   Weight 1–6
   Thrust 1–6
   Drag 1–7
Lift/Drag Ratios 1–10
The V_G Diagram 1–12
Stability 1–12
Turns, Loads and Load Factors 1–15
Stalls and Spins 1–20
Flaps 1–21
Wing Shapes 1–22
Torque 1–23
Ground Effect 1–24
Wake Turbulence 1–25
Glider Aerodynamics 1–27

Chapter 2 Aircraft Systems

Ignition System 2–3
Air/Fuel Mixture 2–4
Carburetor Ice 2–6
Aviation Fuel 2–7
Engine Temperatures 2–8
Propellers 2–9
Cold Weather Operations 2–11
Rotorcraft Systems 2–12
Glider Systems 2–27
Balloon Operations 2–35
Airship Operations 2–41
Airship IFR Operations 2–43

Chapter 3 Flight Instruments

Airspeed Indicator 3–3
Altitude Definitions 3–6
Magnetic Compass 3–6
Gyroscopic Instruments and Systems 3–7
Attitude Instrument Flying 3–8

Chapter 4 Regulations

Pilot Certificate Types and Privileges 4–3
Medical Certificates 4–5
Pilot Logbooks 4–6
High-Performance, Complex, and Tailwheel
   Airplanes 4–7
Recent Flight Experience: Pilot-In-Command 4–9
Change of Address 4–10
Towing 4–10
Responsibility and Authority of the
   Pilot-In-Command 4–11
Preflight Action 4–14
Seatbelts 4–15
Portable Electronic Devices 4–17
Fuel Requirements 4–17
Transponder Requirements 4–18
Supplemental Oxygen 4–19
Instrument and Equipment Requirements 4–19

Continued

Restricted, Limited and Experimental Aircraft:
    Operating Limitations 4–20
Emergency Locator Transmitter (ELT) 4–21
Truth in Leasing 4–22
Operating Near Other Aircraft and
    Right-of-Way Rules 4–22
Speed Limits 4–25
Aerobatic Flight 4–26
Aircraft Lights 4–26
Minimum Altitudes 4–28
Maintenance Responsibility 4–29
Aircraft Inspections 4–30
Maintenance Records 4–31
Maintenance, Preventative Maintenance,
    Rebuilding and Alteration 4–33
NTSB Part 830 4–34
Rotorcraft Regulations 4–36
Glider Regulations 4–37
Lighter-Than-Air Regulations 4–38
LTA Fundamentals of Instructing 4–41

Chapter 5 **Procedures and Airport Operations**

Airspace 5–3
Basic VFR Weather Minimums 5–12
VHF Direction Finder 5–13
Operations on Wet or Slippery Runways 5–14
Land and Hold Short Operations (LAHSO) 5–14
Airport Marking Aids and Signs 5–15
VFR Cruising Altitudes 5–16
Collision Avoidance 5–17
Fitness Physiology 5–19
Aeronautical Decision Making 5–21

Chapter 6 **Weather**

The Earth's Atmosphere 6–3
Temperature 6–6
Wind 6–7
Moisture 6–9
Stable and Unstable Air 6–10
Clouds 6–11
Air Masses and Fronts 6–16
Turbulence 6–17
Icing 6–18
Thunderstorms 6–20
Common IFR Producers 6–23
Wind Shear 6–25
Soaring Weather 6–28

Chapter 7 **Weather Services**

Aviation Routine Weather Report (METAR) 7–3
Pilot Report (UA) 7–5
Terminal Aerodrome Forecast (TAF) 7–8
Aviation Area Forecast (FA) 7–9
Winds and Temperatures Aloft Forecast (FD) 7–10
Inflight Weather Advisories (WA, WS, WST) 7–10
Transcribed Weather Broadcast (TWEB) and
    En Route Flight Advisory Service (EFAS) 7–12
Radar Weather Report 7–14
Surface Analysis Chart 7–15
Weather Depiction Chart 7–16
Constant Pressure Chart 7–18
Tropopause Height/Vertical Wind
    Shear Prognostic Chart 7–18
Significant Weather Prognostics 7–19
Composite Moisture Stability Chart 7–21

Chapter 8 Aircraft Performance

Weight and Balance 8–3
   Computing Weight and Balance 8–4
   Graph Weight and Balance Problems 8–4
   Weight Change 8–5
   Weight Shift 8–6
Rotorcraft Weight and Balance 8–12
Glider Weight and Balance 8–14
Ground Operations During Windy Conditions 8–16
Headwind and Crosswind Components 8–17
Density Altitude 8–19
Takeoff and Landing Considerations 8–20
Takeoff and Landing Distance 8–22
Fuel Consumption vs. Brake Horsepower 8–26
Time, Fuel and Distance to Climb 8–27
   Table Method 8–27
   Graph Method 8–28
Cruise Performance Table 8–32
Cruise and Range Performance Table 8–35
Maximum Rate of Climb 8–36
Rotorcraft Performance 8–37
Glider Performance 8–40
Balloon Performance 8–43

Chapter 9 Navigation

The Flight Computer 9–3
   Finding True Course, Time, Rate,
   Distance, and Fuel 9–3
   Finding Density Altitude 9–10
   Finding Wind Direction and Velocity 9–12
Off-Course Correction 9–13
VHF Omni-Directional Range (VOR) 9–15
Estimating Time and Distance To Station
   Using VOR 9–22
Horizontal Situation Indicator (HSI) 9–25
Automatic Direction Finder (ADF) 9–27
   Determining Bearings with ADF 9–27
   Intercepting Bearings with ADF 9–28
   Estimated Time and Distance to
     Station Using ADF 9–29
     Wing-Tip Bearing Change Method 9–29
     Isosceles Triangle Technique 9–29

Chapter 10 IFR Operations

Instrument Approach Procedures 10–3
Departure Procedures (DPs) 10–6
Enroute Procedures 10–6
Standard Terminal Arrivals (STARs) 10–7

Cross-References

A: Answer, Subject Matter Knowledge Code,
    Category & Page Number ...... A–1 through A–10
B: Subject Matter Knowledge Code
    & Question Number .... B–1 through B–12

Preface

Welcome to ASA's Test Prep Series. ASA's test books have been helping pilots prepare for the FAA Knowledge Tests since 1984 with great success. We are confident that with proper use of this book, you will score very well on any of the commercial pilot certificate tests.

All of the questions in the FAA Commercial Pilot Test Question Bank are included here, and have been arranged into chapters based on subject matter. Topical study, in which similar material is covered under a common subject heading, promotes better understanding, aids recall, and thus provides a more efficient study guide. We suggest you begin by reading the book cover-to-cover. Then go back to the beginning and place emphasis on those questions most likely to be included in your test (identified by the aircraft category above each question). For example: a pilot preparing for the Commercial Airplane test would focus on the questions marked "ALL" and "AIR," and a pilot preparing for the Commercial Helicopter test would focus on the questions marked "ALL" and "RTC."

It is important to answer every question assigned on your FAA Knowledge Test. If in their ongoing review, the FAA authors decide a question has no correct answer, is no longer applicable, or is otherwise defective, your answer will be marked correct no matter which one you chose. However, you will not be given the automatic credit unless you have marked an answer. Unlike some other exams you may have taken, there is no penalty for "guessing" in this instance.

The FAA does not supply the correct answers to questions reproduced in this book, is not responsible for answers contained herein, and will not reveal what they consider the correct answers to be. In order to maintain the integrity of each test, the FAA may rearrange the answer stems around to appear in a different order on your test than you see in this book. For this reason, be careful to fully understand the intent of each question and corresponding answer while studying, rather than memorize the A, B, C answer choice. Our answers are based on research from the reference sources provided with each question. If your study leads you to question an answer choice, we recommend you seek the assistance of a local ground or flight instructor. If you still believe the answer needs review, please forward your questions, recommendations, or concerns to:

Aviation Supplies & Academics, Inc.
7005 132nd Place SE
Newcastle, WA 98059-3153

Voice (425) 235-1500
Fax (425) 235-0128
www.asa2fly.com

Technical Editor:
Jackie Spanitz, ASA

Editor:
Jennie Trerise, ASA

Update Information

Free Test Updates for the One-Year Lifecycle of Test Prep Books

The FAA releases a new test database each spring, and makes amendments to this database approximately twice a year. However, a small number of questions may be withheld from the public for a period of time while the FAA gathers statistics and validates these questions. This means the questions are not available to the public via the internet-posted databases, but they are being issued at the FAA testing centers. In each of these cases, ASA has worded the question to the best of our knowledge, basing it on current figure books, regulations, and procedures, as well as the type of question asked in previous tests.

The questions described above make up a very small percentage of the overall database and are identified by the symbol ^ (printed after the explanation and prior to the subject matter knowledge code— *see* Page xx, "ASA Test Prep Layout"). You can feel confident that you will be prepared for your FAA Knowledge Exam by using the ASA test prep products. ASA publishes test books each July and stays abreast of all changes to the tests, as well as the new questions that have been validated, and posts these changes on the ASA website as a Test Update. Visit the ASA website before taking your test to be certain you have all the current information: www.asa2fly.com

Description of the Tests

Instructions

All test questions are the objective, multiple-choice type, with three choices of answers. Each question can be answered by the selection of a single response. Each test question is independent of other questions, that is, a correct response to one does not depend upon, or influence the correct response to another. The minimum passing score is 70 percent.

The following tests each contain 100 questions and 3 hours are allowed for taking each test:

Test Code
- CAX — Commercial Pilot—Airplane: Focus on the questions marked ALL and AIR
- CRH — Commercial Pilot—Rotorcraft–Helicopter: Focus on the questions marked ALL and RTC
- CRG — Commercial Pilot—Rotorcraft–Gyroplane: Focus on the questions marked ALL and RTC
- CGX — Commercial Pilot—Glider: Focus on the questions marked ALL and GLI
- CBH — Commercial Pilot—Lighter-Than-Air Balloon-Hot Air: Focus on the questions marked ALL and LTA
- CLA — Commercial Pilot—Lighter-Than-Air Airship: Focus on the questions marked ALL and LTA

The following test contains 60 questions and 2.5 hours are allowed for taking the test:

Test Code
- CBG — Commercial Pilot—Balloon–Gas: Focus on the questions marked ALL and LTA

The following tests each contain 50 questions and 2 hours are allowed for taking each test:

Test Code
- MCA — Military Competency—Airplane: Focus on the questions marked MIL
- MCH — Military Competency—Helicopter: Focus on the questions marked MIL

As stated in 14 CFR §61.63, an applicant need not take an additional knowledge test provided the applicant holds an airplane, rotorcraft, powered-lift, or airship rating at that pilot certificate level. For example, an applicant transitioning from gliders to airplanes or helicopters will need to take the 100-question test. An applicant transitioning from airplanes to gliders, or airplanes to helicopters, will not be required to take the Knowledge Exam.

Military Competency Exam

The FAA does not provide a list of questions for the Military Competency Exam. The reference "MIL" provided above questions deemed appropriate for the Military Competency Exam is based on research and history. If you find you were asked additional questions not marked as MIL, please call ASA at **1-800-ASA-2-FLY** so we may update our database to help future candidates.

Please take some time to look at *Federal Aviation Regulation Part 61.73, Military pilots or former military pilots: Special rules*. This part covers the requirements for military pilots or former military pilots who are interested in acquiring their private or commercial pilot certificate.

Process for Taking a Knowledge Test

The Federal Aviation Administration (FAA) has available hundreds of computer testing centers worldwide. These testing centers offer the full range of airman knowledge tests including military competence, instrument foreign pilot, and pilot examiner screening tests. Refer to the list of computer testing designees (CTDs) at the end of this section.

The first step in taking a knowledge test is the registration process. You may either call the testing centers' 1-800 numbers or simply take the test on a walk-in basis. If you choose to use the 1-800 number to register, you will need to select a testing center, schedule a test date, and make financial arrangements for test payment. You may register for tests several weeks in advance, and you may cancel your appointment according to the CTD's cancellation policy. If you do not follow the CTD's cancellation policies, you could be subject to a cancellation fee.

The next step in taking a knowledge test is providing proper identification. You should determine what knowledge test prerequisites are necessary before going to the computer testing center. Your instructor or local Flight Standards District Office (FSDO) can assist you with what documentation to take to the testing facility. Testing center personnel will not begin the test until your identification is verified. A limited number of tests do not require authorization. An endorsement from an authorized instructor is not required for military competency.

Acceptable forms of authorization:

- A certificate of graduation or a statement of accomplishment certifying the satisfactory completion of the ground school portion of a course from an FAA-certificated pilot school.
- A certificate of graduation or a statement of accomplishment certifying the satisfactory completion of the ground school portion of a course from an agency such as a high school, college, adult education program, U.S. Armed Force, ROTC Flight Training School, or Civil Air Patrol.
- A written statement or logbook endorsement from an authorized instructor certifying that you have accomplished a ground-training or home-study course required for the rating sought and you are prepared for the knowledge test.
- Failed Airman Test Report, passing Airman Test Report, or expired Airman Test Report (pass or fail), provided that you still have the original Airman Test Report in your possession.

Before you take the actual test, you will have the option to take a sample test. The actual test is time limited; however, you should have sufficient time to complete and review your test.

When taking a test, keep the following points in mind:

- Answer each question in accordance with the latest regulations and guidance publications.
- Read each question carefully before looking at the possible answers. You should clearly understand the problem before attempting to solve it.
- After formulating an answer, determine which test answer corresponds with your answer. The answer chosen should completely resolve the problem.
- From the answers given, it may appear that there is more than one possible answer; however, there is only one answer that is correct and complete. The other answers are either incomplete, erroneous, or represent common misconceptions.
- If a certain question is difficult for you, it is best to mark it for review and proceed to the next question. After you answer the less difficult questions, return to those which you marked for review and answer them. The review procedure will be explained to you prior to starting the test. Although the computer should alert you to unanswered questions, make sure every question has an answer recorded. This procedure will enable you to use the available time to the maximum advantage.
- When solving a calculation problem, select the answer closest to your solution. The problems have been checked manually and with various types of calculators. If you have solved it correctly, your answer will be closer to the correct answer than any of the other choices.

Upon completion of the knowledge test, you will receive your Airman Test Report, with the testing center's embossed seal, which reflects your score.

The Airman Test Report lists the subject matter knowledge codes for questions answered incorrectly. The total number of subject matter knowledge codes shown on the Airman Test Report is not necessarily an indication of the total number of questions answered incorrectly. Study these knowledge areas to improve your understanding of the subject matter. *See* the *Subject Matter Knowledge Code/Question Number Cross-Reference* in the back of this book for a complete list of which questions apply to each subject matter knowledge code.

Your instructor is required to provide instruction on each of the knowledge areas listed on your Airman Test Report and to complete an endorsement of this instruction. You must present the Airman Test Report to the examiner prior to taking the practical test. During the oral portion of the practical test, the examiner is required to evaluate the noted areas of deficiency.

Should you require a duplicate Airman Test Report due to loss or destruction of the original, send a signed request accompanied by a check or money order for $1 payable to the FAA. Your request should be sent to the Federal Aviation Administration, Airmen Certification Branch, AFS-760, P.O. Box 25082, Oklahoma City, OK 73125.

Computer Testing Designees

The following is a list of the computer testing designees authorized to give FAA knowledge tests. This list should be helpful in case you choose to register for a test or simply want more information. The latest listing of computer testing center locations may be obtained through the FAA website: http://afs600.faa.gov, then select AFS630, Airman Certification, Computer Testing Sites.

Computer Assisted Testing Service (CATS)

1849 Old Bayshore Highway
Burlingame, CA 94010
Applicant inquiry and test registration: 1-800-947-4228
From outside the U.S.: (650) 259-8550

LaserGrade Computer Testing

16209 S.E. McGillivray, Suite L
Vancouver, WA 98683
Applicant inquiry and test registration: 1-800-211-2753 or 1-800-211-2754
From outside the U.S.: (360) 896-9111

AvTest 2000

P.O. Box 26716
Fort Worth, TX 76126
Applicant inquiry and test registration: (817) 249-5564
http://www.avtest2000.com

International pilots who want to apply for an FAA certificate based on their ICAO foreign certificates should go to the nearest FAA Flight Standards District Office (FSDO). Phone numbers for these offices will be found in the blue pages of the local telephone book. This will allow you to fly a U.S.-registered aircraft while in the U.S. If you hold instrument privileges on your foreign license, you can take a 50-question knowledge test (Instrument Rating—Foreign Pilot), and if you pass it, have instrument privileges added to this FAA private pilot certificate.

If you are outside of the U.S., you will have to go in person to an FAA International Field Office (IFO) and apply for an FAA private pilot certificate. When outside of the U.S. you will only be authorized to fly a U.S.-registered aircraft.

International Field Offices (IFO)

1. Brussels, Belgium (32-2) 508.2721
   FAA C/O American Embassy, PSC 82 Box 002, APO AE 09710
2. Frankfurt, Germany (49-69) 69.705.111
   FAA C/O IFO EA-33 Unit 7580, APO AE 09050
3. London, England (44-181) 754.88.19
   FAA C/O American Embassy, PSC 801 Box 63, FPO AE 09498-4063
4. Singapore (65) 543-1466
   FAA C/O American Embassy, PSC 470 AP 96507-0001

These special certificates will not allow you to fly for hire in the U.S. To qualify for a "clean" FAA commercial pilot certificate or higher, you must meet the full certification requirements of 14 CFR Part 61, for the level of certificate you are requesting. Your current, logged flying time will count towards the required experience. However, all required training, knowledge, and practical tests must be completed.

Use of Test Aids and Materials

Airman knowledge tests require applicants to analyze the relationship between variables needed to solve aviation problems, in addition to testing for accuracy of a mathematical calculation. The intent is that all applicants are tested on concepts rather than rote calculation ability. It is permissible to use certain calculating devices when taking airman knowledge tests, provided they are used within the following guidelines. The term "calculating devices" is interchangeable with such items as calculators, computers, or any similar devices designed for aviation-related activities.

1. Guidelines for use of test aids and materials. The applicant may use test aids and materials within the guidelines listed below, if actual test questions or answers are not revealed.

    a. Applicants may use test aids, such as scales, straightedges, protractors, plotters, navigation computers, log sheets, and all models of aviation-oriented calculating devices that are directly related to the test. In addition, applicants may use any test materials provided with the test.

    b. Manufacturer's permanently inscribed instructions on the front and back of such aids listed in 1(a), e.g., formulas, conversions, regulations, signals, weather data, holding pattern diagrams, frequencies, weight and balance formulas, and air traffic control procedures are permissible.

    c. The test proctor may provide calculating devices to applicants and deny them use of their personal calculating devices if the applicant's device does not have a screen that indicates all memory has been erased. The test proctor must be able to determine the calculating device's erasure capability. The use of calculating devices incorporating permanent or continuous type memory circuits without erasure capability is prohibited.

    d. The use of magnetic cards, magnetic tapes, modules, computer chips, or any other device upon which prewritten programs or information related to the test can be stored and retrieved is prohibited. Printouts of data will be surrendered at the completion of the test if the calculating device used incorporates this design feature.

    e. The use of any booklet or manual containing instructions related to the use of the applicant's calculating device is not permitted.

    f. Dictionaries are not allowed in the testing area.

    g. The test proctor makes the final determination relating to test materials and personal possessions that the applicant may take into the testing area.

2. Guidelines for dyslexic applicant's use of test aids and materials. A dyslexic applicant may request approval from the local Flight Standards District Office (FSDO) to take an airman knowledge test using one of the three options listed in preferential order:

    a. Option One. Use current testing facilities and procedures whenever possible.

    b. Option Two. Applicants may use Franklin Speaking Wordmaster® to facilitate the testing process. The Wordmaster® is a self-contained electronic thesaurus that audibly pronounces typed in words and presents them on a display screen. It has a built-in headphone jack for private listening. The headphone feature will be used during testing to avoid disturbing others.

    c. Option Three. Applicants who do not choose to use the first or second option may request a test proctor to assist in reading specific words or terms from the test questions and supplement material. In the interest of preventing compromise of the testing process, the test proctor should be someone who is non-aviation oriented. The test proctor will provide reading assistance only, with no explanation of words or terms. The Airman Testing Standards Branch, AFS-630, will assist in the selection of a test site and test proctor.

Cheating or Other Unauthorized Conduct

Computer testing centers must follow strict security procedures to avoid test compromise. These procedures are established by the FAA and are covered in FAA Order 8080.6, Conduct of Airman Knowledge Tests. The FAA has directed testing centers to terminate a test at any time a test proctor suspects a cheating incident has occurred. An FAA investigation will then be conducted. If the investigation determines that cheating or unauthorized conduct has occurred, then any airman certificate or rating that you hold may be revoked, and you will be prohibited for 1 year from applying for or taking any test for a certificate or rating under 14 CFR Part 61.

Validity of Airman Test Reports

Airman Test Reports are valid within the 24-calendar month period preceding the month you complete the practical test. If the Airman Test Report expires before completion of the practical test, you must retake the knowledge test.

Retesting Procedures

If you receive a grade lower than a 70 percent and wish to retest, you must present the following:
- failed Airman Test Report; and
- a written endorsement from an authorized instructor certifying that additional instruction has been given, and the instructor finds you competent to pass the test.

If you decide to retake the test in anticipation of a better score, you may retake the test after 30 days from the date your last test was taken. The FAA will not allow you to retake a passed test before the 30-day period has lapsed. Prior to retesting, you must give your current Airman Test Report to the test administrator. The last test taken will reflect the official score.

Eligibility Requirements for the Commercial Pilot Certificate

The general prerequisites for a Commercial Pilot Certificate require that the applicant have a combination of experience, knowledge, and skill. For specific information pertaining to certification, an applicant should carefully review the appropriate sections of Federal Aviation Regulations Part 61 for commercial pilot certification.

Additionally, to be eligible for a Commercial Pilot Certificate, applicants must:

1. Be at least 18 years of age (16 to take the knowledge test).
2. Be able to read, speak, write, and understand English or have a limitation placed on the certificate.
3. Hold a current Medical Certificate issued under 14 CFR Part 67. No medical certificate is required for a glider or balloon rating.
4. Pass a knowledge test appropriate to the aircraft rating sought on the subjects in which ground instruction is required. Applicants for a knowledge test must show evidence of completing ground training or a home study course and be prepared for the knowledge test.
5. Pass an oral and flight test appropriate to the rating they seek, covering items selected by the inspector or examiner from those on which training is required.
6. Hold at least a private pilot certificate.
7. Comply with the provisions of 14 CFR Part 61 which apply to the rating they seek.

Certificate and Logbook Endorsement

When you go to take your FAA Knowledge Test, you will be required to show proper identification and have certification of your preparation for the examination, signed by an appropriately certified Flight or Ground Instructor. Ground Schools will have issued the endorsements as you complete the course.

If you choose a home-study for your Knowledge Test, you can either get an endorsement from your instructor or submit your home-study materials to an FAA Office for review and approval prior to taking the test.

Commercial Pilot Endorsement

I certify that Mr./Ms. _____
has received the ground instruction or completed home study required by 14 CFR §61.35 and §61.125.
I have determined he/she is prepared for the _____ knowledge test.

Signed _____ Date _____

CFI Number _____ Expires _____

Test-Taking Tips

Follow these time-proven tips, which will help you develop a skillful, smooth approach to test-taking:

1. In order to maintain the integrity of each test, the FAA may rearrange the answer stems to appear in a different order on your test than you see in this book. For this reason, be careful to fully understand the intent of each question and corresponding answer while studying, rather than memorize the A, B, C answer choice.

2. Take with you to the testing center a sign-off from an instructor, photo I.D., the testing fee, calculator, flight computer (ASA's E6-B or CX-1a Pathfinder), plotter, magnifying glass, and a sharp pointer, such as a safety pin.

3. Your first action when you sit down should be to write on the scratch paper the weight and balance and any other formulas and information you can remember from your study. Remember, some of the formulas may be on your E6-B.

4. Answer each question in accordance with the latest regulations and guidance publications.

5. Read each question carefully before looking at the possible answers. You should clearly understand the problem before attempting to solve it.

6. After formulating an answer, determine which answer choice corresponds the closest with your answer. The answer chosen should completely resolve the problem.

7. From the answer choices given, it may appear that there is more than one possible answer. However, there is only one answer that is correct and complete. The other answers are either incomplete, erroneous, or represent popular misconceptions.

8. If a certain question is difficult for you, it is best to mark it for REVIEW and proceed to the other questions. After you answer the less difficult questions, return to those which you marked for review and answer them. The review marking procedure will be explained to you prior to starting the test. Although the computer should alert you to unanswered questions, make sure every question has an answer recorded. This procedure will enable you to use the available time to the maximum advantage.

9. Perform each math calculation twice to confirm your answer. If adding or subtracting a column of numbers, reverse your direction the second time to reduce the possibility of error.

10. When solving a calculation problem, select the answer nearest to your solution. The problem has been checked with various types of calculators; therefore, if you have solved it correctly, your answer will be closer to the correct answer than any of the other choices.

11. Remember that information is provided in the FAA Legends and FAA Figures.

12. Remember to answer every question, even the ones with no completely correct answer, to ensure the FAA gives you credit for a bad question.

13. Take your time and be thorough but relaxed. Take a minute off every half-hour or so to relax the brain and the body. Get a drink of water halfway through the test.

Suggested Materials for the Commercial Pilot Certificate

The following are some of the publications and products recommended for the Commercial Pilot certificates. All are reprinted by ASA and available from authorized ASA dealers and distributors.

ASA-PM-1	*Flight Training* by Trevor Thom
ASA-PM-2	*Private & Commercial* by Trevor Thom
ASA-PM-3	*Instrument Flying,* by Trevor Thom
ASA-PM-S-I	*Instrument Pilot Syllabus* (for Trevor Thom series)
ASA-PM-S-C	*Commercial Pilot Syllabus* (for Trevor Thom series)
ASA-CAP-2	*The Complete Advanced Pilot* by Bob Gardner
ASA-MPT	*The Complete Multi-Engine Pilot* by Bob Gardner
ASA-SAP	*Say Again, Please: Guide to Radio Communications*
ASA-OEG-BFR	*Guide to the Biennial Flight Review*
ASA-OEG-C	*Commercial Oral Exam Guide*
ASA-8081-12A	*Commercial Pilot Practical Test Standards*
ASA-8081-HD	*Rotorcraft/Helicopter Practical Test Standards*
ASA-VFM-HI	*Visualized Flight Maneuvers for High-Wing Aircraft*
ASA-VFM-LO	*Visualized Flight Maneuvers for Low-Wing Aircraft*
ASA-TP-CFI-01	*Certified Flight Instructor Test Prep*
ASA-TW-C-01	*Prepware™ Exam Software for Commercial Pilot*
ASA-FR-AM-BK	*Federal Aviation Regulations and Aeronautical Information Manual* (combined)
ASA-FAR-FC	*Federal Aviation Regulations: Flight Crew* (Parts 1, 25, 63, 65, 121)
ASA-ANA	*Aerodynamics for Naval Aviators*
ASA-AC00-6A	*Aviation Weather*
ASA-AC00-45E	*Aviation Weather Services*
ASA-8083-21	*Rotorcraft Flying Handbook*
ASA-8083-3	*Airplane Flying Handbook*
ASA-AC61-23C	*Pilot's Handbook of Aeronautical Knowledge*
ASA-8083-1	*Aircraft Weight and Balance Handbook*
ASA-DAT-3	*Dictionary of Aeronautical Terms*
ASA-SP-6	Pilot Master Log
ASA-FL-1	Flightlight™
ASA-KB-3L	Long Tri-Fold Kneeboard
ASA-E6B	E6-B Flight Computer
ASA-CP-RLX	Plotter
ASA-CW-10	Accordion-Fold Chart Wallet (holds 10 charts)
ASA-CW-8	Book-Style Chart Wallet (holds 8 charts)
ASA-ON-TOP	On Top IFR Proficiency Platform Software

ASA Test Prep Layout

The FAA questions have been sorted into chapters according to subject matter. Within each chapter, the questions have been further classified and all similar questions grouped together with a concise discussion of the material covered in each group. This discussion material of "Chapter text" is printed in a larger font and spans the entire width of the page. Immediately following the FAA Question is ASA's Explanation in *italics*. The last line of the Explanation contains the Subject Matter Knowledge Code and further reference (if applicable). *See* the EXAMPLE below.

Figures referenced by the Chapter text only are numbered with the appropriate chapter number, i.e., "Figure 1-1" is Chapter 1's first chapter-text figure.

Some FAA Questions refer to Figures or Legends immediately following the question number, i.e., "5201. (Refer to Figure 14.)." These are FAA Figures and Legends which can be found in the separate booklet: *Computerized Testing Supplement* (CT-8080-XX). This supplement is bundled with the Test Prep and is the exact material you will have access to when you take your computerized test. We provide it separately, so you will become accustomed to referring to the FAA Figures and Legends as you would during the test.

Figures referenced by the Explanation and pertinent to the understanding of that particular question are labeled by their corresponding Question number. For example: the caption "Questions 5245 and 5248" means the figure accompanies the Explanations for both Question 5245 and 5248.

Answers to each question are found at the bottom of each page, and in the Cross-Reference at the back of this book.

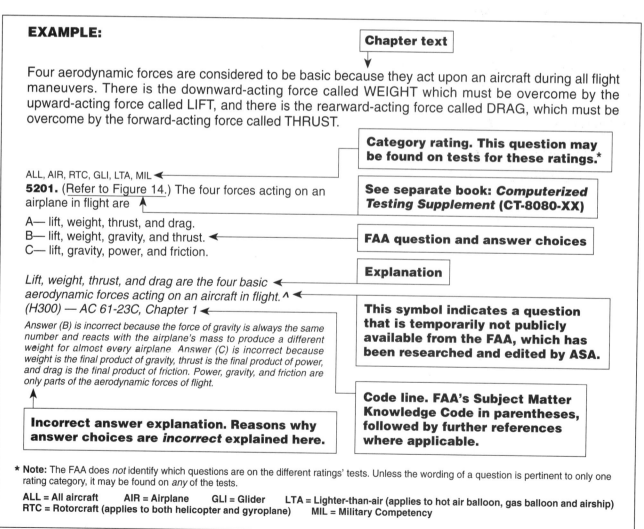

Chapter 1
Basic Aerodynamics

Aerodynamic Terms   *1–3*

Axes of Rotation and the Four Forces Acting in Flight   *1–5*

   Lift   *1–6*

   Weight   *1–6*

   Thrust   *1–6*

   Drag   *1–7*

Lift/Drag Ratios   *1–10*

The V_G Diagram   *1–12*

Stability   *1–12*

Turns, Loads and Load Factors   *1–15*

Stalls and Spins   *1–20*

Flaps   *1–21*

Wing Shapes   *1–22*

Torque   *1–23*

Ground Effect   *1–24*

Wake Turbulence   *1–25*

Glider Aerodynamics   *1–27*

Aerodynamic Terms

An airfoil is a structure of body which produces a useful reaction to air movement. Airplane wings, helicopter rotor blades, and propellers are airfoils. *See* Figure 1-1.

The chord line is a straight reference line from the leading edge to the trailing edge of an airfoil. *See* Figure 1-2.

Changing the shape of an airfoil (by lowering flaps, for example) will change the chord line. *See* Figure 1-3.

In aerodynamics, relative wind is the wind felt by an airfoil. It is created by the movement of air past an airfoil, by the motion of an airfoil through the air, or by a combination of the two. Relative wind is parallel to and in the opposite direction of the flight path of the airfoil. *See* Figure 1-4.

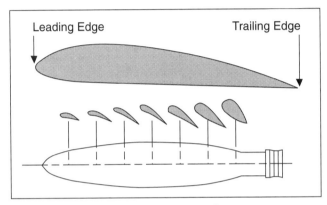

Figure 1-1. A typical airfoil cross-section

Figure 1-2. Chord line

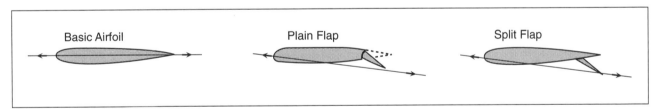

Figure 1-3. Changing shape of wing changes the chord line

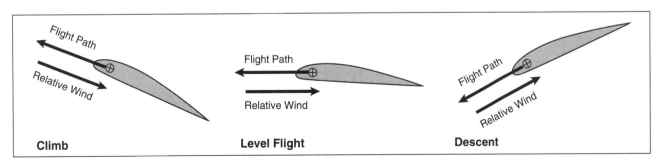

Figure 1-4. Relative wind

Chapter 1 Basic Aerodynamics

The angle of attack is the angle between the chord line of the airfoil and the relative wind. The pilot can vary the angle of attack by manipulating aircraft controls. *See* Figure 1-5. When the angle of attack of a symmetrical airfoil is increased, the center of pressure movement is very limited.

The angle of incidence is the angle between the wing chord line and the center line of the fuselage. The pilot has no control over the angle of incidence. *See* Figure 1-6.

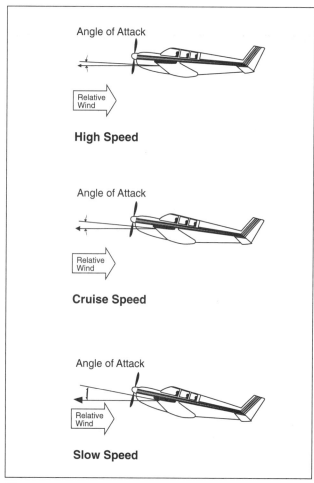

Figure 1-5. Angle of attack

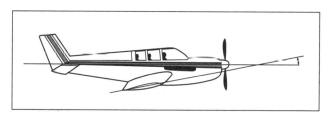

Figure 1-6. Angle of incidence

AIR, GLI

5198. By changing the angle of attack of a wing, the pilot can control the airplane's

A—lift, airspeed, and drag.
B—lift, airspeed, and CG.
C—lift and airspeed, but not drag.

By changing the angle of attack, the pilot can control lift, airspeed, and drag. (H66) — AC 61-23C, Chapter 1

Answer (B) is incorrect because the angle of attack does not affect the CG. Answer (C) is incorrect because the angle of attack also determines drag.

AIR, GLI

5199. The angle of attack of a wing directly controls the

A—angle of incidence of the wing.
B—amount of airflow above and below the wing.
C—distribution of pressures acting on the wing.

The angle of attack of an airfoil directly controls the distribution of pressure below and above it. By changing the angle of attack, the pilot can control lift, airspeed, and drag. (H66) — AC 61-23C, Chapter 1

Answer (A) is incorrect because the angle of incidence is a fixed angle between the chordline of the aircraft and the aircraft's longitudinal axis. Answer (B) is incorrect because the amount of airflow above and below the wing stays constant.

RTC

5239. When the angle of attack of a symmetrical airfoil is increased, the center of pressure will

A—have very limited movement.
B—move aft along the airfoil surface.
C—remain unaffected.

On a symmetrical airfoil, center of pressure movement is very limited. (H70) — FAA-H-8083-21

Answers

5198 [A]   5199 [C]   5239 [A]

Axes of Rotation and the Four Forces Acting in Flight

An airplane has three axes of rotation: the lateral axis, longitudinal axis, and the vertical axis. *See* Figure 1-7.

The lateral axis is an imaginary line from wing tip to wing tip. The rotation about this axis is called pitch. Pitch is controlled by the elevators, and this type of rotation is referred to as longitudinal control, or longitudinal stability. *See* Figure 1-8.

The longitudinal axis is an imaginary line from the nose to the tail. Rotation about the longitudinal axis is called roll. Roll is controlled by the ailerons, and this type of rotation is referred to as lateral control, or lateral stability. *See* Figure 1-9.

The vertical axis is an imaginary line extending vertically through the intersection of the lateral and longitudinal axes. Rotation about the vertical axis is called yaw. Yaw is controlled by the rudder, and this type of rotation is referred to as directional control or directional stability. *See* Figure 1-10.

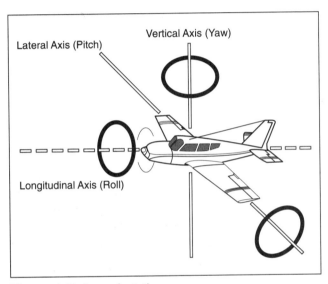

Figure 1-7. Axes of rotation

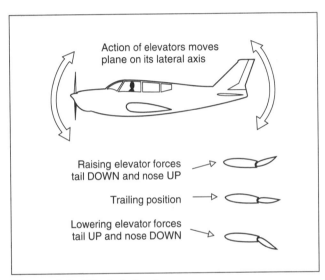

Figure 1-8. Effect of elevators

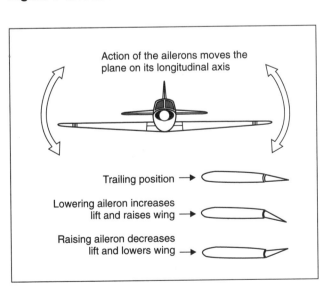

Figure 1-9. Effect of ailerons

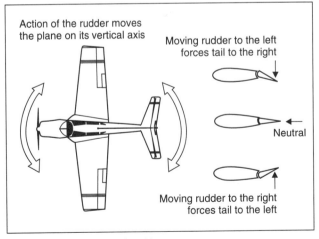

Figure 1-10. Effect of rudder

Chapter 1 **Basic Aerodynamics**

The center of gravity (the imaginary point where all the weight is concentrated) is the point at which an airplane would balance if suspended from that point. The three axes intersect at the center of gravity. Movement of the center of gravity can affect the stability of the airplane.

Four aerodynamic forces are considered basic because they act upon an aircraft during all flight maneuvers. The downward-acting force called weight is counteracted by the upward-acting force called lift. The rearward-acting force called drag is counteracted by the forward-acting force called thrust. *See* Figure 1-11.

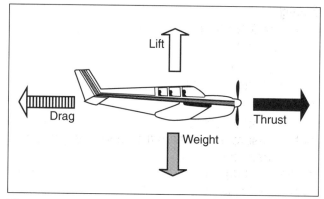

Figure 1-11. Relationship of forces in flight

Lift

Air is a gas that can be compressed or expanded. When compressed, more air can occupy a given volume and air density is increased. When allowed to expand, air occupies a greater space and density is decreased. Temperature, atmospheric pressure, and humidity all affect air density. Air density has significant effects on an aircraft's performance.

As the velocity of a fluid (gas or liquid) increases, its pressure decreases. This is known as Bernoulli's Principle. *See* Figure 1-12.

Lift is the result of a pressure difference between the top and the bottom of the wing. A wing is designed to accelerate air over the top camber of the wing, thereby decreasing the pressure on the top and producing lift. *See* Figure 1-13.

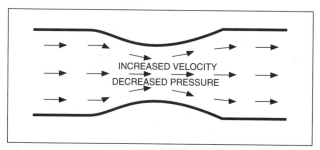

Figure 1-12. Flow of air through a constriction

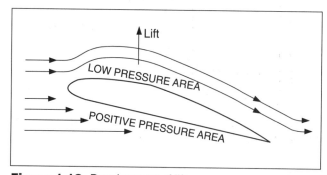

Figure 1-13. Development of lift

Several factors are involved in the creation of lift: angle of attack, wing area and shape (planform), air velocity, and air density. All of these factors have an effect on the amount of lift produced at any given moment. The pilot can actively control the angle of attack and the airspeed, and increasing either of these will result in an increase in lift.

Weight

Weight is the force with which gravity attracts all bodies (masses) vertically toward the center of the Earth.

Thrust

Thrust is the forward force produced by the propeller acting as an airfoil to displace a large mass of air rearward.

Drag

Drag, the force acting parallel to the flight path, resists the forward movement of an airplane through the air. Drag may be classified into two main types: parasite drag and induced drag.

Parasite drag is the resistance of the air produced by any part of an airplane that does not produce lift (antennae, landing gear, etc.). Parasite drag will increase as airspeed increases. If the airspeed of an airplane is doubled, parasite drag will be quadrupled.

Induced drag is a by-product of lift. In other words, this drag is generated as the wing develops lift. The high-pressure air beneath the wing trying to flow around and over the wing tips into the area of low pressure causes a vortex behind the wing tip. This vortex causes a spanwise flow and creates vortices along the trailing edge of the wing. As angle of attack is increased (up to the critical angle), lift will increase and so will the vortices and downwash. This downwash redirects the lift vector rearward, causing a rearward component of lift (induced drag). Induced drag will increase as airspeed decreases. *See* Figure 1-14.

During unaccelerated (straight-and-level) flight, the four aerodynamic forces which act on an airplane are said to be in equilibrium, or: Lift = Weight, and Thrust = Drag.

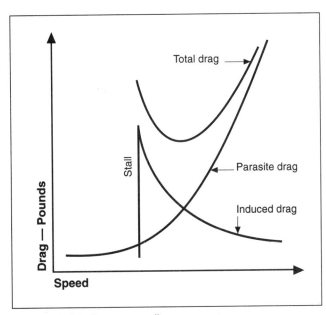

Figure 1-14. Drag curve diagram

AIR, GLI

5158. Lift on a wing is most properly defined as the

A—force acting perpendicular to the relative wind.
B—differential pressure acting perpendicular to the chord of the wing.
C—reduced pressure resulting from a laminar flow over the upper camber of an airfoil, which acts perpendicular to the mean camber.

Lift opposes the downward force of weight. It is produced by the dynamic effect of the air acting on the wing, and acts perpendicular to the flight path (relative wind) through the wing's center of lift. (H300) — AC 61-23C, Chapter 1

AIR

5200. In theory, if the angle of attack and other factors remain constant and the airspeed is doubled, the lift produced at the higher speed will be

A—the same as at the lower speed.
B—two times greater than at the lower speed.
C—four times greater than at the lower speed.

Lift is proportional to the square of the airplane's velocity. For example, an airplane traveling at 200 knots has four times the lift as the same airplane traveling at 100 knots, if the angle of attack and other factors remain constant. (H66) — AC 61-23C, Chapter 1

Answers (A) and (B) are incorrect because as airspeed doubles, lift will be four times greater than at the lower speed.

AIR, GLI

5201. An aircraft wing is designed to produce lift resulting from a difference in the

A—negative air pressure below and a vacuum above the wing's surface.
B—vacuum below the wing's surface and greater air pressure above the wing's surface.
C—higher air pressure below the wing's surface and lower air pressure above the wing's surface.

The wing is designed to provide actions greater than its weight by shaping it to develop a relatively positive (high)-pressure lifting action from the air mass below the wing and a negative (low)-pressure lifting action from lowered pressure above the wing. The increased speed of the air over the top of the airfoil produces the drop in

Continued

Answers

5158 [A]   5200 [C]   5201 [C]

pressure. The pressure difference between the upper and lower surface does not account for all the lift produced. Air also strikes the lower surface of the wing, and the reaction of this downward-backward flow results in an upward-forward force on the wing. (H66) — AC 61-23C, Chapter 1

AIR
5203. Which statement is true, regarding the opposing forces acting on an airplane in steady-state level flight?

A—These forces are equal.
B—Thrust is greater than drag and weight and lift are equal.
C—Thrust is greater than drag and lift is greater than weight.

During straight-and-level flight at constant airspeed, thrust and drag are equal and lift and weight are equal. (H66) — AC 61-23C, Chapter 1

AIR, GLI
5219. Which is true regarding the force of lift in steady, unaccelerated flight?

A—At lower airspeeds the angle of attack must be less to generate sufficient lift to maintain altitude.
B—There is a corresponding indicated airspeed required for every angle of attack to generate sufficient lift to maintain altitude.
C—An airfoil will always stall at the same indicated airspeed; therefore, an increase in weight will require an increase in speed to generate sufficient lift to maintain altitude.

To maintain the lift and weight forces in balance, and to keep the airplane straight-and-level in a state of equilibrium, as velocity increases, angle of attack must be decreased. Conversely, as the airplane slows, the decreasing velocity requires the angle of attack be increased enough to create sufficient lift to maintain flight. Therefore, for every angle of attack, there is a corresponding indicated airspeed required to maintain altitude in steady, unaccelerated flight—all other factors being constant. (H66) — AC 61-23C, Chapter 1

Answer (A) is incorrect because to provide sufficient lift, the angle of attack must be increased as the airspeed is reduced. Answer (C) is incorrect because the angle of attack must be increased to compensate for the decreased lift.

AIR
5223. To generate the same amount of lift as altitude is increased, an airplane must be flown at

A—the same true airspeed regardless of angle of attack.
B—a lower true airspeed and a greater angle of attack.
C—a higher true airspeed for any given angle of attack.

In order to maintain its lift at a higher altitude, an airplane must fly at a greater true airspeed for any given angle of attack. (H66) — AC 61-23C, Chapter 1

Answers (A) and (B) are incorrect because true airspeed must be increased as altitude increases to generate the same amount of lift.

AIR
5229. What changes in airplane longitudinal control must be made to maintain altitude while the airspeed is being decreased?

A—Increase the angle of attack to produce more lift than drag.
B—Increase the angle of attack to compensate for the decreasing lift.
C—Decrease the angle of attack to compensate for the increasing drag.

As the airplane slows down, the decreasing airspeed or velocity requires increasing the angle of attack to produce the constant lift needed to maintain altitude. (H66) — AC 61-23C, Chapter 1

Answer (A) is incorrect because if you are generating more lift than drag, an increase in the angle of attack will cause the airplane to climb. Answer (C) is incorrect because the angle of attack must be increased in order to maintain altitude, if airspeed is being decreased.

AIR
5161. In theory, if the airspeed of an airplane is doubled while in level flight, parasite drag will become

A—twice as great.
B—half as great.
C—four times greater.

Parasite drag has more influence at high speed, and induced drag has more influence at low speed. For example, if an airplane in a steady flight condition at 100 knots is then accelerated to 200 knots, the parasite drag becomes four times as great. (H300) — AC 61-23C, Chapter 1

Answers

5203 [A]   5219 [B]   5223 [C]   5229 [B]   5161 [C]

Chapter 1 Basic Aerodynamics

AIR

5162. As airspeed decreases in level flight below that speed for maximum lift/drag ratio, total drag of an airplane

A—decreases because of lower parasite drag.
B—increases because of increased induced drag.
C—increases because of increased parasite drag.

Parasite drag is greatest at higher airspeeds. Induced drag is the by-product of lift and becomes a greater influence at higher angles of attack and slower airspeeds. It increases in direct proportion to increases in the angle of attack. Any angle of attack lower or higher than that for L/D$_{MAX}$ reduces the lift-drag ratio and consequently increases the total drag. (H300) — AC 61-23C, Chapter 1

Answer (A) is incorrect because total drag increases when airspeed decreases below L/D$_{MAX}$ due to increased induced drag. Answer (C) is incorrect because parasite drag decreases with speeds below L/D$_{MAX}$.

AIR, GLI

5167. Which statement is true relative to changing angle of attack?

A—A decrease in angle of attack will increase pressure below the wing, and decrease drag.
B—An increase in angle of attack will increase drag.
C—An increase in angle of attack will decrease pressure below the wing, and increase drag.

Air striking the underside of the wing is deflected downward, producing an opposite reaction which pushes (lifts) the wing upward. To increase lift, the wing is tilted upward, increasing the angle of attack and deflecting more air downward. The larger the angle of attack, the more the lift force tilts toward the rear of the aircraft, increasing drag. (H300) — AC 61-23C, Chapter 1

AIR, GLI

5202. On a wing, the force of lift acts perpendicular to and the force of drag acts parallel to the

A—chord line.
B—flightpath.
C—longitudinal axis.

Lift acts upward and perpendicular to the relative wind. Drag acts parallel to, and in the same direction as the relative wind, which is parallel to the flightpath. (H66) — AC 61-23C, Chapter 1

Answers (A) and (C) are incorrect because there is not a fixed relationship between lift and drag with respect to the airplane's chord line or longitudinal axis.

GLI

5280. Which is true regarding aerodynamic drag?

A—Induced drag is created entirely by air resistance.
B—All aerodynamic drag is created entirely by the production of lift.
C—Induced drag is a by-product of lift and is greatly affected by changes in airspeed.

When an aircraft develops lift, the upward force on the aircraft results in an equal and downward force on the air which is then set in motion, generally downward. It takes energy to do this, and the transfer of energy from the aircraft to the air implies the existence of drag on the aircraft. This lift-related part of the total drag is called induced drag. The induced drag varies inversely with the square of the velocity. (N20) — Soaring Flight Manual

Answer (A) is incorrect because skin friction drag (a form of parasite drag), is created by air resistance. Answer (B) is incorrect because aerodynamic drag (a form of parasite drag), is created by friction between the air and the surface over which it is flowing.

AIR, GLI

5218. Which is true regarding the forces acting on an aircraft in a steady-state descent? The sum of all

A—upward forces is less than the sum of all downward forces.
B—rearward forces is greater than the sum of all forward forces.
C—forward forces is equal to the sum of all rearward forces.

In steady-state flight, the sum of the opposing forces is equal to zero. The sum of all upward forces equals the sum of all downward forces. The sum of all forward forces equals the sum of all backward forces. (H66) — AC 61-23C, Chapter 1

Answers

5162 [B]   5167 [B]   5202 [B]   5280 [C]   5218 [C]

Chapter 1 **Basic Aerodynamics**

AIR
5220. During the transition from straight-and-level flight to a climb, the angle of attack is increased and lift

A—is momentarily decreased.
B—remains the same.
C—is momentarily increased.

When transitioning from level flight to a climb, the forces acting on the airplane go through certain changes. The first change, an increase in lift, occurs when back pressure is applied to the elevator control. This initial change is a result of the increase in the angle of attack which occurs when the airplane's pitch attitude is raised. This results in a climbing attitude. When the inclined flight path and the climb speed are established, the angle of attack and the corresponding lift stabilize at approximately the original value. (H66) — AC 61-23C, Chapter 1

Lift/Drag Ratios

The lift-to-drag ratio is the lift required for level flight (weight) divided by the drag produced at the airspeed and angle of attack required to produce that lift. The L/D ratio for a particular angle of attack is equal to the power-off glide ratio.

Problem:

Refer to FAA Figure 3. If an airplane glides at an angle of attack of 10°, how much altitude will it lose in 1 mile?

Solution:

1. Enter the L/D chart from the bottom at a 10° angle of attack.
2. Proceed vertically upward until intersecting the L/D curve.
3. Follow the horizontal reference lines to the right to the point of intersection with the glide ratio scale. L/D at 10° angle of attack = 11.0.
4. 5,280 feet ÷ 11 = 480 foot altitude loss.

L/D_{MAX} occurs at the angle of attack that gives maximum glide performance and maximum range in a propeller driven aircraft. At an airspeed slower (or at a higher angle of attack) than needed for L/D_{MAX}, the glide distance will be reduced due to the increase in induced drag.

AIR
5165. (Refer to Figure 1.) At the airspeed represented by point A, in steady flight, the airplane will

A—have its maximum L/D ratio.
B—have its minimum L/D ratio.
C—be developing its maximum coefficient of lift.

At point A, the total drag curve is at its lowest point. When an aircraft is flown at the airspeed and angle of attack that results in the lowest total drag possible, then the resulting L/D ratio is at its maximum. (H300) — AC 61-23C, Chapter 1

Answer (B) is incorrect because the minimum L/D ratio occurs when parasite drag is very high, a result of high airspeeds. Answer (C) is incorrect because the maximum coefficient of lift occurs at slower airspeeds, which results in higher induced drag and a lower L/D ratio.

AIR
5166. (Refer to Figure 1.) At an airspeed represented by point B, in steady flight, the pilot can expect to obtain the airplane's maximum

A—endurance.
B—glide range.
C—coefficient of lift.

Point B represents the airspeed that results in the greatest L/D ratio. At this point the aircraft will have its maximum glide range. (H300) — AC 61-23C, Chapter 1

Answer (A) is incorrect since only jet aircraft will obtain maximum endurance at L/D_{MAX}. Answer (C) is incorrect because the critical angle of attack and the maximum coefficient of lift occur at the same point, where total drag is also high because of an increase in induced drag.

Answers

5220 [C]   5165 [A]   5166 [B]

AIR
5213. (Refer to Figure 3.) If an airplane glides at an angle of attack of 10°, how much altitude will it lose in 1 mile?

A—240 feet.
B—480 feet.
C—960 feet.

To find the glide ratio (L/D) at an angle of attack of 10°, move upward from the angle of attack scale to the L/D curve. Then move horizontally to the right to find the value located on the L/D scale. This gives an 11:1 glide ratio. The question only deals with glide ratio, so that is the only scale needed. With this glide ratio, the airplane will descend 1 foot of altitude for every 11 feet covered horizontally.

$$\frac{L}{D} = \frac{11}{1} = \frac{\text{horizontal distance}}{\text{vertical distance}}$$

$$\frac{11}{1} = \frac{5{,}280 \text{ feet (1 SM)}}{X}$$

$$X = 480 \text{ feet}$$

(H66) — AC 61-23C, Chapter 1

AIR
5214. (Refer to Figure 3.) How much altitude will this airplane lose in 3 miles of gliding at an angle of attack of 8°?

A—440 feet.
B—880 feet.
C—1,320 feet.

To find the glide ratio (L/D) at an angle of attack of 8°, move upward from the angle of attack scale to the L/D curve. Then move horizontally to the right to find the value located on the L/D scale. This gives a 12:1 glide ratio. The question only deals with glide ratio, so that is the only scale needed. With this glide ratio, the airplane will descend 1 foot of altitude for every 12 feet covered horizontally.

$$\frac{L}{D} = \frac{12}{1} = \frac{\text{horizontal distance}}{\text{vertical distance}}$$

$$\frac{12}{1} = \frac{5{,}280 \text{ feet (3 SM)}}{X}$$

$$X = 1{,}320 \text{ feet}$$

(H66) — AC 61-23C, Chapter 1

AIR, GLI
5215. (Refer to Figure 3.) The L/D ratio at a 2° angle of attack is approximately the same as the L/D ratio for a

A—9.75° angle of attack.
B—10.5° angle of attack.
C—16.5° angle of attack.

The glide ratio (L/D) at a 2° angle of attack is about 7.6:1, which is the same glide ratio (L/D) as at a 16.5° angle of attack. (H66) — AC 61-23C, Chapter 1

AIR
5217. What performance is characteristic of flight at maximum lift/drag ratio in a propeller-driven airplane? Maximum

A—gain in altitude over a given distance.
B—range and maximum distance glide.
C—coefficient of lift and minimum coefficient of drag.

Maximum range condition would occur where the proportion between speed and power required is greatest. The maximum range condition (of propeller driven airplanes) is obtained at maximum lift-drag ratio (L/D_{MAX}). The best angle of glide is one that allows the airplane to travel the greatest distance over the ground with the least loss of altitude. This is also the airplane's maximum L/D and is usually expressed as a ratio. This implies that the airplane should be flown at L/D_{MAX} to obtain the greatest glide distance. (H66) — AC 61-23C, Chapter 1

ALL
5505. Which maximum range factor decreases as weight decreases?

A—Altitude.
B—Airspeed.
C—Angle of attack.

The maximum range condition is obtained at maximum L/D ratio which occurs at a particular angle of attack and lift coefficient. As gross weight decreases, the airspeed for maximum L/D decreases. (H66) — AC 61-23C, Chapter 1

Answer (A) is incorrect because as weight decreases, the maximum range altitude may increase. Answer (C) is incorrect because angle of attack does not play a role in determining maximum range factor.

The V_G Diagram

FAA Figure 5 is a V_G diagram which plots load factor against indicated airspeed and shows the pilot the limits within which the aircraft will safely handle structural loads. Point C is maneuvering speed (V_A). Any plotted combination of load factor and airspeed which falls in the shaded area may result in structural damage.

V_{NO}, the maximum speed for normal operations, is shown by the vertical line from point D to point G on the V_G diagrams, and is marked as the upper limit of the green arc on the airspeed indicator. The red line on the airspeed indicator is represented by the line from point E to point F. The line connecting points C, D, and E represents the limit load factor above which structural damage may occur.

AIR, GLI
5231. (Refer to Figure 5.) The horizontal dashed line from point C to point E represents the

A—ultimate load factor.
B—positive limit load factor.
C—airspeed range for normal operations.

C to E is the maximum positive load limit. In this case it is 3.8 Gs, which is appropriate for normal category airplanes. (H66) — AC 61-23C, Chapter 1

AIR, GLI
5232. (Refer to Figure 5.) The vertical line from point E to point F is represented on the airspeed indicator by the

A—upper limit of the yellow arc.
B—upper limit of the green arc.
C—blue radial line.

V_{NE} (never exceed airspeed), the vertical line from point E to F, is marked on airspeed indicators with a red radial line, the upper limit of the yellow arc. (H66) — AC 61-23C, Chapter 1

Stability

Stability is the inherent ability of an airplane to return, or not return, to its original flight condition after being disturbed by an outside force, such as rough air.

Positive static stability is the initial tendency of an aircraft to return to its original position. *See* Figure 1-15.

Positive dynamic stability is the tendency of an oscillating airplane (with positive static stability) to return to its original position relative to time. *See* Figure 1-16.

Aircraft design normally assures that the aircraft will be stable in pitch. The pilot can adversely affect this longitudinal stability by allowing the center of gravity (CG) to move forward or aft of specified CG limits through improper loading procedures. One undesirable flight characteristic a pilot might experience in an airplane loaded with the CG located behind the aft CG limit would be the inability to recover from a stalled condition.

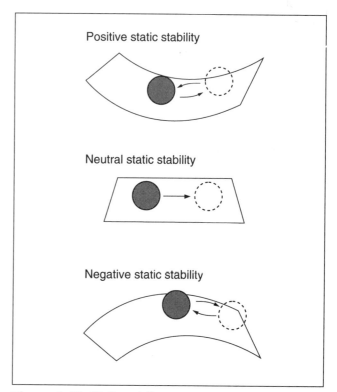

Figure 1-15. Static stability

Answers
5231 [B]   5232 [A]

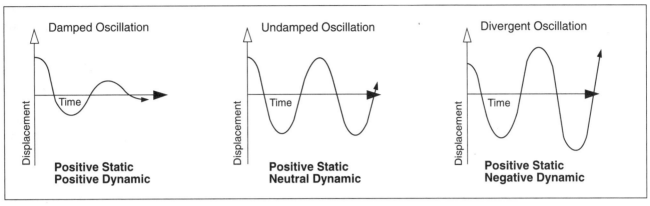

Figure 1-16. Positive static stability relative to dynamic stability

The location of the CG with respect to the center of lift (CL) will determine the longitudinal stability of an airplane. *See* Figure 1-17.

An airplane will be less stable at all airspeeds if it is loaded to the most aft CG. An advantage of an airplane said to be inherently stable is that it will require less effort to control.

Changes in pitch can also be experienced with changes in power setting (except in T-tail airplanes). When power is reduced, there is a corresponding reduction in downwash on the tail, which results in the nose "pitching" down.

Effects of Forward CG

1. Increased longitudinal stability.
2. Lower cruise speed. The wing flies at a higher angle of attack to create more lift to counter the added downward forces produced by the tail, therefore the wing also produces more induced drag.
3. Higher stall speed. The wing flies at a higher angle of attack to create more lift to counter the added downward forces produced by the tail, therefore the wing also produces more induced drag.

Effects of Aft CG

1. Decreased longitudinal stability.
2. Higher cruise speed (for just the opposite reason listed above).
3. Lower stall speed.
4. Poor stall/spin recovery.

Figure 1-17. Effects of CG on aircraft stability

5205. In small airplanes, normal recovery from spins may become difficult if the

A—CG is too far rearward and rotation is around the longitudinal axis.
B—CG is too far rearward and rotation is around the CG.
C—spin is entered before the stall is fully developed.

The recovery from a stall in any airplane becomes progressively more difficult as its center of gravity moves aft. This is particularly important in spin recovery, as there is a point in rearward loading of any airplane at which a "flat" spin will develop. (H66) — AC 61-23C, Chapter 1

Answer (A) is incorrect because rotation is around the CG in a spin. Answer (C) is incorrect because an airplane must first stall in order to spin.

Answers
5205 [B]

AIR

5207. If an airplane is loaded to the rear of its CG range, it will tend to be unstable about its

A—vertical axis.
B—lateral axis.
C—longitudinal axis.

An airplane becomes less longitudinally stable as the CG is moved rearward. Longitudinal stability (pitching) is stability about the lateral axis. (H66) — AC 61-23C, Chapter 1

Answer (A) is incorrect because the CG has little to do with the vertical axis. Answer (C) is incorrect because lateral stability is not greatly affected by the CG location.

AIR

5212. An airplane will stall at the same

A—angle of attack regardless of the attitude with relation to the horizon.
B—airspeed regardless of the attitude with relation to the horizon.
C—angle of attack and attitude with relation to the horizon.

The definition of a stall is when the airplane exceeds the critical angle of attack. This happens because the smooth airflow over the airplane's wing is disrupted and the lift degenerates rapidly. This can occur at any airspeed, in any attitude, with any power setting. (H66) — AC 61-23C, Chapter 1

AIR, GLI

5226. If the airplane attitude remains in a new position after the elevator control is pressed forward and released, the airplane displays

A—neutral longitudinal static stability.
B—positive longitudinal static stability.
C—neutral longitudinal dynamic stability.

Neutral static stability is the initial tendency of the airplane to remain in the new condition after its equilibrium has been disturbed. When an airplane's attitude is momentarily displaced and it remains at the new attitude, it is displaying neutral longitudinal static stability. Longitudinal stability makes an airplane stable about its lateral axis (pitch). (H66) — AC 61-23C, Chapter 1

Answer (B) is incorrect because positive longitudinal static stability is the initial tendency of the airplane to return to its original attitude after the elevator control is pressed forward and released. Answer (C) is incorrect because neutral longitudinal dynamic stability is the overall tendency for the airplane to remain in the new condition over a period of time.

AIR, GLI

5227. Longitudinal dynamic instability in an airplane can be identified by

A—bank oscillations becoming progressively steeper.
B—pitch oscillations becoming progressively steeper.
C—Trilatitudinal roll oscillations becoming progressively steeper.

Longitudinal stability, or pitching about the lateral axis, is considered to be the most affected by certain variables in various flight conditions. A longitudinally unstable airplane has a tendency to dive or climb progressively into a steep dive or climb which may result in a stall. A longitudinally unstable airplane is difficult and sometimes dangerous to fly. (H66) — AC 61-23C, Chapter 1

Answers (A) and (C) are incorrect because roll oscillations refer to lateral stability.

AIR

5228. Longitudinal stability involves the motion of the airplane controlled by its

A—rudder.
B—elevator.
C—ailerons.

Longitudinal stability or pitching is the motion around the lateral axis. Pitch is controlled by the elevator. (H66) — AC 61-23C, Chapter 1

Answer (A) is incorrect because the rudder affects directional stability. Answer (C) is incorrect because the ailerons affect lateral stability.

AIR

5230. If the airplane attitude initially tends to return to its original position after the elevator control is pressed forward and released, the airplane displays

A—positive dynamic stability.
B—positive static stability.
C—neutral dynamic stability.

Static stability deals with initial tendencies. Positive static stability is the initial tendency of the airplane to return to its original state after being disturbed. (H66) — AC 61-23C, Chapter 1

Answers

5207 [B]   5212 [A]   5226 [A]   5227 [B]   5228 [B]   5230 [B]

Turns, Loads and Load Factors

When an airplane is banking into a turn, a portion of the lift developed is diverted into a horizontal component of lift. It is this horizontal (sideward) force that forces the airplane from straight-and-level flight and causes it to turn. The reduced vertical lift component results in a loss of altitude unless total lift is increased by increasing the angle of attack, increasing airspeed or both.

In aerodynamics, load is the force, or imposed stress, that must be supported by an airplane structure in flight. The loads imposed on the wings in flight are stated in terms of load factor.

In straight-and-level flight, the wings of an airplane support a load equal to the sum of the weight of the airplane plus its contents. This particular load factor is equal to "one G," where "G" refers to the pull of gravity.

However, a force which acts toward the outside of the curve, called centrifugal force, is generated any time an airplane is flying a curved path (turns, climbs, or descents).

Whenever the airplane is flying in a curved flight path with a positive load, the load that the wings must support will be equal to the weight of the airplane plus the load imposed by centrifugal force; therefore, it can be said that turns increase the load factor on an airplane.

As the angle of bank of a turn increases, the load factor increases, as shown in Figure 1-18.

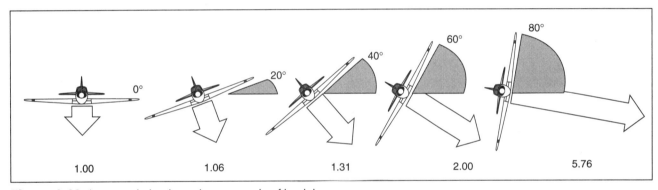

Figure 1-18. Increase in load on wings as angle of bank increases

The amount of excess load that can be imposed on the wing of an airplane depends on the speed of the airplane. An example of this would be a change in direction made at high speed with forceful control movement, which results in a high load factor being imposed.

An increased load factor (weight) will cause an airplane to stall at a higher airspeed, as shown in Figure 1-19.

Some conditions that increase the weight (load) of an aircraft are: overloading the airplane, too steep an angle of bank, turbulence and abrupt movement of the controls.

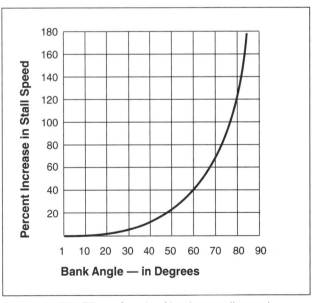

Figure 1-19. Effect of angle of bank on stall speed

Chapter 1 Basic Aerodynamics

Because different types of operations require different maneuvers (and therefore varying bank angles and load factors), aircraft are separated into categories determined by the loads that their wing structures can support:

Category	Positive Limit Load
Normal (nonacrobatic) (N)	3.8 times gross weight
Utility (normal operations and limited acrobatic maneuvers)	4.4 times gross weight
Acrobatic (A)	6.0 times gross weight

The limit loads should not be exceeded in actual operation, even though a safety factor of 50% above limit loads is incorporated into the strength of an airplane.

AIR, GLI
5192. To increase the rate of turn and at the same time decrease the radius, a pilot should

A—maintain the bank and decrease airspeed.
B—increase the bank and increase airspeed.
C—increase the bank and decrease airspeed.

The horizontal component of lift will equal the centrifugal force of steady, turning flight. To increase the rate of turn, the angle of bank may be increased and the airspeed may be decreased. (H55) — AC 61-23C, Chapter 1

Answer (A) is incorrect because although at a given angle of bank a decrease in airspeed will increase the rate of turn and decrease the radius, it will not be as effective as steepening the bank and decreasing the airspeed. Answer (B) is incorrect because the pilot should decrease airspeed to decrease the turn radius.

AIR
5193. Which is correct with respect to rate and radius of turn for an airplane flown in a coordinated turn at a constant altitude?

A—For a specific angle of bank and airspeed, the rate and radius of turn will not vary.
B—To maintain a steady rate of turn, the angle of bank must be increased as the airspeed is decreased.
C—The faster the true airspeed, the faster the rate and larger the radius of turn regardless of the angle of bank.

At a specific angle of bank and a specific airspeed, the radius of the turn and the rate of turn would remain constant if the altitude were maintained. Rate of turn varies with airspeed, or bank angle. If the angle of bank is held constant and the airspeed is increased, the rate of turn will decrease, and the radius of turn will increase. To maintain a constant rate of turn as the airspeed is increased, the angle of bank must be increased. (H55) — AC 61-23C, Chapter 1

Answer (B) is incorrect because you must decrease the angle of bank when the airspeed is decreased if you are to maintain a steady rate of turn. Answer (C) is incorrect because for a given bank angle a faster airspeed gives a slower rate of turn.

AIR
5194. Why is it necessary to increase back elevator pressure to maintain altitude during a turn? To compensate for the

A—loss of the vertical component of lift.
B—loss of the horizontal component of lift and the increase in centrifugal force.
C—rudder deflection and slight opposite aileron throughout the turn.

Lift during a bank is divided into two components, one vertical and the other horizontal. The vertical component of lift must be equal to the weight to maintain altitude. Since the vertical component of lift decreases as the bank angle increases, the angle of attack must be progressively increased to produce sufficient vertical lift to support the airplane's weight. The increased back elevator pressure provides the increased angle of attack. (H55) — AC 61-23C, Chapter 1

Answer (B) is incorrect because this describes a skidding turn. Answer (C) is incorrect because slight opposite aileron pressure may be required to compensate for the overbanking tendency, not to maintain altitude.

Answers

5192 [C]   5193 [A]   5194 [A]

Chapter 1 Basic Aerodynamics

AIR

5195. To maintain altitude during a turn, the angle of attack must be increased to compensate for the decrease in the

A—forces opposing the resultant component of drag.
B—vertical component of lift.
C—horizontal component of lift.

Lift during a bank is divided into two components, one vertical and the other horizontal. The vertical component of lift must be equal to the weight to maintain altitude. Since the vertical component of lift decreases as the bank angle increases, the angle of attack must be progressively increased to produce sufficient vertical lift to support the airplane's weight. The increased back elevator pressure provides the increased angle of attack. (H55) — AC 61-23C, Chapter 1

Answer (A) is incorrect because the phrase "the resultant component of drag" has no meaning. Answer (C) is incorrect because as the horizontal component of lift decreases, the vertical component of lift increases, therefore the angle of attack must be decreased.

AIR

5210. If airspeed is increased during a level turn, what action would be necessary to maintain altitude? The angle of attack

A—and angle of bank must be decreased.
B—must be increased or angle of bank decreased.
C—must be decreased or angle of bank increased.

To compensate for added lift which would result if the airspeed were increased during a turn, the angle of attack must be decreased, or the angle of bank increased, if a constant altitude is to be maintained. (H66) — AC 61-23C, Chapter 1

Answer (A) is incorrect because either the angle of attack can be decreased or the angle of bank increased to maintain altitude as airspeed is increased. Answer (B) is incorrect because to maintain constant altitude in a turn as the airspeed increases, the angle of bank must decrease.

AIR, GLI

5225. As the angle of bank is increased, the vertical component of lift

A—decreases and the horizontal component of lift increases.
B—increases and the horizontal component of lift decreases.
C—decreases and the horizontal component of lift remains constant.

Lift during a bank is divided into two components, one vertical and the other horizontal. The vertical component of lift must be equal to the weight to maintain altitude. Since the vertical component of lift decreases as the bank angle increases, the angle of attack must be progressively increased to produce sufficient vertical lift to support the airplane's weight. The increased back elevator pressure provides the increased angle of attack. (H66) — AC 61-23C, Chapter 1

AIR, GLI

5181. Which is true regarding the use of flaps during level turns?

A—The lowering of flaps increases the stall speed.
B—The raising of flaps increases the stall speed.
C—Raising flaps will require added forward pressure on the yoke or stick.

If flaps are raised, the stall speed increases. (H305) — AC 61-23C, Chapter 1

Answer (A) is incorrect because flaps decrease the stall speed. Answer (C) is incorrect because raising the flaps decreases the lift provided, therefore, back pressure is required to maintain altitude.

AIR, MIL

5017. If an airplane category is listed as utility, it would mean that this airplane could be operated in which of the following maneuvers?

A—Limited acrobatics, excluding spins.
B—Limited acrobatics, including spins (if approved).
C—Any maneuver except acrobatics or spins.

Utility category airplanes can do all normal category maneuvers plus limited acrobatics, including spins (if approved). (A150) — 14 CFR §23.3(b)

Answer (A) is incorrect because the utility category includes spins. Answer (C) is incorrect because the normal category prohibits acrobatics and spins.

Answers

5195 [B]　　　5210 [C]　　　5225 [A]　　　5181 [B]　　　5017 [B]

Chapter 1 Basic Aerodynamics

AIR, GLI

5151. The ratio between the total airload imposed on the wing and the gross weight of an aircraft in flight is known as

A—load factor and directly affects stall speed.
B—aspect load and directly affects stall speed.
C—load factor and has no relation with stall speed.

The load factor is the ratio between the total airload imposed on the wings of an airplane and the gross weight of the airplane. An increased load factor increases stalling speed and a decreased load factor decreases stalling speed. (H303) — AC 61-23C, Chapter 1

AIR, GLI

5152. Load factor is the lift generated by the wings of an aircraft at any given time

A—divided by the total weight of the aircraft.
B—multiplied by the total weight of the aircraft.
C—divided by the basic empty weight of the aircraft.

Load factor is the ratio between the total airload supported by the wing to the total weight of the airplane; i.e., the total airload supported by the wings divided by the total weight of the airplane. (H303) — AC 61-23C, Chapter 1

Answer (B) is incorrect because load factor multiplied by airplane weight equals required lift. Answer (C) is incorrect because load factor is lift divided by the total weight of the airplane.

AIR

5153. For a given angle of bank, in any airplane, the load factor imposed in a coordinated constant-altitude turn

A—is constant and the stall speed increases.
B—varies with the rate of turn.
C—is constant and the stall speed decreases.

In an airplane at any airspeed, if a constant altitude is maintained during the turn, the load factor for a given degree of bank is the same, which is the resultant of gravity and centrifugal force. Load supported by the wings increases as the angle of bank increases. Stall speeds increase in proportion to the square root of the load factor. (H303) — AC 61-23C, Chapter 1

Answer (B) is incorrect because rate of turn does not affect the load factor. Answer (C) is incorrect because stall speed will increase with an increase in bank.

AIR

5154. Airplane wing loading during a level coordinated turn in smooth air depends upon the

A—rate of turn.
B—angle of bank.
C—true airspeed.

For any given angle of bank, the rate of turn varies with the airspeed. If the angle of bank is held constant and the airspeed is increased, the rate of turn will decrease. Because of this, there is no change in centrifugal force for any given bank angle. Therefore, the load factor remains the same. Load factor varies with changing bank angle and increases at a rapid rate after the angle of bank reaches 50°. (H303) — AC 61-23C, Chapter 1

Answers (A) and (C) are incorrect because rate of turn and true airspeed do not have an impact on wing loading in a coordinated turn.

AIR, GLI

5155. In a rapid recovery from a dive, the effects of load factor would cause the stall speed to

A—increase.
B—decrease.
C—not vary.

There is a direct relationship between the load factor imposed upon the wing and its stalling characteristics. The stalling speed increases in proportion to the square root of the load factor. (H303) — AC 61-23C, Chapter 1

AIR, GLI

5156. If an aircraft with a gross weight of 2,000 pounds was subjected to a 60° constant-altitude bank, the total load would be

A—3,000 pounds.
B—4,000 pounds.
C—12,000 pounds.

The load factor in a 60° bank is 2 Gs. Load Factor = G Load x Aircraft Weight. Therefore, 2,000 x 2 = 4,000 pounds. (H303) — AC 61-23C, Chapter 1

Answers

5151 [A]   5152 [A]   5153 [A]   5154 [B]   5155 [A]   5156 [B]

AIR, GLI
5157. While maintaining a constant angle of bank and altitude in a coordinated turn, an increase in airspeed will

A—decrease the rate of turn resulting in a decreased load factor.
B—decrease the rate of turn resulting in no change in load factor.
C—increase the rate of turn resulting in no change in load factor.

For any given angle of bank the rate of turn varies with the airspeed. In other words, if the angle of bank is held constant and the airspeed is increased, the rate of turn will decrease, or if the airspeed is decreased, the rate of turn will increase. Because of this, there is no change in centrifugal force for any given bank. Therefore, the load factor remains the same. (H303) — AC 61-23C, Chapter 1

Answer (A) is incorrect because load factor remains the same at a constant angle of bank. Answer (C) is incorrect because the rate of turn in a constant angle of bank will decrease with an increase in airspeed, and the load factor will remain the same.

AIR, GLI
5159. While holding the angle of bank constant, if the rate of turn is varied the load factor would

A—remain constant regardless of air density and the resultant lift vector.
B—vary depending upon speed and air density provided the resultant lift vector varies proportionately.
C—vary depending upon the resultant lift vector.

For any given angle of bank the rate of turn varies with the airspeed. In other words, if the angle of bank is held constant and the airspeed is increased, the rate of turn will decrease, or if the airspeed is decreased, the rate of turn will increase. Because of this, there is no change in centrifugal force for any given bank. Therefore, the load factor remains the same. (H303) — AC 61-23C, Chapter 1

Answer (B) is incorrect because rate of turn will vary based on airspeed with a constant angle of bank. Answer (C) is incorrect because load factor varies based on the resultant load vector.

AIR
5163. If the airspeed is increased from 90 knots to 135 knots during a level 60° banked turn, the load factor will

A—increase as well as the stall speed.
B—decrease and the stall speed will increase.
C—remain the same but the radius of turn will increase.

At a given angle of bank, a higher airspeed will make the radius of the turn larger and the airplane will be turning at a slower rate. This compensates for added centrifugal force, allowing the load factor to remain the same. (H303) — AC 61-23C, Chapter 1

AIR
5179. (Refer to Figure 2.) Select the correct statement regarding stall speeds.

A—Power-off stalls occur at higher airspeeds with the gear and flaps down.
B—In a 60° bank the airplane stalls at a lower airspeed with the gear up.
C—Power-on stalls occur at lower airspeeds in shallower banks.

Power-on stalls occur at lower airspeeds than power-off stalls because of increased airflow over the wing and because some lift is produced by the vertical component of thrust, reducing the lift needed to be produced by velocity. Power-on or -off stalls occur at a lower airspeed in a shallower bank. (H303) — AC 61-23C, Chapter 1

Answer (A) is incorrect because stall speed is lower with power-off stalls, with gear and flaps down. Answer (B) is incorrect because the gear position alone will not affect stall speed in a 60° bank.

AIR
5180. (Refer to Figure 2.) Select the correct statement regarding stall speeds. The airplane will stall

A—10 knots higher in a power-on 60° bank with gear and flaps up than with gear and flaps down.
B—35 knots lower in a power-off, flaps-up, 60° bank, than in a power-off, flaps-down, wings-level configuration.
C—10 knots higher in a 45° bank, power-on stall than in a wings-level stall.

The stalling speed (in knots) for a power-on, 60° bank, with gear and flaps up is 76 knots. For a power-on, 60° bank, and gear and flaps down, the stalling speed is 66 knots, which is 10 knots slower. (H303) — AC 61-23C, Chapter 1

Answers

5157 [B]   5159 [A]   5163 [C]   5179 [C]   5180 [A]

Chapter 1 **Basic Aerodynamics**

AIR
5221. (Refer to Figure 4.) What is the stall speed of an airplane under a load factor of 2 Gs if the unaccelerated stall speed is 60 knots?

A—66 knots.
B—74 knots.
C—84 knots.

From a load factor of 2, go horizontally to the curved line labeled "Load Factor." From that point of intersection, go up to the curve labeled "Stall Speed Increase." From there, go to the left and read the increase: 40%. The new accelerated stall speed will be 140%, or 1.4 times, the original value. 60 x 1.4 = 84 knots. (H66) — AC 61-23C, Chapter 1

AIR, GLI
5222. (Refer to Figure 4.) What increase in load factor would take place if the angle of bank were increased from 60° to 80°?

A—3 Gs.
B—3.5 Gs.
C—4 Gs.

Proceed vertically along the line above 60° angle of bank to where it intersects the curve labeled "Load Factor." Next, proceed along this line to the left to the corresponding load factor or "G" unit, which is 2 Gs in this case. Now, repeat the procedure using 80° of bank, which should yield a load factor of 6, which is a difference of 4 Gs. (H66) — AC 61-23C, Chapter 1

Stalls and Spins

As the angle of attack is increased (to increase lift), the air will no longer flow smoothly over the upper wing surface, but instead will become turbulent or "burble" near the trailing edge. A further increase in the angle of attack will cause the turbulent area to expand forward. At an angle of attack of approximately 18° to 20° (for most wings), turbulence over the upper wing surface decreases lift so drastically that flight cannot be sustained and the wing stalls. *See* Figure 1-20.

The angle at which a stall occurs is called the critical angle of attack. An airplane can stall at any airspeed or any attitude, but will always stall at the same critical angle of attack. The indicated airspeed at which a given airplane will stall in a particular configuration, however, will remain the same regardless of altitude. Because air density decreases with an increase in altitude, the airplane has to be flown faster at higher altitudes to cause the same pressure difference between pitot impact pressure and static pressure. The recovery from a stall in any airplane becomes progressively more difficult as its center of gravity moves aft.

An aircraft will spin only after it has stalled, and will continue to spin as long as the outside wing continues to provide more lift than the inside wing, and the aircraft remains stalled.

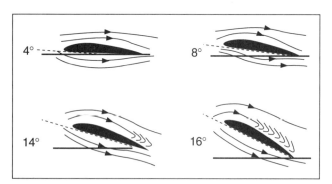

Figure 1-20. Flow of air over wing at various angles of attack

AIR, GLI
5160. The need to slow an aircraft below V_A is brought about by the following weather phenomenon:

A—High density altitude which increases the indicated stall speed.
B—Turbulence which causes an increase in stall speed.
C—Turbulence which causes a decrease in stall speed.

When severe turbulence is encountered, the airplane should be flown at or below maneuvering speed. This is the speed least likely to result in structural damage to the airplane, even if full control travel is used, and yet allows a sufficient margin of safety above stalling speed in turbulent air. When an airplane is flying at a high speed with a low angle of attack, and suddenly encounters a vertical current of air moving upward, the relative wind changes to an upward direction as it meets the airfoil. This increases the angle of attack on the wings and has

Answers

5221 [C]   5222 [C]   5160 [B]

the same effect as applying a sharp back pressure on the elevator control. It increases the load factor which in turn increases the stall speed. (H303) — AC 61-23C, Chapter 1

Answer (A) is incorrect because indicated stall speed is not affected by changes in density altitude. Answer (C) is incorrect because the higher load factors imposed on the aircraft by turbulence increases stall speed.

AIR
5196. Stall speed is affected by

A—weight, load factor, and power.
B—load factor, angle of attack, and power.
C—angle of attack, weight, and air density.

The indicated stalling speed is affected by:

1. *Weight. As the weight is increased, the stall speed also increases.*
2. *Bank angle (load factor). As the bank angle increases, so does the stalling speed.*
3. *Power. An increase in power will decrease the stalling speed.*

A change in density altitude (air density) or angle of attack has no effect on the indicated stalling speed, since for a given airplane, the stalling or critical angle of attack remains constant. (H66) — AC 61-23C, Chapter 1

AIR, GLI
5204. The angle of attack at which a wing stalls remains constant regardless of

A—weight, dynamic pressure, bank angle, or pitch attitude.
B—dynamic pressure, but varies with weight, bank angle, and pitch attitude.
C—weight and pitch attitude, but varies with dynamic pressure and bank angle.

When the angle of attack becomes so great that the air can no longer flow smoothly over the top wing surface, it becomes impossible for the air to follow the contour of the wing. This is the stalling or critical angle of attack. For any given airplane, the stalling or critical angle of attack remains constant regardless of weight, dynamic pressure, bank angle, or pitch attitude. These factors will affect the speed at which the stall occurs, but not the angle. (H66) — AC 61-23C, Chapter 1

Answers (B) and (C) are incorrect because the stall speed varies with weight and bank angle.

AIR
5211. The stalling speed of an airplane is most affected by

A—changes in air density.
B—variations in flight altitude.
C—variations in airplane loading.

The indicated stalling speed is most affected by load factor. The airplane's stalling speed increases in proportion to the square root of the load factor, whereas a change in altitude (air density) has no effect on the indicated stalling speed. (H66) — AC 61-23C, Chapter 1

AIR
5206. Recovery from a stall in any airplane becomes more difficult when its

A—center of gravity moves aft.
B—center of gravity moves forward.
C—elevator trim is adjusted nosedown.

The recovery from a stall in any airplane becomes progressively more difficult as its center of gravity moves aft. (H66) — AC 61-23C, Chapter 1

Flaps

Extending the flaps increases the wing camber and the angle of attack of a wing. This increases wing lift and also increases induced drag. The increased lift enables the pilot to make steeper approaches to a landing without an increase in airspeed. Spoilers, unlike flaps, do not change the wing camber. Their primary purpose is to decrease or "spoil" the lift (increase drag) of the wing. See Figure 1-21.

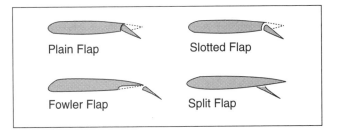

Figure 1-21. Use of flaps increases lift and drag

Answers

5196 [A] 5204 [A] 5211 [C] 5206 [A]

Chapter 1 **Basic Aerodynamics**

AIR, GLI
5182. One of the main functions of flaps during the approach and landing is to

A—decrease the angle of descent without increasing the airspeed.
B—provide the same amount of lift at a slower airspeed.
C—decrease lift, thus enabling a steeper-than-normal approach to be made.

Extending the flaps increases wing lift and also increases induced drag. The increased drag enables the pilot to make steeper approaches without an increase in airspeed. (H305) — AC 61-23C, Chapter 1

Answer (A) is incorrect because the flaps will increase the angle of descent without increasing the airspeed. Answer (C) is incorrect because the flaps increase lift and induced drag.

GLI
5276. The primary purpose of wing spoilers is to decrease

A—the drag.
B—landing speed.
C—the lift of the wing.

Deployable spoilers added to the upper wing surface "spoil" lift. These devices, when deployed, also increase drag. (N20) — Soaring Flight Manual

GLI
5282. Both lift and drag would be increased when which of these devices are extended?

A—Flaps.
B—Spoilers.
C—Slats.

Spoilers decrease a wing's lift whereas flaps increase lift. Both spoilers and flaps increase the drag. (N20) — Soaring Flight Manual

Wing Shapes

It is desirable for a wing to stall in the root area before it stalls near the tips. This provides a warning of an impending stall and allows the ailerons to be effective during the stall. A rectangular wing stalls in the root area first. The stall begins near the tip and progresses inboard on a highly tapered wing, or a wing with sweepback. *See* Figure 1-22.

Aspect ratio is the ratio of the span of an aircraft wing to its mean, or average, chord. Generally speaking, the higher the aspect ratio, the more efficient the wing. But for practical purposes, structural considerations normally limit the aspect ratio for all except high-performance sailplanes. A high-aspect-ratio wing has a low stall speed, and at a constant air velocity and high angle of attack, it has less drag than a low-aspect-ratio wing.

AIR, GLI
5197. A rectangular wing, as compared to other wing planforms, has a tendency to stall first at the

A—wingtip, with the stall progression toward the wing root.
B—wing root, with the stall progression toward the wing tip.
C—center trailing edge, with the stall progression outward toward the wing root and tip.

The rectangular wing has a tendency to stall first at the wing root with the stall pattern progressing outward to the tip. This type of stall pattern decreases undesirable rolling tendencies and increases lateral control when approaching a stall. (H66) — AC 61-23C, Chapter 1

Answers

5182 [B]   5276 [C]   5282 [A]   5197 [B]

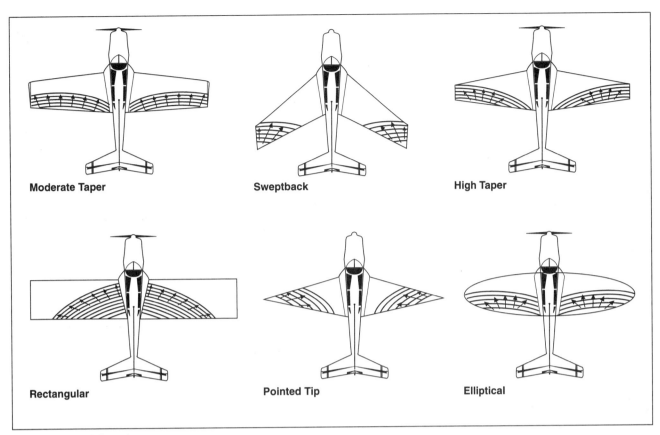

Figure 1-22. Wing shapes

Torque

In an airplane of standard configuration there is an inherent tendency for the airplane to turn to the left. This tendency, called "torque," is a combination of the following four forces:

- reactive force
- spiraling slipstream
- gyroscopic precession
- "P" factor

Spiraling slipstream is the only one addressed in this test. The spiraling slipstream is the reaction of the air to a rotating propeller, which forces the air to spiral in a clockwise direction around the fuselage. This spiraling slipstream tends to rotate the airplane right around the longitudinal axis.

Chapter 1 Basic Aerodynamics

AIR
5238. A propeller rotating clockwise as seen from the rear, creates a spiraling slipstream that tends to rotate the airplane to the

A—right around the vertical axis, and to the left around the longitudinal axis.
B—left around the vertical axis, and to the right around the longitudinal axis.
C—left around the vertical axis, and to the left around the longitudinal axis.

The slipstream strikes the vertical fin on the left causing a yaw to the left, at the same time it causes a rolling moment to the right. (H66) — AC 61-23C, Chapter 1

Ground Effect

Ground effect occurs when flying within one wingspan or less above the surface. The airflow around the wing and wing tips is modified and the resulting pattern reduces the downwash which reduces the induced drag. These changes can result in an aircraft either becoming airborne before reaching recommended takeoff speed or floating during an approach to land. *See* Figure 1-23.

An airplane leaving ground effect after takeoff will require an increase in angle of attack to maintain the same lift coefficient, which in turn will cause an increase in induced drag and therefore, require increased thrust.

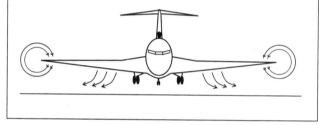

Figure 1-23. Ground effect phenomenon

AIR
5209. An airplane leaving ground effect will

A—experience a reduction in ground friction and require a slight power reduction.
B—experience an increase in induced drag and require more thrust.
C—require a lower angle of attack to maintain the same lift coefficient.

As the wing encounters ground effect and is maintained at a constant lift coefficient, there is a reduction in the upwash, downwash, and wing-tip vortices. This causes a reduction in induced drag. While in ground effect, the airplane requires less thrust to maintain lift. It will also require a lower angle of attack. When an airplane leaves ground effect, there is an increase in drag which will require a higher angle of attack. Additional thrust will be required to compensate for the loss. (H66) — AC 61-23C, Chapter 4

Answer (A) is incorrect because ground friction is reduced when breaking ground. Answer (C) is incorrect because a higher angle of attack is required to maintain the same lift coefficient when leaving the ground.

AIR, GLI
5216. If the same angle of attack is maintained in ground effect as when out of ground effect, lift will

A—increase, and induced drag will decrease.
B—decrease, and parasite drag will increase.
C—increase, and induced drag will increase.

If the airplane is brought into ground effect with a constant angle of attack, it will experience an increase in lift coefficient and a reduction in the thrust required. The reduction of the wing-tip vortices due to ground effect alters the spanwise lift distribution and reduces the induced flow. The reduction in induced flow causes a significant reduction in induced drag, but has no direct effect on parasite drag. (H66) — AC 61-23C, Chapter 4

Answers

5238 [B]   5209 [B]   5216 [A]

AIR
5224. To produce the same lift while in ground effect as when out of ground effect, the airplane requires

A—a lower angle of attack.
B—the same angle of attack.
C—a greater angle of attack.

If the airplane is brought into ground effect with a constant angle of attack, it will experience an increase in lift coefficient and a reduction in the thrust required. The reduction of the wing-tip vortices due to ground effect alters the spanwise lift distribution and reduces the induced flow. The reduction in induced flow causes a significant reduction in induced drag, but has no direct effect on parasite drag. (H66) — AC 61-23C, Chapter 4

Wake Turbulence

All aircraft leave behind two types of wake turbulence: Prop or jet blast, and wing-tip vortices.

Prop or jet blast could be hazardous to light aircraft on the ground behind large aircraft that are either taxiing or running-up their engines. In the air, prop or jet blast dissipates rapidly.

Wing-tip vortices are a by-product of lift. When a wing is flown at a positive angle of attack, a pressure differential is created between the upper and lower wing surfaces, and the pressure above the wing will be lower than the pressure below the wing. In attempting to equalize the pressure, air moves outward, upward, and around the wing tip, setting up a vortex which trails behind each wing. *See* Figure 1-24.

The strength of a vortex is governed by the weight, speed, and the shape of the wing of the generating aircraft. Maximum vortex strength occurs when the generating aircraft is heavy, clean, and slow.

Vortices generated by large aircraft in flight tend to sink below the flight path of the generating aircraft. A pilot should fly at or above the larger aircraft's flight path in order to avoid the wake turbulence created by the wing-tip vortices. *See* Figure 1-25.

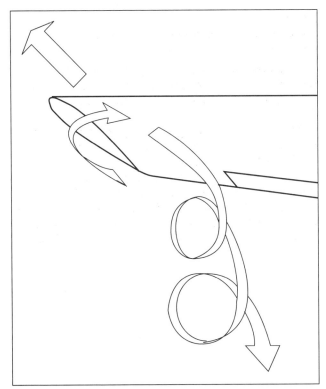

Figure 1-24. Wing-tip vortices

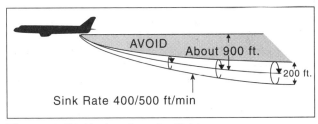

Figure 1-25. Vortices in cruise flight

Answers
5224 [A]

Chapter 1 **Basic Aerodynamics**

Close to the ground, vortices tend to move laterally. A crosswind will tend to hold the upwind vortex over the landing runway, while a tailwind may move the vortices of a preceding aircraft forward into the touchdown zone.

To avoid wake turbulence when landing, a pilot should note the point where a preceding large aircraft touched down and then land past that point. *See* Figure 1-26.

On takeoff, a pilot should lift off prior to reaching the rotation point of a preceding large aircraft; the flight path should then remain upwind and above the preceding aircraft's flight path. *See* Figure 1-27.

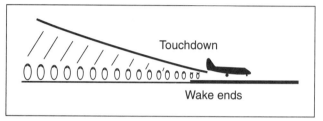

Figure 1-26. Touchdown and wake end

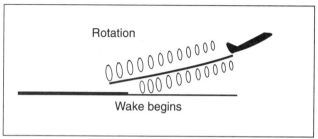

Figure 1-27. Rotation and wake beginning

ALL, MIL
5750. Choose the correct statement regarding wake turbulence.

A—Vortex generation begins with the initiation of the takeoff roll.
B—The primary hazard is loss of control because of induced roll.
C—The greatest vortex strength is produced when the generating airplane is heavy, clean, and fast.

Vortices sink at 400-500 fpm. Vortex generation begins when lift is being produced at takeoff. Greatest vortex strength is produced when the airplane is heavy, clean, and slow. The primary hazard is loss of control due to induced roll caused by the spinning vortices. (J27) — AIM ¶7-3-3

Answer (A) is incorrect because vortex generation begins at the rotation point when the airplane takes off. Answer (C) is incorrect because the greatest vortex strength is when the generating aircraft is heavy, clean, and slow.

ALL, MIL
5751. During a takeoff made behind a departing large jet airplane, the pilot can minimize the hazard of wingtip vortices by

A—being airborne prior to reaching the jet's flightpath until able to turn clear of its wake.
B—maintaining extra speed on takeoff and climbout.
C—extending the takeoff roll and not rotating until well beyond the jet's rotation point.

Vortices begin to form when the jet rotates. Plan to be off the runway prior to reaching the jet's point of rotation, then fly above or turn away from the jet's flight path. (J27) — AIM ¶7-3-4

ALL, MIL
5752. Which procedure should you follow to avoid wake turbulence if a large jet crosses your course from left to right approximately 1 mile ahead and at your altitude?

A—Make sure you are slightly above the path of the jet.
B—Slow your airspeed to V_A and maintain altitude and course.
C—Make sure you are slightly below the path of the jet and perpendicular to the course.

Fly above the jet's flight path whenever possible, because the vortices descend. Avoid flight below and behind a large aircraft's path. (J27) — AIM ¶7-3-3

Answers

5750 [B]   5751 [A]   5752 [A]

Chapter 1 Basic Aerodynamics

ALL, MIL
5753. To avoid possible wake turbulence from a large jet aircraft that has just landed prior to your takeoff, at which point on the runway should you plan to become airborne?

A—Past the point where the jet touched down.
B—At the point where the jet touched down, or just prior to this point.
C—Approximately 500 feet prior to the point where the jet touched down.

Vortices cease to be generated when the aircraft lands. Plan to become airborne beyond this point. (J27) — AIM ¶7-3-3

ALL, MIL
5754. When landing behind a large aircraft, which procedure should be followed for vortex avoidance?

A—Stay above its final approach flightpath all the way to touchdown.
B—Stay below and to one side of its final approach flightpath.
C—Stay well below its final approach flightpath and land at least 2,000 feet behind.

Stay at or above the large aircraft's final approach flight path. Note the touchdown point and land beyond it. (J27) — AIM ¶7-3-3

Glider Aerodynamics

GLI
5053. GIVEN:

Glider's maximum certificated
operating weight .... 1,140 lb
Towline breaking strength .... 3,050 lb

Which meets the requirement for one of the safety links? A breaking strength of

A—812 pounds installed where the towline is attached to the towplane.
B—912 pounds installed where the towline is attached to the glider.
C—2,300 pounds installed where the towline is attached to the glider.

The breaking strength of the safety link between the towline and the glider must be between 80% of the maximum operating weight of the glider and twice that weight (or it must be between 912 and 2,280 pounds). A towplane-towline safety link strength must be greater than the glider-towline link which, in this case, cannot be less than 912 pounds. (B12) — 14 CFR §91.309(a)(3)

GLI
5054. During aerotow of a glider that weighs 940 pounds, which towrope tensile strength would require the use of safety links at each end of the rope?

A—752 pounds.
B—1,500 pounds.
C—2,000 pounds.

A towline of up to 1,880 pounds breaking strength may be used without safety links in this case since this is twice the weight of the glider. (B12) — 14 CFR §91.309(a)(3)

GLI
5277. That portion of the glider's total drag created by the production of lift is called

A—induced drag, and is not affected by changes in airspeed.
B—induced drag, and is greatly affected by changes in airspeed.
C—parasite drag, and is greatly affected by changes in airspeed.

When a sailplane develops lift, the upward force on the sailplane results in an equal and downward force on the air which is then set in motion, generally downward. It takes energy to do this, and the transfer of energy from the sailplane to the air implies the existence of drag on the sailplane. This part of the total drag is called induced drag. The induced drag varies inversely with the square of the velocity. (N20) — Soaring Flight Manual

Answers

5753 [A]   5754 [A]   5053 [B]   5054 [C]   5277 [B]

Chapter 1 **Basic Aerodynamics**

GLI
5278. The best L/D ratio of a glider occurs when parasite drag is

A—equal to induced drag.
B—less than induced drag.
C—greater than induced drag.

L/D_{MAX} is the airspeed where parasite drag and induced drag are equal. (N20) — Soaring Flight Manual

GLI
5279. A glider is designed for an L/D ratio of 22:1 at 50 MPH in calm air. What would the approximate GLIDE RATIO be with a direct headwind of 25 MPH?

A—44:1.
B—22:1.
C—11:1.

$$\text{Glide ratio} = \frac{88 (V_C + V_W)}{R_S}$$

Where V_C = gliding airspeed
V_W = wind speed (+ = tailwind, − = head wind)
R_S = rate of sink

Thus, if $V_W = 0$ in calm air, $V_W = -25$ MPH, V_C is fixed at 50 MPH. Glide ratio would be reduced in the ratio of:

$$\frac{V_C + V_W}{V_C + V_W} = \frac{50 - 25}{50 - 0} = \frac{1}{2}$$

and would become 11:1.

(N20) — Soaring Flight Manual

GLI
5281. At a given airspeed, what effect will an increase in air density have on lift and drag of a glider?

A—Lift and drag will decrease.
B—Lift will increase but drag will decrease.
C—Lift and drag will increase.

Lift and drag vary directly with the density of the air. As air density increases, lift and drag increase. (N20) — Soaring Flight Manual

GLI
5283. If the airspeed of a glider is increased from 45 MPH to 90 MPH, the parasite drag will be

A—two times greater.
B—four times greater.
C—six times greater.

As speed increases, the amount of parasite drag increases as the square of the velocity. If the speed is doubled, four times as much drag is produced. (N20) — Soaring Flight Manual

GLI
5284. If the indicated airspeed of a glider is decreased from 90 MPH to 45 MPH, the induced drag will be

A—four times less.
B—two times greater.
C—four times greater.

The induced drag varies inversely with the square of the velocity. Airspeed is decreased, so induced drag will increase. Velocity is changed by a factor of two; induced drag will vary as two squared or a factor of four. (N20) — Soaring Flight Manual

GLI
5285. Which is true regarding wing camber of a glider's airfoil? The camber is

A—the same on both the upper and lower wing surface.
B—less on the upper wing surface than it is on the lower wing surface.
C—greater on the upper wing surface than it is on the lower wing surface.

Camber is the curvature of the upper and lower surfaces of an airfoil from the leading edge to the trailing edge. The upper surface normally has a greater camber than the lower wing surface. (N20) — Soaring Flight Manual

GLI
5286. If the glider's radius of turn is 175 feet at 40 MPH, what would the radius of turn be if the TAS is increased to 80 MPH while maintaining a constant angle of bank?

A—350 feet.
B—525 feet.
C—700 feet.

Answers

5278 [A]   5279 [C]   5281 [C]   5283 [B]   5284 [C]   5285 [C]
5286 [C]

Chapter 1 Basic Aerodynamics

At a given angle of bank, the radius of the turn increases in proportion to the square of the velocity. If the velocity is doubled, the radius will increase four times.

175 x 4 = 700 feet

(N20) — Soaring Flight Manual

GLI
5290. With regard to the effects of spoilers and wing flaps, which is true if the glider's pitch attitude is held constant when such devices are being operated? (Disregard negative flap angles above neutral position.) Retracting flaps

A—will reduce the glider's stall speed.
B—or extending spoilers will increase the glider's rate of descent.
C—or extending spoilers will decrease the glider's rate of descent.

Opening the spoilers causes the glider to sink faster while decelerating, whereas raising the flaps will increase the rate of descent without a speed increase. (N20) — Soaring Flight Manual

GLI
5291. If the angle of attack is increased beyond the critical angle of attack, the wing will no longer produce sufficient lift to support the weight of the glider

A—regardless of airspeed or pitch attitude.
B—unless the airspeed is greater than the normal stall speed.
C—unless the pitch attitude is on or below the natural horizon.

A glider can be stalled in any attitude and at any airspeed. (N20) — Soaring Flight Manual

GLI
5292. What force causes the glider to turn in flight?

A—Vertical component of lift.
B—Horizontal component of lift.
C—Positive yawing movement of the rudder.

The horizontal component of lift is the force that pulls the glider from a straight flight path to make it turn. (N20) — Soaring Flight Manual

GLI
5293. GIVEN:

Glider A
Wingspan .... 51 ft
Average wing chord .... 4 ft

Glider B
Wingspan .... 48 ft
Average wing chord .... 3.5 ft

Determine the correct aspect ratio and its effect on performance at low speeds.

A—Glider A has an aspect ratio of 13.7, and will generate less lift with greater drag than glider B.
B—Glider B has an aspect ratio of 13.7, and will generate greater lift with less drag than glider A.
C—Glider B has an aspect ratio of 12.7, and will generate less lift with greater drag than glider A.

Aspect ratio is defined as the ratio between the glider's span and the mean chord of its wings. High-aspect ratio in a glider is associated with a high glide ratio (higher lift/lower drag), other factors being equal.

Wingspan ÷ Average Wing Chord = Aspect Ratio
51 ÷ 4 = 12.75 (Glider A)
48 ÷ 3.5 = 13.7 (Glider B)

(N20) — Soaring Flight Manual

GLI
5294. GIVEN:

Glider A
Wingspan .... 48 ft
Average wing chord .... 4.5 ft

Glider B
Wingspan .... 54 ft
Average wing chord .... 3.7 ft

Determine the correct aspect ratio and its effect on performance at low speeds.

A—Glider A has an aspect ratio of 10.6, and will generate greater lift with less drag than will glider B.
B—Glider B has an aspect ratio of 14.5, and will generate greater lift with less drag than will glider A.
C—Glider B has an aspect ratio of 10.6, and will generate less lift with greater drag than will glider A.

Aspect ratio is defined as the ratio between the glider's span and the mean chord of its wings. High-aspect ratio in a glider is associated with a high glide ratio (higher lift/lower drag), other factors being equal.

Wingspan ÷ Average Wing Chord = Aspect Ratio
48 ÷ 4.5 = 10.67 (Glider A)
54 ÷ 3.7 = 14.59 (Glider B)

(N20) — Soaring Flight Manual

Answers

5290 [B]   5291 [A]   5292 [B]   5293 [B]   5294 [B]

Chapter 1 Basic Aerodynamics

GLI
5295. The best L/D ratio of a glider is a value that

A—varies depending upon the weight being carried.
B—remains constant regardless of airspeed changes.
C—remains constant and is independent of the weight being carried.

A heavily loaded glider goes forward and down faster than when lightly loaded. The glide ratios are the same for both loading conditions, but occur at different airspeeds. (N21) — Soaring Flight Manual

GLI
5296. A glide ratio of 22:1 with respect to the air mass will be

A—11:1 in a tailwind and 44:1 in a headwind.
B—22:1 regardless of wind direction and speed.
C—11:1 in a headwind and 44:1 in a tailwind.

Glide ratio may be taken as the ratio of forward to downward motion and is numerically the same as the lift to drag ratio (L/D). Within the air mass this value remains unchanged. However, glide ratio, when expressed in terms of motion over the ground, will vary as a function of wind velocity. (N21) — Soaring Flight Manual

Answers

5295 [C]   5296 [B]

Chapter 2
Aircraft Systems

Ignition System   *2–3*

Air/Fuel Mixture   *2–4*

Carburetor Ice   *2–6*

Aviation Fuel   *2–7*

Engine Temperatures   *2–8*

Propellers   *2–9*

Cold Weather Operations   *2–11*

Rotorcraft Systems   *2–12*

Glider Systems   *2–27*

Balloon Operations   *2–35*

Airship Operations   *2–41*

Airship IFR Operations   *2–43*

Ignition System

Most reciprocating engines used to power small aircraft incorporate two separate magneto ignition systems. The primary advantages of the dual ignition system are increased safety and improved engine performance.

A magneto ("mag") is a self-contained source of electrical energy, so even if an aircraft loses total electric power, the engine will continue to run. For electrical energy, magnetos depend upon a rotating magnet and a coil.

When checking for magneto operation prior to flight, the engine should run smoothly when operating with the magneto selector set on BOTH, and should experience a slight drop in revolutions per minute (RPM) when running on only one or the other magneto. The drop in RPM is caused by reduced efficiency of a single spark plug, as opposed to two.

If the ground wire between the magneto and the ignition switch becomes disconnected or broken, the engine cannot be shut down by turning off the ignition switch.

AIR, RTC
5169. Before shutdown, while at idle, the ignition key is momentarily turned OFF. The engine continues to run with no interruption; this

A—is normal because the engine is usually stopped by moving the mixture to idle cut-off.
B—should not normally happen. Indicates a magneto not grounding in OFF position.
C—is an undesirable practice, but indicates that nothing is wrong.

If the magneto switch ground wire is disconnected, the magneto is ON even though the ignition switch is in the OFF position. The engine could fire if the propeller is moved from outside the airplane. (H307) — AC 61-23C, Chapter 2

Answer (A) is incorrect because the engine should stop when the ignition key is turned to the OFF position. Answer (C) is incorrect because this indicates there is a faulty ground wire.

AIR, RTC
5171. A way to detect a broken magneto primary grounding lead is to

A—idle the engine and momentarily turn the ignition off.
B—add full power, while holding the brakes, and momentarily turn off the ignition.
C—run on one magneto, lean the mixture, and look for a rise in manifold pressure.

If the magneto switch ground wire is disconnected, the magneto is ON even though the ignition switch is in the OFF position. The engine could fire if the propeller is moved from outside the airplane. (H307) — AC 61-23C, Chapter 2

Answer (B) is incorrect because it is not necessary to add full power when performing the check. Answer (C) is incorrect because the way to detect a broken magneto ground wire is to turn the ignition to the OFF position; if the engine continues to run, the problem is confirmed.

AIR, RTC
5173. The most probable reason an engine continues to run after the ignition switch has been turned off is

A—carbon deposits glowing on the spark plugs.
B—a magneto ground wire is in contact with the engine casing.
C—a broken magneto ground wire.

If the magneto switch ground wire is disconnected, the magneto is ON even though the ignition switch is in the OFF position. The engine could fire if the propeller is moved from outside the airplane. (H307) — AC 61-23C, Chapter 2

Answer (A) is incorrect because glowing carbon deposits is a result of preignition. Answer (B) is incorrect because a magneto ground wire should be in contact with the engine casing to provide grounding.

Answers

5169 [B]　　　5171 [A]　　　5173 [C]

Chapter 2 Aircraft Systems

AIR, RTC
5174. If the ground wire between the magneto and the ignition switch becomes disconnected, the engine

A—will not operate on one magneto.
B—cannot be started with the switch in the BOTH position.
C—could accidentally start if the propeller is moved with fuel in the cylinder.

If the magneto switch ground wire is disconnected, the magneto is ON even though the ignition switch is in the OFF position. The engine could fire if the propeller is moved from outside the airplane. (H307) — AC 61-23C, Chapter 2

Answer (A) is incorrect because both magnetos remain on when the ground wire is disconnected. Answer (B) is incorrect because the engine can still be started, and the magnetos cannot be turned off.

Air/Fuel Mixture

Carburetors are normally set to deliver the correct air/fuel mixture (air/fuel ratio) at sea level. This air/fuel ratio is the ratio of the weight of fuel to the weight of air entering the cylinder. This ratio is determined by the setting of the mixture control in both fuel injection and carburetor-equipped engines.

When climbing, the mixture control allows the pilot to decrease the fuel flow as altitude increases (air density decreases), thus maintaining the correct mixture (air/fuel ratio). If the fuel flow is allowed to remain constant by not leaning the mixture, the fuel/air ratio becomes too rich, as the density (weight per unit volume) of air decreases with increased altitude, resulting in a loss of efficiency. Operating with an excessively rich mixture may cause fouling of spark plugs.

When descending, air density increases. Unless fuel flow is increased, the mixture may become excessively lean; i.e., the weight of fuel is too low for the weight of air reaching the cylinders. This may result in the creation of high temperatures and pressures.

The best power mixture is the air/fuel ratio from which the most power can be obtained for any given throttle setting.

AIR, RTC, LTA
5172. Fouling of spark plugs is more apt to occur if the aircraft

A—gains altitude with no mixture adjustment.
B—descends from altitude with no mixture adjustment.
C—throttle is advanced very abruptly.

If the fuel/air mixture is too rich, excessive fuel consumption, rough engine operation, and appreciable loss of power will occur. Because of excessive fuel, a cooling effect takes place which causes below normal temperatures in the combustion chambers. This cooling results in spark plug fouling. Unless the mixture is leaned with a gain in altitude, the mixture becomes excessively rich. (H307) — AC 61-23C, Chapter 2

Answer (B) is incorrect because descending without a mixture adjustment (operating with an excessively lean mixture) would result in overheating, rough engine operation, a loss of power, and detonation. Answer (C) is incorrect because advancing the throttle abruptly may cause the engine to hesitate or stop.

AIR, RTC, LTA
5176. The pilot controls the air/fuel ratio with the

A—throttle
B—manifold pressure
C—mixture control

The fuel/air ratio of the combustible mixture delivered to the engine is controlled by the mixture control. (H307) — AC 61-23C, Chapter 2

Answer (A) is incorrect because the throttle regulates the total volume of fuel and air entering the combustion chamber. Answer (B) is incorrect because the manifold pressure indicates the engine's power output.

Answers
5174 [C]   5172 [A]   5176 [C]

AIR, RTC, LTA
5187. Fuel/air ratio is the ratio between the

A— volume of fuel and volume of air entering the cylinder.
B— weight of fuel and weight of air entering the cylinder.
C— weight of fuel and weight of air entering the carburetor.

The mixture control is used to change the fuel to air mixture entering the combustion chamber (cylinder). Fuel-to-air ratio is the weight of fuel to a given weight of air. (H51) — AC 61-23C, Chapter 2

Answer (A) is incorrect because, as altitude increases, the amount of air in a fixed volume decreases. Answer (C) is incorrect because the carburetor is where the fuel/air ratio is established prior to entering the cylinders.

AIR, RTC, LTA
5188. The mixture control can be adjusted, which

A— prevents the fuel/air combination from becoming too rich at higher altitudes.
B— regulates the amount of air flow through the carburetor's venturi.
C— prevents the fuel/air combination from becoming lean as the airplane climbs.

As the aircraft climbs, the fuel/air mixture becomes richer and the excessive fuel causes the engine to lose power and to run rougher. The mixture control provides a means for the pilot to decrease fuel to compensate for this imbalance in mixture as altitude increases. (H307) — AC 61-23C, Chapter 2

Answer (B) is incorrect because the throttle regulates the airflow through the carburetor's venturi. Answer (C) is incorrect because the fuel/air ratio becomes richer as the aircraft climbs.

AIR, RTC, LTA
5298. The best power mixture is that fuel/air ratio at which

A— cylinder head temperatures are the coolest.
B— the most power can be obtained for any given throttle setting.
C— a given power can be obtained with the highest manifold pressure or throttle setting.

The throttle setting determines the amount of air flowing into the engine. The mixture control is then adjusted to get the best fuel/air ratio, resulting in the best power the engine can develop at this particular throttle setting. (H51) — AC 61-23C, Chapter 2

Answer (A) is incorrect because the cylinder heads will be the coolest when mixture is richest. Answer (C) is incorrect because this describes the highest power setting.

AIR, RTC, LTA
5608. What will occur if no leaning is made with the mixture control as the flight altitude increases?

A— The volume of air entering the carburetor decreases and the amount of fuel decreases.
B— The density of air entering the carburetor decreases and the amount of fuel increases.
C— The density of air entering the carburetor decreases and the amount of fuel remains constant.

Fuel flow remains constant if no adjustments are made. The same volume of air goes into the carburetor, but the weight and density of the air is less, causing an excessively rich mixture, which causes spark plug fouling and decreased power. (H307) — AC 61-23C, Chapter 2

AIR, RTC, LTA
5609. Unless adjusted, the fuel/air mixture becomes richer with an increase in altitude because the amount of fuel

A— decreases while the volume of air decreases.
B— remains constant while the volume of air decreases.
C— remains constant while the density of air decreases.

Fuel flow remains constant if no adjustments are made. The same volume of air goes into the carburetor, but the weight and density of the air is less, causing an excessively rich mixture, which causes spark plug fouling and decreased power. (H307) — AC 61-23C, Chapter 2

AIR, RTC, LTA
5610. The basic purpose of adjusting the fuel/air mixture control at altitude is to

A— decrease the fuel flow to compensate for decreased air density.
B— decrease the amount of fuel in the mixture to compensate for increased air density.
C— increase the amount of fuel in the mixture to compensate for the decrease in pressure and density of the air.

Fuel flow remains constant if no adjustments are made. The same volume of air goes into the carburetor, but the weight and density of the air is less, causing an excessively rich mixture, which causes spark plug fouling and decreased power. (H307) — AC 61-23C, Chapter 2

Answers
5187 [B]   5188 [A]   5298 [B]   5608 [C]   5609 [C]   5610 [A]

Chapter 2 Aircraft Systems

AIR, RTC, LTA
5611. At high altitudes, an excessively rich mixture will cause the

A—engine to overheat.
B—fouling of spark plugs.
C—engine to operate smoother even though fuel consumption is increased.

Fuel flow remains constant if no adjustments are made. The same volume of air goes into the carburetor, but the weight and density of the air is less, causing an excessively rich mixture, which causes spark plug fouling and decreased power. (H307) — AC 61-23C, Chapter 2

Answer (A) is incorrect because a lean mixture will cause the engine to overheat. Answer (C) is incorrect because an engine runs smoother when the mixture is adjusted for the altitude.

Carburetor Ice

As air flows through a carburetor, it expands rapidly. At the same time, fuel entering the airstream is vaporized. Expansion of the air and vaporization of the fuel causes a sudden cooling of the mixture which may cause ice to form inside the carburetor. The possibility of icing should always be considered when operating in conditions where the outside air temperature is between 20°F and 70°F and the relative humidity is high.

Carburetor heat preheats the air before it enters the carburetor and either prevents carburetor ice from forming or melts any ice which may have formed. When carburetor heat is applied, the heated air that enters the carburetor is less dense. This causes the air/fuel mixture to become enriched, and this in turn decreases engine output (less engine horsepower) and increases engine operating temperatures.

During engine run-up, prior to departure from a high-altitude airport, the pilot may notice a slight engine roughness which is not affected by the magneto check but grows worse during the carburetor heat check. In this case the air/fuel mixture may be too rich due to the lower air density at the high altitude, and applying carburetor heat decreases the air density even more. A leaner setting of the mixture control may correct this problem.

In an airplane with a fixed-pitch propeller, the first indication of carburetor ice will likely be a decrease in RPM as the air supply is choked off. Application of carburetor heat will decrease air density, causing the RPM to drop even lower. Then, as the carburetor ice melts, the RPM will rise gradually.

Fuel injection systems, which do not utilize a carburetor, are generally considered to be less susceptible to icing than carburetor systems are.

AIR, RTC, LTA
5170. Leaving the carburetor heat on while taking off

A—leans the mixture for more power on takeoff.
B—will decrease the takeoff distance.
C—will increase the ground roll.

Use of carburetor heat enriches the mixture, which tends to reduce the output of the engine and also increases the operating temperature. Therefore, the heat should not be used when full power is required (such as during takeoff) or during normal engine operations except to check for the presence of, or removal of carburetor ice. A decrease in engine output will increase distance required to reach lift off speed. Therefore, it will increase ground roll. (H307) — AC 61-23C, Chapter 2

AIR, RTC, LTA
5189. Which statement is true concerning the effect of the application of carburetor heat?

A—It enriches the fuel/air mixture.
B—It leans the fuel/air mixture.
C—It has no effect on the fuel/air mixture.

Use of carburetor heat enriches the mixture which tends to reduce the output of the engine and also increases the operating temperature. (H307) — AC 61-23C, Chapter 2

Answers

5611 [B]   5170 [C]   5189 [A]

Chapter 2 **Aircraft Systems**

AIR, RTC, LTA
5606. Applying carburetor heat will

A—not affect the mixture.
B—lean the fuel/air mixture.
C—enrich the fuel/air mixture.

Use of carburetor heat enriches the mixture which tends to reduce the output of the engine and also increases the operating temperature. (H307) — AC 61-23C, Chapter 2

Aviation Fuel

Fuel does two things for the engine; it acts both as an agent for combustion and as an agent for cooling (based on the mixture setting of the engine).

Aviation fuel is available in several grades. The proper grade for a specific engine will be listed in the aircraft flight manual. If the proper grade of fuel is not available, it is possible to use the next higher grade. A lower grade of fuel should never be used.

The use of low-grade fuel or a too lean air/fuel mixture may cause detonation, which is the uncontrolled spontaneous explosion of the mixture in the cylinder, instead of burning progressively and evenly. Detonation produces extreme heat.

Preignition is the premature uncontrolled firing of the fuel/air mixture. It is caused by an incandescent area (such as a carbon or lead deposit heated to a red hot glow) serving as an ignitor in advance of normal ignition.

Fuel can be contaminated by water and/or dirt. The air inside the aircraft fuel tanks can cool at night, which allows formation of water droplets (through condensation) on the insides of the fuel tanks. These droplets then fall into the fuel. To avoid this problem, always fill the tanks completely when parking overnight.

Thoroughly drain all of the aircraft's sumps, drains, and strainers before a flight to get rid of any water that may have collected.

Dirt can get into the fuel if refueling equipment is poorly maintained or if the refueling operation is sloppy. Use care when refueling an aircraft.

On aircraft equipped with fuel pumps, the practice of running a fuel tank dry before switching tanks is considered unwise because the engine-driven fuel pump or electric fuel boost pump may draw air into the fuel system and cause vapor lock.

AIR, RTC, LTA
5185-1. Detonation may occur at high-power settings when

A—the fuel mixture ignites instantaneously instead of burning progressively and evenly.
B—an excessively rich fuel mixture causes an explosive gain in power.
C—the fuel mixture is ignited too early by hot carbon deposits in the cylinder.

Detonation or knock is a sudden explosion or shock to a small area of the piston top, rather than the normal smooth burn in the combustion chamber. (H307) — AC 61-23C, Chapter 2

Answer (B) is incorrect because detonation may occur with an excessively lean fuel mixture and a loss in power. Answer (C) is incorrect because this describes preignition.

AIR, RTC, LTA
5185-2. Detonation can be caused by

A—a "rich" mixture.
B—low engine temperatures.
C—using a lower grade of fuel than recommended.

Detonation can be caused by low grade fuel or a too lean air/fuel mixture. (H307) — AC 61-23C, Chapter 2

Answers

5606 [C]   5185-1 [A]   5185-2 [C]

Chapter 2 **Aircraft Systems**

AIR, RTC, LTA
5186. The uncontrolled firing of the fuel/air charge in advance of normal spark ignition is known as

A—instantaneous combustion.
B—detonation.
C—pre-ignition.

When the cylinder head gets too hot, it can ignite the fuel/air mixture before the spark. This condition is called preignition. (H51) — AC 61-23C, Chapter 2

Answers (A) and (B) are incorrect because detonation is an instantaneous combustion of the fuel/air mixture, which is caused by using too lean a mixture, using too low a grade of fuel, or operating with temperatures that are too high.

AIR, RTC, LTA
5190. Detonation occurs in a reciprocating aircraft engine when

A—there is an explosive increase of fuel caused by too rich a fuel/air mixture.
B—the spark plugs receive an electrical jolt caused by a short in the wiring.
C—the unburned fuel/air charge in the cylinders is subjected to instantaneous combustion.

Detonation is a sudden explosion, or instantaneous combustion, of the fuel/air mixture in the cylinders, producing extreme heat and severe structural stresses on the engine. (H51) — AC 61-23C, Chapter 2

Answer (A) is incorrect because detonation is caused by too lean a mixture. Answer (B) is incorrect because detonation does not have anything to do with the wiring.

AIR, RTC, LTA
5299. Detonation can be caused by

A—a "rich" mixture.
B—low engine temperatures.
C—using a lower grade fuel than recommended.

Detonation is a sudden explosion or shock to a small area of the piston top, rather than the normal smooth burn in the combustion chamber. It can be caused by low grade fuel or a lean mixture. (H51) — AC 61-23C, Chapter 2

Engine Temperatures

Most light aircraft engines are cooled externally by air. For internal cooling and lubrication, an engine depends on circulating oil. Engine lubricating oil not only prevents direct metal-to-metal contact of moving parts, it also absorbs and dissipates part of the engine heat produced by internal combustion. If the engine oil level is too low, an abnormally high engine oil temperature indication may result.

On the ground or in the air, excessively high engine temperatures can cause excessive oil consumption, loss of power, and possible permanent internal engine damage.

If the engine oil temperature and cylinder head temperature gauges have exceeded their normal operating range, or if the pilot suspects that the engine (with a fixed-pitch propeller) is detonating during climb-out, the pilot may have been operating with either too much power and the mixture set too lean, using fuel of too low a grade, or operating the engine with not enough oil in it. Reducing the rate of climb and increasing airspeed, enriching the fuel mixture, or retarding the throttle will help cool an overheating engine. Also, rapid throttle operation can induce detonation, which may detune the crankshaft.

The most important rule to remember in the event of a power failure after becoming airborne is to maintain safe airspeed.

AIR
5175. For internal cooling, reciprocating aircraft engines are especially dependent on

A—a properly functioning cowl flap augmenter.
B—the circulation of lubricating oil.
C—the proper freon/compressor output ratio.

Lubricating oil serves two purposes:

1. *It furnishes a coating of oil over the surfaces of the moving parts, preventing metal-to-metal contact and the generation of heat; and*

Answers
5186 [C]   5190 [C]   5299 [C]   5175 [B]

2. It absorbs and dissipates, through the oil cooling system, part of the engine heat produced by the internal combustion process.

(H307) — AC 61-23C, Chapter 2

Answer (A) is incorrect because although cowl flaps aid internal cooling, they are not the primary cooling source. Answer (C) is incorrect because the proper freon/compressor output ratio controls cabin cooling.

AIR
5271. A detuning of engine crankshaft counterweights is a source of overstress that may be caused by

A—rapid opening and closing of the throttle.
B—carburetor ice forming on the throttle valve.
C—operating with an excessively rich fuel/air mixture.

Rapid throttle operation can induce detonation, which may detune the crankshaft. (K20) — AC 20-103

Answer (B) is incorrect because carburetor ice can cause the engine to stop running, but it will not affect the engine crankshaft counterweights. Answer (C) is incorrect because operating with an excessively rich mixture fouls the spark plugs, but does not affect the crankshaft.

AIR, RTC, LTA
5607. An abnormally high engine oil temperature indication may be caused by

A—a defective bearing.
B—the oil level being too low.
C—operating with an excessively rich mixture.

The oil pressure indication varies inversely with the oil temperature. High temperature and low-pressure usually indicate low oil level. (H307) — AC 61-23C, Chapter 2

Answer (A) is incorrect because a defective bearing will increase metal particles in the oil, but will not significantly affect the oil temperature. Answer (C) is incorrect because a rich mixture results in lower engine operating temperatures; therefore, it would not increase engine oil temperature.

Propellers

The propeller is a rotating airfoil which produces thrust by creating a positive dynamic pressure, usually on the engine side.

When a propeller rotates, the tips travel at a greater speed than the hub. To compensate for the greater speed at the tips, the blades are twisted slightly. The propeller blade angles decrease from the hub to the tips with the greatest angle of incidence, or highest pitch, at the hub and the smallest at the tip. This produces a relatively uniform angle of attack (uniform lift) along the blade's length in cruise flight.

No propeller is 100% efficient. There is always some loss of power when converting engine output into thrust. This loss is primarily due to propeller slippage. A propeller's efficiency is the ratio of thrust horsepower (propeller output) to brake horsepower (engine output). A fixed propeller will have a peak (best) efficiency at only one combination of airspeed and RPM.

A constant-speed (controllable-pitch) propeller allows the pilot to select the most efficient propeller blade angle for each phase of flight. In this system, the throttle controls the power output as registered on the manifold pressure gauge, and the propeller control regulates the engine RPM (propeller RPM). The pitch angle of the blades is changed by governor regulated oil pressure which keeps engine speed at a constant selected RPM. A constant-speed propeller allows the pilot to select a small propeller blade angle (flat pitch) and high RPM to develop maximum power and thrust for takeoff.

To reduce the engine output to climb power after takeoff, a pilot should decrease the manifold pressure. The RPM is decreased by increasing the propeller blade angle. When the throttle is advanced (increased) during cruise, the propeller pitch angle will automatically increase to allow engine RPM to remain the same. A pilot should avoid a high manifold pressure setting with low RPM on engines equipped with a constant-speed propeller to prevent placing undue stress on engine components. To avoid high manifold pressure combined with low RPM, the manifold pressure should be reduced before reducing RPM when decreasing power settings, and the RPM increased before increasing the manifold pressure when increasing power settings.

Answers

5271 [A]   5607 [B]

Chapter 2 Aircraft Systems

AIR
5183. Which statement best describes the operating principle of a constant-speed propeller?

A—As throttle setting is changed by the pilot, the prop governor causes pitch angle of the propeller blades to remain unchanged.
B—A high blade angle, or increased pitch, reduces the propeller drag and allows more engine power for takeoffs.
C—The propeller control regulates the engine RPM and in turn the propeller RPM.

The propeller control regulates the engine RPM and in turn, the propeller RPM. The RPM is registered on the tachometer. (H51) — AC 61-23C, Chapter 2

Answer (A) is incorrect because the prop governor causes the pitch angle of the prop blades to change to help maintain a specified airspeed. Answer (B) is incorrect because a high blade angle will increase propeller drag with less engine power.

AIR
5184. In aircraft equipped with constant-speed propellers and normally-aspirated engines, which procedure should be used to avoid placing undue stress on the engine components? When power is being

A—decreased, reduce the RPM before reducing the manifold pressure.
B—increased, increase the RPM before increasing the manifold pressure.
C—increased or decreased, the RPM should be adjusted before the manifold pressure.

Power change procedure on a constant-speed propeller is to increase RPM before manifold pressure. To decrease power, reduce manifold pressure before reducing RPM. This will help avoid placing undue stress on the engine components. (H308) — AC 61-23C, Chapter 2

Answers (A) and (C) are incorrect because when power is being decreased, the manifold pressure should be reduced before reducing the RPM.

AIR
5235. Propeller efficiency is the

A—ratio of thrust horsepower to brake horsepower.
B—actual distance a propeller advances in one revolution.
C—ratio of geometric pitch to effective pitch.

Since the efficiency of any machine is the ratio of useful power output to actual power input, propeller efficiency is the ratio of thrust horsepower to brake horsepower. (H66) — AC 61-23C, Chapter 17

Answer (B) is incorrect because effective pitch is the actual distance a propeller advances in one revolution. Answer (C) is incorrect because the ratio of geometric pitch to effective pitch is called slippage.

AIR
5236. A fixed-pitch propeller is designed for best efficiency only at a given combination of

A—altitude and RPM.
B—airspeed and RPM.
C—airspeed and altitude.

Fixed-pitch propellers are most efficient only at a given combination of airspeed and RPM. (H66) — AC 61-23C, Chapter 17

Answers (A) and (C) are incorrect because altitude does not affect propeller efficiency.

AIR
5237. The reason for variations in geometric pitch (twisting) along a propeller blade is that it

A—permits a relatively constant angle of incidence along its length when in cruising flight.
B—prevents the portion of the blade near the hub from stalling during cruising flight.
C—permits a relatively constant angle of attack along its length when in cruising flight.

Twisting, or variations in the geometric pitch of the blades, permits the propeller to operate with a relatively constant angle of attack along its length when in cruising flight. (H66) — AC 61-23C, Chapter 17

Answer (A) is incorrect because variations in geometric pitch permit a constant angle of attack along its length. Answer (B) is incorrect because the propeller tips would be stalled during cruising flight if there was no variation in geometric pitch.

AIR
5654. To establish a climb after takeoff in an aircraft equipped with a constant-speed propeller, the output of the engine is reduced to climb power by decreasing manifold pressure and

A—increasing RPM by decreasing propeller blade angle.
B—decreasing RPM by decreasing propeller blade angle.
C—decreasing RPM by increasing propeller blade angle.

Answers
5183 [C]   5184 [B]   5235 [A]   5236 [B]   5237 [C]   5654 [C]

A low-pitch, high RPM setting is utilized to obtain maximum power for takeoff. Then, after the airplane is airborne, an increasing blade angle/pitch will cause lower RPM which provides adequate thrust and better economy while maintaining the proper airspeed. (H308) — AC 61-23C, Chapter 2

Answer (A) is incorrect because RPM is decreased to reduce power, by increasing propeller blade angle. Answer (B) is incorrect because blade angle must be increased to decrease RPM.

AIR
5667. To develop maximum power and thrust, a constant-speed propeller should be set to a blade angle that will produce a

A—large angle of attack and low RPM.
B—small angle of attack and high RPM.
C—large angle of attack and high RPM.

Smaller angle of attack makes the blades take smaller amounts of air, which in turn allows the engine to run at higher RPM, producing more power. (H66) — AC 61-23C, Chapter 17

AIR
5668. For takeoff, the blade angle of a controllable-pitch propeller should be set at a

A—small angle of attack and high RPM.
B—large angle of attack and low RPM.
C—large angle of attack and high RPM.

Smaller angle of attack makes the blades take smaller amounts of air, which in turn allows the engine to run at higher RPM, producing more power. (H66) — AC 61-23C, Chapter 17

Cold Weather Operations

At low temperatures, changes occur in the viscosity of engine oil, batteries can lose a high percentage of their effectiveness, instruments can stick, and warning lights can stick in the pushed position when "pushed to test." Therefore, preheating the engines, as well as the cockpit, before starting is advisable in low temperatures. The pilot should also be aware that at extremely low temperatures, the engine can develop more than rated takeoff power even though the manifold pressure (MAP) and RPM readings are normal.

Overpriming is a frequent cause of difficult starting in cold weather because oil is washed off the cylinder walls and poor compression results. The manufacturer's instructions should be followed for starting an overprimed engine.

During cold weather preflight operations, be sure to check the oil breather lines. The vapors caused by combustion may condense, then freeze, clogging these lines.

Since most aircraft heaters work by using the engine to heat outside air, a pilot should frequently inspect a manifold type heating system to minimize the possibility of hazardous exhaust gases leaking into the cockpit.

AIR, RTC
5653. Frequent inspections should be made of aircraft exhaust manifold-type heating systems to minimize the possibility of

A—exhaust gases leaking into the cockpit.
B—a power loss due to back pressure in the exhaust system.
C—a cold-running engine due to the heat withdrawn by the heater.

Carbon monoxide poisoning from exhaust gases leaking into the cockpit from a faulty exhaust manifold has been linked to several fatal aircraft accidents. (H351) — AC 61-23C, Chapter 9

Answer (B) is incorrect because a leak in the exhaust system would decrease back pressure. Answer (C) is incorrect because engine temperature is not affected by heat withdrawn by the heater.

Answers

5667 [B]   5668 [A]   5653 [A]

Chapter 2 **Aircraft Systems**

AIR, RTC

5766. During preflight in cold weather, crankcase breather lines should receive special attention because they are susceptible to being clogged by

A—congealed oil from the crankcase.
B—moisture from the outside air which has frozen.
C—ice from crankcase vapors that have condensed and subsequently frozen.

The crankcase breather requires special consideration when preparing for cold weather. Frozen breather lines can create numerous problems. When crankcase vapors cool, they may condense in the breather line and subsequently freeze it closed. Special care is recommended during the preflight to ensure that the breather system is free of ice. (L52) — AC 91-13

Answer (A) is incorrect because oil that lays in the bottom of the crankcase never gets into the breather lines. Answer (B) is incorrect because a low air temperature is usually associated with a low moisture content.

AIR, RTC

5767. Which is true regarding preheating an aircraft during cold weather operations?

A—The cabin area as well as the engine should be preheated.
B—The cabin area should not be preheated with portable heaters.
C—Hot air should be blown directly at the engine through the air intakes.

Low temperatures may cause a change in the viscosity of engine oils, batteries may lose a high percentage of their effectiveness, and instruments may stick. Because of the above, preheating the engines as well as the cabin before starting is desirable in low temperatures. Extreme caution should be used in the preheat process to avoid fire. (L52) — AC 91-13

Rotorcraft Systems

The pitch angle of a helicopter rotor blade is the acute angle between the chord line of the blade and the plane of rotor rotation.

In a semirigid rotor system, the rotor blades are rigidly interconnected to the hub, but the hub is free to tilt and rock with respect to the rotor shaft. In this system in which only two-bladed rotors are used, the blades flap as a unit; that is, as one blade flaps up, the other blade flaps down an equal amount. The hinge which permits the flapping or see-saw effect is called a teetering hinge.

A rocking hinge, perpendicular to the teetering hinge and parallel to the rotor blades, allows the head to rock in response to tilting of the swash plate by the cyclic pitch control. This changes the pitch angle an equal amount on each blade—decreasing it on one blade and increasing it on the other.

In a fully articulated rotor system, each blade is attached to the hub by three hinges, oriented at approximately right angles to each other. A horizontal hinge, called the flapping hinge, allows the blades to move up and down independently. A vertical hinge, called a drag, or lag hinge, allows each blade to move back and forth in the plane of the rotor disk. This movement is called dragging, or hunting. The blades can also rotate about their spanwise axis to change their individual blade pitch angle, or feather.

Fully articulated helicopter rotor systems generally use three or more blades, and each blade can flap, drag, and feather independently of the other blades.

The freewheeling unit in a helicopter rotor system allows the engine to automatically disengage from the rotor when the engine stops or slows below the corresponding rotor RPM. This makes autorotation possible.

Because a helicopter rotor system weighs so much more than a propeller, a helicopter must have some way to disconnect the engine from the rotor to relieve the starter load. For this reason, it is necessary to have a clutch between the engine and the rotor. The clutch in a helicopter rotor system allows the engine to be started without the load, and when the engine is running properly, the rotor load can be gradually applied.

Answers

5766 [C]   5767 [A]

High-frequency vibrations are associated with the engine in most helicopters and are impossible to count, because of their high frequency. A high-frequency vibration that suddenly occurs during flight could be an indication of a transmission bearing failure. Such a failure will result in vibrations whose frequencies are directly related to the engine speed. Abnormal low-frequency vibrations in a helicopter are always associated with the main rotor.

RTC
5168. For gyroplanes with constant-speed propellers, the first indication of carburetor icing is usually

A—a decrease in engine RPM.
B—a decrease in manifold pressure.
C—engine roughness followed by a decrease in engine RPM.

For gyroplanes with controllable-pitch (constant speed) propellers, the first indication of carburetor icing is usually a drop in manifold pressure. There will be no reduction in RPM with constant-speed propellers, since propeller pitch is automatically adjusted to compensate for the loss of power, thus maintaining constant RPM. (H307) — AC 61-23C, Chapter 2

RTC
5240. Coning is caused by the combined forces of

A—drag, weight, and translational lift.
B—lift and centrifugal force.
C—flapping and centrifugal force.

Coning is the upward bending of the blades caused by the combined forces of lift and centrifugal force. (H71) — FAA-H-8083-21

RTC
5241. The forward speed of a rotorcraft is restricted primarily by

A—dissymmetry of lift.
B—transverse flow effect.
C—high-frequency vibrations.

A tendency for the retreating blade to stall in forward flight is inherent in all present-day helicopters and is a major factor in limiting their forward airspeed. Retreating blade stall is caused by the dissymmetry of lift created when the airflow over the retreating blade of the helicopter slows down as forward speed of the helicopter increases. (H71) — FAA-H-8083-21

RTC
5242. When hovering, a helicopter tends to move in the direction of tail rotor thrust. This statement is

A—true; the movement is called transverse tendency.
B—true; the movement is called translating tendency.
C—false; the movement is opposite the direction of tail rotor thrust, and is called translating tendency.

Due to translating tendency or drift, the entire helicopter has a tendency to move in the direction of tail rotor thrust (to the right) when hovering. (H71) — FAA-H-8083-21

RTC
5243. The purpose of lead-lag (drag) hinges in a three-bladed, fully articulated helicopter rotor system is to compensate for

A—Coriolis effect.
B—dissymmetry of lift.
C—blade flapping tendency.

When a rotor blade of a three-bladed rotor system flaps upward, the center of mass of the blade moves closer to the axis of rotation and blade acceleration takes place. This acceleration and deceleration action (often referred to as leading, lagging or hunting) in the plane of rotation is due to "Coriolis effect." (H71) — FAA-H-8083-21

RTC
5244. What happens to the helicopter as it experiences translating tendency?

A—It tends to dip slightly to the right as the helicopter approaches approximately 15 knots in takeoff.
B—It gains increased rotor efficiency as air over the rotor system reaches approximately 15 knots.
C—It moves in the direction of tail rotor thrust.

Due to translating tendency or drift, the entire helicopter has a tendency to move in the direction of tail rotor thrust (to the right) when hovering. (H71) — FAA-H-8083-21

Answers

| 5168 | [B] | 5240 | [B] | 5241 | [A] | 5242 | [B] | 5243 | [A] | 5244 | [C] |

RTC
5245. The unequal lift across the rotor disc that occurs in horizontal flight as a result of the difference in velocity of the air over the advancing half of the disc area and the air passing over the retreating half of the disc area is known as

A—coning.
B—disc loading.
C—dissymmetry of lift.

Dissymmetry of lift is created by horizontal flight or by wind during hovering flight, and is the difference in lift that exists between the advancing blade half of the disc area and the retreating blade half. (H71) — FAA-H-8083-21

RTC
5246. The lift differential that exists between the advancing blade and the retreating blade is known as

A—Coriolis effect.
B—translational lift.
C—dissymmetry of lift.

Dissymmetry of lift is created by horizontal flight or by wind during hovering flight, and is the difference in lift that exists between the advancing blade half of the disc area and the retreating blade half. (H71) — FAA-H-8083-21

RTC
5247. Most helicopters, by design tend to drift to the right when hovering in a no-wind condition. This statement is

A—false; helicopters have no tendency to drift, but will rotate in that direction.
B—true; the mast or cyclic pitch system of most helicopters is rigged forward, this with gyroscopic precession will overcome this tendency.
C—true; the mast or cyclic pitch system of most helicopters is rigged to the left to overcome this tendency.

The entire helicopter has a tendency to move in the direction of tail rotor thrust (to the right) when hovering. To counteract this drift, the rotor mast in some helicopters is rigged slightly to the left side. (H71) — FAA-H-8083-21

RTC
5248. When a rotorcraft transitions from straight-and-level flight into a 30° bank while maintaining a constant altitude, the total lift force must

A—increase and the load factor will increase.
B—increase and the load factor will decrease.
C—remain constant and the load factor will decrease.

The steeper the angle of bank, the greater the angle of attack of the rotor blades required to maintain altitude. Thus, with an increase in bank and a greater angle of attack, the resultant lifting force will be increased. The load factor and hence, apparent gross weight increase, is relatively small in banks up to 30°. (H72) — FAA-H-8083-21

RTC
5249. Cyclic control pressure is applied during flight that results in a maximum increase in main rotor blade pitch angle at the "three o'clock" position. Which way will the rotor disc tilt?

A—Aft.
B—Left.
C—Right.

Because of the gyroscopic precession property, the blades do not rise or lower to maximum deflection until a point approximately 90° later in the plane of rotation. (H73) — FAA-H-8083-21

RTC
5250. Cyclic control pressure is applied during flight that results in a maximum decrease in pitch angle of the rotor blades at the "12 o'clock" position. Which way will the rotor disc tilt?

A—Aft.
B—Left.
C—Forward.

Because of the gyroscopic precession property, the blades do not rise or lower to maximum deflection until a point approximately 90° later in the plane of rotation. (H73) — FAA-H-8083-21

RTC
5251. The primary purpose of the tail rotor system is to

A—assist in making coordinated turns.
B—maintain heading during forward flight.
C—counteract the torque effect of the main rotor.

Answers

| 5245 | [C] | 5246 | [C] | 5247 | [C] | 5248 | [A] | 5249 | [A] | 5250 | [B] |
| 5251 | [C] | | | | | | | | | | |

Chapter 2 **Aircraft Systems**

The force that compensates for torque and keeps the fuselage from turning in the direction opposite to the main rotor is produced by means of an auxiliary rotor located on the end of the tail boom. (H73) — FAA-H-8083-21

Carburetor icing is a frequent cause of engine failure. Even a slight accumulation of this deposit will reduce power and may lead to complete engine failure, particularly when the throttle is partly or fully closed. (H74) — FAA-H-8083-21

RTC
5252. Can the tail rotor produce thrust to the left?

A—No; the right thrust can only be reduced, causing tail movement to the left.
B—Yes; primarily so that hovering turns can be accomplished to the right.
C—Yes; primarily to counteract the drag of the transmission during autorotation.

The maximum positive-pitch angle of the tail rotor is generally somewhat greater than the maximum negative pitch angle available. This is because the primary purpose of the tail rotor is to counteract the torque of the main rotor. The capability for tail rotors to produce thrust to the left (negative-pitch angle) is necessary because, during autorotation, the drag of the transmission tends to yaw the nose to the left — in the same direction that the main rotor is turning. (H73) — FAA-H-8083-21

RTC
5253. The main rotor blades of a fully-articulated rotor system can

A—flap and feather collectively.
B—flap, drag, and feather independently.
C—feather independently, but cannot flap or drag.

Each blade of a fully articulated rotor system can flap, drag, and feather independently of the other blades. (H74) — FAA-H-8083-21

RTC
5254. A reciprocating engine in a helicopter is more likely to stop due to in-flight carburetor icing than will the same type engine in an airplane. This statement

A—has no basis in fact. The same type engine will run equally well in either aircraft.
B—is true. The freewheeling unit will not allow windmilling (flywheel) effect to be exerted on a helicopter engine.
C—is false. The clutch will immediately release the load from the helicopter engine under engine malfunctioning conditions.

RTC
5255. What is the primary purpose of the clutch?

A—It allows the engine to be started without driving the main rotor system.
B—It provides disengagement of the engine from the rotor system for autorotation.
C—It transmits engine power to the main rotor, tail rotor, generator/alternator, and other accessories.

Because of the much greater weight of a helicopter rotor in relation to the power of the engine, than the weight of a propeller in relation to the power of the engine in an airplane, it is necessary to have the rotor disconnected from the engine to relieve the starter load. For this reason, a clutch is needed between the engine and rotor. The clutch allows the engine to be started and gradually assume the load of driving the heavy rotor. (H74) — FAA-H-8083-21

RTC
5256. What is the primary purpose of the freewheeling unit?

A—It allows the engine to be started without driving the main rotor system.
B—It provides speed reduction between the engine, main rotor system, and tail rotor system.
C—It provides disengagement of the engine from the rotor system for autorotation purposes.

The freewheeling coupling provides for autorotative capabilities by automatically disconnecting the rotor system from the engine when the engine stops or slows below the equivalent of rotor RPM. When the engine is disconnected from the rotor system through the automatic action of the freewheeling coupling, the transmission continues to rotate with the main rotor thereby enabling the tail rotor to continue turning at its normal rate. This permits the pilot to maintain directional control during autorotation. (H74) — FAA-H-8083-21

Answers

5252 [C]   5253 [B]   5254 [B]   5255 [A]   5256 [C]

Chapter 2 Aircraft Systems

RTC
5257. The main rotor blades of a semirigid rotor system can

A—flap and feather as a unit.
B—flap, drag, and feather independently.
C—feather independently, but cannot flap or drag.

A semi-rigid rotor system can flap and feather as a unit. (H74) — FAA-H-8083-21

RTC
5258. Rotorcraft climb performance is most adversely affected by

A—higher than standard temperature and low relative humidity.
B—lower than standard temperature and high relative humidity.
C—higher than standard temperature and high relative humidity.

High elevations, high temperatures, and high moisture content, all of which contribute to a high density altitude condition, lessen helicopter performance. (H77) — FAA-H-8083-21

RTC
5259. The most unfavorable combination of conditions for rotorcraft performance is

A—low density altitude, low gross weight, and calm wind.
B—high density altitude, high gross weight, and calm wind.
C—high density altitude, high gross weight, and strong wind.

The most adverse conditions for helicopter performance are the combination of a high density altitude, heavy gross weight, and calm or no wind. (H77) — FAA-H-8083-21

RTC
5260. How does high density altitude affect rotorcraft performance?

A—Engine and rotor efficiency is reduced.
B—Engine and rotor efficiency is increased.
C—It increases rotor drag, which requires more power for normal flight.

Helicopter performance is reduced because the thinner air at high density altitudes reduces the amount of lift of the rotor blades. Also, the (unsupercharged) engine does not develop as much power because of the thinner air and the decreased atmospheric pressure. (H77) — FAA-H-8083-21

RTC
5261. A medium-frequency vibration that suddenly occurs during flight could be indicative of a defective

A—main rotor system.
B—tail rotor system.
C—transmission system.

Medium-frequency vibrations are a result of trouble with the tail rotor in most helicopters. Improper rigging, imbalance, defective blades, or bad bearings in the tail rotor are all sources of these vibrations. If the vibration occurs only during turns, the trouble may be caused by insufficient tail rotor flapping action. (H78) — FAA-H-8083-21

RTC
5262. In most helicopters, medium-frequency vibrations indicate a defective

A—engine.
B—main rotor system.
C—tail rotor system.

Medium-frequency vibrations are a result of trouble with the tail rotor in most helicopters. Improper rigging, imbalance, defective blades, or bad bearings in the tail rotor are all sources of these vibrations. If the vibration occurs only during turns, the trouble may be caused by insufficient tail rotor flapping action. (H78) — FAA-H-8083-21

RTC
5263. Abnormal helicopter vibrations in the low-frequency range are associated with which system or component?

A—Tail rotor.
B—Main rotor.
C—Transmission.

Abnormal vibrations in this category are always associated with the main rotor. The vibration will be some frequency related to the rotor RPM and the number of blades of the rotor, such as one vibration per revolution, two per revolution, or three per revolution. Low-frequency vibrations are slow enough that they can be counted. (H78) — FAA-H-8083-21

Answers

| 5257 | [A] | 5258 | [C] | 5259 | [B] | 5260 | [A] | 5261 | [B] | 5262 | [C] |
| 5263 | [B] | | | | | | | | | | |

Chapter 2 Aircraft Systems

RTC
5264. Helicopter low-frequency vibrations are always associated with the

A—main rotor.
B—tail rotor.
C—transmission.

Abnormal vibrations in this category are always associated with the main rotor. The vibration will be some frequency related to the rotor RPM and the number of blades of the rotor, such as one vibration per revolution, two per revolution, or three per revolution. Low-frequency vibrations are slow enough that they can be counted. (H78) — FAA-H-8083-21

RTC
5265. A high-frequency vibration that suddenly occurs during flight could be an indication of a defective

A—transmission.
B—freewheeling unit.
C—main rotor system.

High-frequency vibrations are associated with the engine in most helicopters, and will be impossible to count due to the high rate of vibration. However, they could be associated with the tail rotor for helicopters in which the tail rotor RPM is approximately equal to or greater than the engine RPM. A defective clutch, or missing or bent fan blades will cause vibrations which should be corrected. Any bearings in the engine or in the transmission or the tail rotor drive shaft that go bad will result in vibrations with frequencies directly related to the speed of the engine. (H78) — FAA-H-8083-21

RTC
5266. Ground resonance is more likely to occur with helicopters that are equipped with

A—rigid rotor systems.
B—semi-rigid rotor systems.
C—fully articulated rotor systems.

In general, if ground resonance occurs, it will occur only in helicopters possessing three-bladed, fully articulated rotor systems and landing wheels. (H78) — FAA-H-8083-21

RTC
5267. The proper action to initiate a quick stop is to apply

A—forward cyclic, while raising the collective and applying right antitorque pedal.
B—aft cyclic, while raising the collective and applying left antitorque pedal.
C—aft cyclic, while lowering the collective and applying right antitorque pedal.

The deceleration is initiated by applying aft cyclic to reduce forward speed. Simultaneously, the collective pitch should be lowered as necessary to counteract any climbing tendency. The timing must be exact. If too little down collective is applied for the amount of aft cyclic applied, a climb will result. If too much down collective is applied for the amount of aft cyclic applied, a descent will result. A rapid application of aft cyclic requires an equally rapid application of down collective. As collective pitch is lowered, right pedal should be increased to maintain heading and throttle should be adjusted to maintain RPM. (H80) — FAA-H-8083-21

RTC
5671. During the flare portion of a power-off landing, the rotor RPM tends to

A—remain constant.
B—increase initially.
C—decrease initially.

Forward speed during autorotation descent permits a pilot to incline the rotor disc rearward, thus causing a flare. The additional induced lift created by the greater volume of air momentarily checks forward speed as well as descent. The greater volume of air acting on the rotor disc will normally increase rotor RPM during the flare. (H71) — FAA-H-8083-21

RTC
5672. Which would produce the slowest rotor RPM?

A—A vertical descent with power.
B—A vertical descent without power.
C—Pushing over after a steep climb.

A pushover out of a steep climb will produce the lowest rotor RPM. (H71) — FAA-H-8083-21

Answers

5264 [A]   5265 [A]   5266 [C]   5267 [C]   5671 [B]   5672 [C]

Chapter 2 **Aircraft Systems**

RTC
5673. If the RPM is low and the manifold pressure is high, what initial corrective action should be taken?

A—Increase the throttle.
B—Lower the collective pitch.
C—Raise the collective pitch.

Problem: RPM low, manifold pressure high.

Solution: Lowering the collective pitch will reduce the manifold pressure, decrease drag on the rotor, and therefore, increase the RPM.

(H73) — FAA-H-8083-21

RTC
5674. During climbing flight, the manifold pressure is low and the RPM is high. What initial corrective action should be taken?

A—Increase the throttle.
B—Decrease the throttle.
C—Raise the collective pitch.

Problem: RPM high, manifold pressure low.

Solution: Raising the collective pitch will increase the manifold pressure, increase drag on the rotor, and therefore decrease the RPM.

(H73) — FAA-H-8083-21

RTC
5675. During level flight, if the manifold pressure is high and the RPM is low, what initial corrective action should be made?

A—Decrease the throttle.
B—Increase the throttle.
C—Lower the collective pitch.

Problem: RPM low, manifold pressure high.

Solution: Lowering the collective pitch will reduce the manifold pressure, decrease drag on the rotor, and therefore, increase the RPM.

(H73) — FAA-H-8083-21

RTC
5676. When operating a helicopter in conditions favorable for carburetor icing, the carburetor heat should be

A—adjusted to keep the carburetor air temperature gauge indicating in the green arc at all times.
B—OFF for takeoffs, adjusted to keep the carburetor air temperature gauge indicating in the green arc at all other times.
C—OFF during takeoffs, approaches, and landings; adjusted to keep the carburetor air temperature gauge indicating in the green arc at all other times.

When a carburetor temperature gauge is used, the carburetor heat should be adjusted to keep the temperature in the green band. (H75) — FAA-H-8083-21

RTC
5686. As altitude increases, the V_{NE} of a helicopter will

A—increase.
B—decrease.
C—remain the same.

As the altitude increases, the never-exceed airspeed (red line) for most helicopters decreases. (H77) — FAA-H-8083-21

RTC
5695. The antitorque system fails during cruising flight and a powered approach landing is commenced. If the helicopter yaws to the right just prior to touchdown, what could the pilot do to help swing the nose to the left?

A—Increase the throttle.
B—Decrease the throttle.
C—Increase collective pitch.

Directional control should be maintained primarily with cyclic control and secondarily, by gently applying throttle momentarily, with needles joined, to swing the nose to the right, or decreasing throttle to swing the nose to the left. (H78) — FAA-H-8083-21

RTC
5696. If antitorque failure occurred during cruising flight, what could be done to help straighten out a left yaw prior to touchdown?

A—A normal running landing should be made.
B—Make a running landing using partial power and left cyclic.
C—Apply available throttle to help swing the nose to the right just prior to touchdown.

Answers

| 5673 | [B] | 5674 | [C] | 5675 | [C] | 5676 | [A] | 5686 | [B] | 5695 | [B] |
| 5696 | [C] | | | | | | | | | | |

Directional control should be maintained primarily with cyclic control and secondarily, by gently applying throttle momentarily, with needles joined, to swing the nose to the right. (H78) — FAA-H-8083-21

RTC
5697. Should a helicopter pilot ever be concerned about ground resonance during takeoff?

A—No; ground resonance occurs only during an autorotative touchdown.
B—Yes; although it is more likely to occur on landing, it can occur during takeoff.
C—Yes; but only during slope takeoffs.

Ground resonance occurs when the helicopter makes contact with the surface during landing or while in contact with the surface during an attempted takeoff. (H78) — FAA-H-8083-21

RTC
5698. An excessively steep approach angle and abnormally slow closure rate should be avoided during an approach to a hover, primarily because

A—the airspeed indicator would be unreliable.
B—a go-around would be very difficult to accomplish.
C—settling with power could develop, particularly during the termination.

Situations that are conducive to a settling-with-power condition are:

1. *Attempting to hover out of ground effect at altitudes above the hovering ceiling of the helicopter;*
2. *Attempting to hover out of ground effect without maintaining precise altitude control; or*
3. *A steep power approach in which airspeed is permitted to drop nearly to zero.*

(H78) — FAA-H-8083-21

RTC
5699. During a near-vertical power approach into a confined area with the airspeed near zero, what hazardous condition may develop?

A—Ground resonance.
B—Settling with power.
C—Blade stall vibration.

Situations that are conducive to a settling-with-power condition are:

1. *Attempting to hover out of ground effect at altitudes above the hovering ceiling of the helicopter;*
2. *Attempting to hover out of ground effect without maintaining precise altitude control; or*
3. *A steep power approach in which airspeed is permitted to drop nearly to zero.*

(H78) — FAA-H-8083-21

RTC
5700. Which procedure will result in recovery from settling with power?

A—Increase collective pitch and power.
B—Maintain constant collective pitch and increase throttle.
C—Increase forward speed and partially lower collective pitch.

In recovering from a settling-with-power condition, the tendency on the part of the pilot is to first try to stop the descent by increasing collective pitch which will result in increasing the stalled area of the rotor and increasing the rate of descent. Since inboard portions of the blades are stalled, cyclic control will be reduced. Recovery can be accomplished by increasing forward speed, and/or partially lowering collective pitch. (H78) — FAA-H-8083-21

RTC
5701. The addition of power in a settling with power situation produces an

A—increase in airspeed.
B—even greater rate of descent.
C—increase in cyclic control effectiveness.

In recovering from a settling-with-power condition, the tendency on the part of the pilot is to first try to stop the descent by increasing collective pitch which will result in increasing the stalled area of the rotor and increasing the rate of descent. Since inboard portions of the blades are stalled, cyclic control will be reduced. Recovery can be accomplished by increasing forward speed, and/or partially lowering collective pitch. (H78) — FAA-H-8083-21

Answers

| 5697 | [B] | 5698 | [C] | 5699 | [B] | 5700 | [C] | 5701 | [B] |

Chapter 2 Aircraft Systems

RTC
5702. Under which situation is accidental settling with power likely to occur?

A—A steep approach in which the airspeed is permitted to drop to nearly zero.
B—A shallow approach in which the airspeed is permitted to drop below 10 MPH.
C—Hovering in ground effect during calm wind, high-density altitude conditions.

Situations that are conducive to a settling-with-power condition are:

1. *Attempting to hover out of ground effect at altitudes above the hovering ceiling of the helicopter;*
2. *Attempting to hover out of ground effect without maintaining precise altitude control; or*
3. *A steep power approach in which airspeed is permitted to drop nearly to zero.*

(H78) — FAA-H-8083-21

RTC
5703. Which is true with respect to recovering from an accidental settling with power situation?

A—Antitorque pedals should not be utilized during the recovery.
B—Recovery can be accomplished by increasing rotor RPM, reducing forward airspeed, and minimizing maneuvering.
C—Since the inboard portions of the main rotor blades are stalled, cyclic control effectiveness will be reduced during the initial portion of the recovery.

In recovering from a settling-with-power condition, the tendency on the part of the pilot is to first try to stop the descent by increasing collective pitch which will result in increasing the stalled area of the rotor and increasing the rate of descent. Since inboard portions of the blades are stalled, cyclic control will be reduced. Recovery can be accomplished by increasing forward speed, and/or partially lowering collective pitch. (H78) — FAA-H-8083-21

RTC
5704. When operating at high forward airspeed, retreating blade stall is more likely to occur under conditions of

A—low gross weight, high density altitude, and smooth air.
B—high gross weight, low density altitude, and smooth air.
C—high gross weight, high density altitude, and turbulent air.

When operating at high forward airspeeds, stalls are more likely to occur under conditions of:

1. *High gross weight.*
2. *Low RPM.*
3. *High density altitude.*
4. *Steep or abrupt turns.*
5. *Turbulent air.*

(H78) — FAA-H-8083-21

RTC
5705. What are the major indications of an incipient retreating blade stall situation, in order of occurrence?

A—Low-frequency vibration, pitchup of the nose, and a tendency for the helicopter to roll.
B—Slow pitchup of the nose, high-frequency vibration, and a tendency for the helicopter to roll.
C—Slow pitchup of the nose, tendency for the helicopter to roll, followed by a medium-frequency vibration.

The major warnings of approaching retreating blade stall conditions, in the order in which they will generally be experienced, are:

1. *Abnormal two-per-revolution vibration in two-bladed rotors or three-per-revolution vibration in three-bladed rotors.*
2. *Pitchup of the nose.*
3. *Tendency for the helicopter to roll.*

(H78) — FAA-H-8083-21

RTC
5706. How should a pilot react at the onset of retreating blade stall?

A—Reduce collective pitch, rotor RPM, and forward airspeed.
B—Reduce collective pitch, increase rotor RPM, and reduce forward airspeed.
C—Increase collective pitch, reduce rotor RPM, and reduce forward airspeed.

At the onset of blade stall vibration, the pilot should take the following corrective measures:

1. *Reduce collective pitch.*
2. *Increase rotor RPM.*
3. *Reduce forward airspeed.*
4. *Minimize maneuvering.*

(H78) — FAA-H-8083-21

Answers

| 5702 [A] | 5703 [C] | 5704 [C] | 5705 [A] | 5706 [B] |

Chapter 2 Aircraft Systems

RTC
5707. The most power will be required to hover over which surface?

A—High grass.
B—Concrete ramp.
C—Rough/uneven ground.

Tall grass will tend to disperse or absorb the ground effect. More power will be required to hover, and takeoff may be very difficult. (H79) — FAA-H-8083-21

RTC
5708. Which flight technique is recommended for use during hot weather?

A—During takeoff, accelerate quickly into forward flight.
B—During takeoff, accelerate slowly into forward flight.
C—Use minimum allowable RPM and maximum allowable manifold pressure during all phases of flight.

Flight technique in hot weather:

1. *Make full use of wind and translational lift.*
2. *Hover as low as possible and no longer than necessary.*
3. *Maintain maximum allowable engine RPM.*
4. *Accelerate very slowly into forward flight.*
5. *Employ running takeoffs and landings when necessary.*
6. *Use caution in maximum performance takeoffs and steep approaches.*

(H79) — FAA-H-8083-21

RTC
5709. To taxi on the surface in a safe and efficient manner, helicopter pilots should use the

A—cyclic pitch to control starting, taxi speed, and stopping.
B—collective pitch to control starting, taxi speed, and stopping.
C—antitorque pedals to correct for drift during crosswind conditions.

The collective pitch controls starting, stopping and rate of speed while taxiing. The higher the collective pitch, the faster will be the taxi speed. Taxi at a speed no greater than that of a normal walk. (H80) — FAA-H-8083-21

RTC
5710. During surface taxiing, the cyclic pitch stick is used to control

A—heading.
B—ground track.
C—forward movement.

Move the cyclic slightly forward of the neutral position and apply a gradual upward pressure on the collective pitch to move the helicopter forward along the surface. Use pedals to maintain heading and cyclic to maintain ground track. (H80) — FAA-H-8083-21

RTC
5711. To taxi on the surface in a safe and efficient manner, one should use the cyclic pitch to

A—start and stop aircraft movement.
B—maintain heading during crosswind conditions.
C—correct for drift during crosswind conditions.

During crosswind taxi, the cyclic should be held into the wind a sufficient amount to eliminate any drifting movement. (H80) — FAA-H-8083-21

RTC
5712. A pilot is hovering during calm wind conditions. The greatest amount of engine power will be required when

A—ground effect exists.
B—making a left-pedal turn.
C—making a right-pedal turn.

During a hovering turn to the left, the RPM will decrease if throttle is not added. In a hovering turn to the right, RPM will increase if throttle is not reduced slightly. This is due to the amount of engine power that is being absorbed by the tail rotor which is dependent upon the pitch angle at which the tail rotor blades are operating. Avoid making large corrections in RPM while turning, since the throttle adjustment will result in erratic nose movements due to torque changes. (H80) — FAA-H-8083-21

Answers

| 5707 | [A] | 5708 | [B] | 5709 | [B] | 5710 | [B] | 5711 | [C] | 5712 | [B] |

Chapter 2 **Aircraft Systems**

RTC
5713. Which statement is true about an autorotative descent?

A—Generally, only the cyclic control is used to make turns.
B—The pilot should use the collective pitch control to control the rate of descent.
C—The rotor RPM will tend to decrease if a tight turn is made with a heavily loaded helicopter.

When making turns during an autorotative descent, generally use cyclic control only. Use of antitorque pedals to assist or speed the turn causes loss of airspeed and downward pitching of the nose, especially when the left pedal is used. When the autorotation is initiated, sufficient right pedal pressure should be used to maintain straight flight and prevent yawing to the left. This pressure should not be changed to assist the turn. (H80) — FAA-H-8083-21

RTC
5714. Using right pedal to assist a right turn during an autorotative descent will probably result in what actions?

A—A decrease in rotor RPM, pitch up of the nose, decrease in sink rate, and increase in indicated airspeed.
B—An increase in rotor RPM, pitch up of the nose, decrease in sink rate, and increase in indicated airspeed.
C—An increase in rotor RPM, pitch down of the nose, increase in sink rate, and decrease in indicated airspeed.

Using right pedal to assist a right turn during an autorotative descent will probably result in an increase in rotor RPM, pitch down of the nose, increase in sink rate, and decrease in indicated airspeed. (H80) — FAA-H-8083-21

RTC
5715. Using left pedal to assist a left turn during an autorotative descent will probably cause the rotor RPM to

A—increase and the airspeed to decrease.
B—decrease and the aircraft nose to pitch down.
C—increase and the aircraft nose to pitch down.

Use of antitorque pedals to assist or speed the turn causes loss of airspeed and downward pitching of the nose, especially when the left pedal is used. (H80) — FAA-H-8083-21

RTC
5716. When planning slope operations, only slopes of 5° gradient or less should be considered, primarily because

A—ground effect is lost on slopes of steeper gradient.
B—downwash turbulence is more severe on slopes of steeper gradient.
C—most helicopters are not designed for operations on slopes of steeper gradient.

As collective pitch is lowered, continue to move the cyclic stick toward the slope to maintain a fixed position and use cyclic as necessary to stop forward or aft movement of the helicopter. The slope must be shallow enough to allow the pilot to hold the helicopter against it with the cyclic stick during the entire landing. A slope of 5° is considered maximum for normal operation of most helicopters. Each make of helicopter will generally have its own peculiar way of indicating to the pilot when lateral cyclic stick travel is about to run out: i.e., the rotor hub hitting the rotor mast, vibrations felt through the cyclic stick, and others. A landing should not be made in these instances since this indicates to the pilot that the slope is too steep. (H80) — FAA-H-8083-21

RTC
5717. When making a slope landing, the cyclic pitch control should be used to

A—lower the downslope skid to the ground.
B—hold the upslope skid against the slope.
C—place the rotor disc parallel to the slope.

As collective pitch is lowered, continue to move the cyclic stick toward the slope to maintain a fixed position and use cyclic as necessary to stop forward or aft movement of the helicopter. The slope must be shallow enough to allow the pilot to hold the helicopter against it with the cyclic stick during the entire landing. A slope of 5° is considered maximum for normal operation of most helicopters. Each make of helicopter will generally have its own peculiar way of indicating to the pilot when lateral cyclic stick travel is about to run out: i.e., the rotor hub hitting the rotor mast, vibrations felt through the cyclic stick, and others. A landing should not be made in these instances since this indicates to the pilot that the slope is too steep. (H80) — FAA-H-8083-21

Answers

5713 [A]   5714 [C]   5715 [C]   5716 [C]   5717 [B]

RTC
5718. Takeoff from a slope is normally accomplished by

A—making a downslope running takeoff if the surface is smooth.
B—simultaneously applying collective pitch and downslope cyclic control.
C—bringing the helicopter to a level attitude before completely leaving the ground.

As the downslope skid is rising and the helicopter approaches a level attitude, move the cyclic stick back to the neutral position keeping the rotor disc parallel to the true horizon. Continue to apply collective pitch and take the helicopter straight up to a hover before moving away from the slope. In moving away from the slope, the tail should not be turned upslope because of the danger of the tail rotor striking the surface. (H80) — FAA-H-8083-21

RTC
5719. What is the procedure for a slope landing?

A—Use maximum RPM and maximum manifold pressure.
B—If the slope is 10° or less, the landing should be made perpendicular to the slope.
C—When parallel to the slope, slowly lower the upslope skid to the ground prior to lowering the downslope skid.

A downward pressure on the collective pitch will start the helicopter descending. As the upslope skid touches the ground, apply cyclic stick in the direction of the slope. This will hold the skid against the slope while the downslope skid is continuing to be let down with the collective pitch. (H80) — FAA-H-8083-21

RTC
5720. You are hovering during calm wind conditions and decide to make a right-pedal turn. In most helicopters equipped with reciprocating engines, the engine RPM will tend to

A—increase.
B—decrease.
C—remain unaffected.

During a hovering turn to the left, the RPM will decrease if throttle is not added. In a hovering turn to the right, RPM will increase if throttle is not reduced slightly. This is due to the amount of engine power that is being absorbed by the tail rotor which is dependent upon the pitch angle at which the tail rotor blades are operating. Avoid making large corrections in RPM while turning, since the throttle adjustment will result in erratic nose movements due to torque changes. (H80) — FAA-H-8083-21

RTC
5721. During calm wind conditions, in most helicopters, which of these flight operations would require the most power?

A—A left-pedal turn.
B—A right-pedal turn.
C—Hovering in ground effect.

During a hovering turn to the left, the RPM will decrease if throttle is not added. In a hovering turn to the right, RPM will increase if throttle is not reduced slightly. This is due to the amount of engine power that is being absorbed by the tail rotor which is dependent upon the pitch angle at which the tail rotor blades are operating. Avoid making large corrections in RPM while turning, since the throttle adjustment will result in erratic nose movements due to torque changes. (H80) — FAA-H-8083-21

RTC
5722. If complete power failure should occur while cruising at altitude, the pilot should

A—partially lower the collective pitch, close the throttle, then completely lower the collective pitch.
B—lower the collective pitch as necessary to maintain proper rotor RPM, and apply right pedal to correct for yaw.
C—close the throttle, lower the collective pitch to the full-down position, apply left pedal to correct for yaw, and establish a normal power-off glide.

By immediately lowering collective pitch (which must be done in case of engine failure), lift and drag will be reduced and the helicopter will begin an immediate descent, thus producing an upward flow of air through the rotor system. (H80) — FAA-H-8083-21

Answers

| 5718 | [C] | 5719 | [C] | 5720 | [A] | 5721 | [A] | 5722 | [B] |

Chapter 2 Aircraft Systems

RTC
5723. When making an autorotation to touchdown, what action is most appropriate?

A—A slightly nose-high attitude at touchdown is the proper procedure.
B—The skids should be in a longitudinally level attitude at touchdown.
C—Aft cyclic application after touchdown is desirable to help decrease ground run.

As the helicopter approaches normal hovering altitude, maintain a landing attitude with cyclic control, maintain heading with pedals, apply sufficient collective pitch (while holding the throttle in the closed position) to cushion the touchdown and be sure the helicopter is landing parallel to its direction of motion upon contact with the surface. Avoid landing on the heels of the skid gear. (H80) — FAA-H-8083-21

RTC
5724. During the entry into a quick stop, how should the collective pitch control be used? It should be

A—lowered as necessary to prevent ballooning.
B—raised as necessary to prevent a rotor overspeed.
C—raised as necessary to prevent a loss of altitude.

The deceleration is initiated by applying aft cyclic to reduce forward speed. Simultaneously, the collective pitch should be lowered as necessary to counteract any climbing tendency. (H80) — FAA-H-8083-21

RTC
5725. During a normal approach to a hover, the collective pitch control is used primarily to

A—maintain RPM.
B—control the rate of closure.
C—control the angle of descent.

The angle of descent is primarily controlled by collective pitch. The airspeed is primarily controlled by the cyclic control. Heading on final approach is maintained with pedal control. However, the approach can only be accomplished successfully by the coordination of all controls. (H80) — FAA-H-8083-21

RTC
5726. During a normal approach to a hover, the cyclic pitch is used primarily to

A—maintain heading.
B—control rate of closure.
C—control angle of descent.

The angle of descent is primarily controlled by collective pitch. The airspeed, or rate of closure, is primarily controlled by the cyclic control. Heading on final approach is maintained with pedal control. However, the approach can only be accomplished successfully by the coordination of all controls. (H80) — FAA-H-8083-21

RTC
5727. Normal RPM should be maintained during a running landing primarily to ensure

A—adequate directional control until the helicopter stops.
B—that sufficient lift is available should an emergency develop.
C—longitudinal and lateral control, especially if the helicopter is heavily loaded or high density altitude conditions exist.

After surface contact, the cyclic control should be placed slightly forward of neutral to tilt the main rotor away from the tail boom, antitorque pedals should be used to maintain heading, throttle should be used to maintain RPM, and cyclic stick should be used to maintain surface track. Normally, the collective pitch is held stationary after touchdown until the helicopter comes to a complete stop. However, if braking is desired or required, the collective pitch may be lowered cautiously. To ensure directional control, normal rotor RPM must be maintained until the helicopter stops. (H80) — FAA-H-8083-21

RTC
5728. Which is true concerning a running takeoff?

A—If a helicopter cannot be lifted vertically, a running takeoff should be made.
B—One advantage of a running takeoff is that the additional airspeed can be converted quickly to altitude.
C—A running takeoff may be possible when gross weight or density altitude prevents a sustained hover at normal hovering altitude.

Answers

| 5723 | [B] | 5724 | [A] | 5725 | [C] | 5726 | [B] | 5727 | [A] | 5728 | [C] |

Chapter 2 **Aircraft Systems**

A running takeoff is used when conditions of load and/ or density altitude prevent a sustained hover at normal hovering altitude. It is often referred to as a high-altitude takeoff. (H80) — FAA-H-8083-21

RTC
5729. When conducting a confined area-type operation, the primary purpose of the high reconnaissance is to determine the

A—type of approach to be made.
B—suitability of the area for landing.
C—height of the obstructions surrounding the area.

The primary purpose of the high reconnaissance is to determine the suitability of an area for a landing. In a high reconnaissance, the following items should be accomplished:

1. *Determine wind direction and speed;*
2. *Select the most suitable flight paths into and out of the area, with particular consideration being given to forced landing areas;*
3. *Plan the approach and select a point for touchdown; and*
4. *Locate and determine the size of barriers, if any, immediately around the area.*

(H81) — FAA-H-8083-21

RTC
5730. During a pinnacle approach to a rooftop heliport under conditions of high wind and turbulence, the pilot should make a

A—shallow approach, maintaining a constant line of descent with cyclic applications.
B—normal approach, maintaining a slower-than-normal rate of descent with cyclic applications.
C—steeper-than-normal approach, maintaining the desired angle of descent with collective applications.

A steep approach is used primarily when there are obstacles in the approach path that are too high to allow a normal approach. A steep approach will permit entry into most confined areas and is sometimes used to avoid areas of turbulence around a pinnacle. (H81) — FAA-H-8083-21

RTC
5731. What type approach should be made to a rooftop heliport under conditions of relatively high wind and turbulence?

A—A normal approach.
B—A steeper-than-normal approach.
C—A shallower-than-normal approach.

A steep approach is used primarily when there are obstacles in the approach path that are too high to allow a normal approach. A steep approach will permit entry into most confined areas and is sometimes used to avoid areas of turbulence around a pinnacle. (H81) — FAA-H-8083-21

RTC
5732. If turbulence and downdrafts are expected during a pinnacle approach to a rooftop heliport, plan to make a

A—steeper-than-normal approach.
B—normal approach, maintaining a lower-than-normal airspeed.
C—shallow approach, maintaining a higher-than-normal airspeed.

A steep approach is used primarily when there are obstacles in the approach path that are too high to allow a normal approach. A steep approach will permit entry into most confined areas and is sometimes used to avoid areas of turbulence around a pinnacle. (H81) — FAA-H-8083-21

RTC
5733. If ground resonance is experienced during rotor spin-up, what action should you take?

A—Taxi to a smooth area.
B—Make a normal takeoff immediately.
C—Close the throttle and slowly raise the spin-up lever.

A corrective action for ground resonance is an immediate takeoff if RPM is in proper range (for helicopters) or an immediate closing of the throttle and placing the blades in low pitch if the RPM is low. "During spin-up" implies low RPM, so closing the throttle is appropriate. (H767) — FAA-H-8083-21, Chapter 21

Answers

5729 [B]　　5730 [C]　　5731 [B]　　5732 [A]　　5733 [C]

Chapter 2 Aircraft Systems

RTC
5734. The principal factor limiting the never-exceed speed (V_{NE}) of a gyroplane is

A—turbulence and altitude.
B—blade-tip speed, which must remain below the speed of sound.
C—lack of sufficient cyclic stick control to compensate for dissymmetry of lift or retreating blade stall, depending on which occurs first.

Retreating blade stall is the principal factor limiting the never-exceed speed. (H762) — FAA-H-8083-21, Chapter 16

RTC
5735. Why should gyroplane operations within the cross-hatched portion of a Height vs. Velocity chart be avoided?

A—The rotor RPM may build excessively high if it is necessary to flare at such low altitudes.
B—Sufficient airspeed may not be available to ensure a safe landing in case of an engine failure.
C—Turbulence near the surface can dephase the blade dampers causing geometric unbalanced conditions on the rotor system.

A rotorcraft or gyroplane pilot must become familiar with this chart for the particular gyroplane he or she is flying. From it, the pilot can determine what altitudes and airspeeds are required to safely make an autorotative landing in case of an engine failure. The chart can be used to determine altitude-airspeed combinations from which it would be nearly impossible to successfully complete an autorotative landing. The altitude-airspeed combinations that should be avoided are represented by the shaded areas of the chart. (H765) — FAA-H-8083-21, Chapter 19

RTC
5736. The principal reason the shaded area of a Height vs. Velocity chart should be avoided is

A—rotor RPM may decay before ground contact is made if an engine failure should occur.
B—rotor RPM may build excessively high if it is necessary to flare at such low altitudes.
C—insufficient airspeed would be available to ensure a safe landing in case of an engine failure.

A rotorcraft or gyroplane pilot must become familiar with this chart for the particular gyroplane he or she is flying. From it, the pilot can determine what altitudes and airspeeds are required to safely make an autorotative landing in case of an engine failure. The chart can be used to determine altitude-airspeed combinations from which it would be nearly impossible to successfully complete an autorotative landing. The altitude-airspeed combinations that should be avoided are represented by the shaded areas of the chart. (H765) — FAA-H-8083-21, Chapter 19

RTC
5737. During the transition from pre-rotation to flight, all rotor blades change pitch

A—simultaneously to the same angle of incidence.
B—simultaneously but to different angles of incidence.
C—to the same degree at the same point in the cycle of rotation.

Compensation for dissymmetry of lift requires constant change in the blade angle of incidence, with one increasing as another simultaneously decreases. During the transition from prerotation to flight (or any time there is dissymmetry of lift) all rotor blades change pitch simultaneously, but to different angles of incidence. (H766) — FAA-H-8083-21, Chapter 20

RTC
5738. Select the true statement concerning gyroplane taxi procedures.

A—Avoid abrupt control movements when blades are turning.
B—The cyclic stick should be held in the neutral position at all times.
C—The cyclic stick should be held slightly aft of neutral at all times.

Avoid abrupt control motions while taxiing. (H766) — FAA-H-8083-21, Chapter 20

Answers

5734 [C]	5735 [B]	5736 [C]	5737 [B]	5738 [A]

RTC
5755. With respect to vortex circulation, which is true?

A—Helicopters generate downwash turbulence, not vortex circulation.
B—The vortex strength is greatest when the generating aircraft is flying fast.
C—Vortex circulation generated by helicopters in forward flight trail behind in a manner similar to wingtip vortices generated by airplanes.

In forward flight, departing or landing helicopters produce a pair of high velocity trailing vortices similar to wing-tip vortices of large fixed-wing aircraft. Pilots of small aircraft should use caution when operating behind or crossing behind landing and departing helicopters. (J27) — AIM ¶7-3-7

RTC
5756. Which is true with respect to vortex circulation?

A—Helicopters generate downwash turbulence only, not vortex circulation.
B—The vortex strength is greatest when the generating aircraft is heavy, clean, and slow.
C—When vortex circulation sinks into ground effect, it tends to dissipate rapidly and offer little danger.

The strength of the vortex is governed by the weight, speed, and shape of the airfoil of the generating aircraft. The greatest vortex strength occurs when the generating aircraft is heavy, clean, and slow. (J27) — AIM ¶7-3-7

Glider Systems

Variometers used in sailplanes are so sensitive that they indicate climbs and descents as a result of changes in airspeed. A total energy compensator for a variometer reduces the climb and dive errors that are caused by airspeed changes and cancels out errors caused by "stick thermals" and changes in airspeed. The variometer shows only when the sailplane is climbing in rising air currents.

GLI
5273. Which is true regarding electric variometers?

A—Are generally considered to be less sensitive and has a slower response time than a vertical-speed indicator.
B—The sensitivity can be adjusted in flight to suit existing air conditions.
C—They do not utilize outside air static pressure lines.

Variometers are so sensitive they indicate climbs and descents as a result of changes in airspeed. A total energy compensator is used to cancel out errors caused by thermals and changes in airspeed. (N22) — Soaring Flight Manual

GLI
5274. Which is true regarding variometers?

A—An electric variometer does not utilize outside air static pressure lines.
B—One of the advantages of the pellet variometer over the vane variometer, is that dirt, moisture, or static electricity will not affect its operation.
C—A total energy system senses airspeed changes and tends to cancel out the resulting variometer climb or dive indications.

A total energy variometer has been compensated so it only responds to changes in total energy, there by canceling out false indications of climbs or descents. (N22) — Soaring Flight Manual

GLI
5275. Which is true concerning total energy compensators? The instrument

A—responds to up and down air currents only.
B—will register climbs that result from stick thermals.
C—reacts to climbs and descents like a conventional rate-of-climb indicator.

A total energy variometer has been compensated so it only responds to changes in total energy of the sailplane; thus, a change in altitude and airspeed due to stick deflection does not register as lift or sink on the variometer. If the system is properly designed, the false climb or descent will be canceled out. (N22) — Soaring Flight Manual

Answers

| 5755 [C] | 5756 [B] | 5273 [B] | 5274 [C] | 5275 [A] |

Chapter 2 Aircraft Systems

GLI

5297. The advantage of total energy compensators is that this system

A—includes a speed ring around the rim of the variometer.
B—adds the effect of stick thermals to the total energy produced by thermals.
C—reduces climb and dive errors on variometer indications caused by airspeed changes.

A total energy variometer has been compensated so it responds only to changes in total energy of the sailplane; thus, a change in altitude and airspeed due to stick deflection does not register as lift or sink on the variometer. If the system is properly designed, the false climb or descent will be canceled out. (N22) — Soaring Flight Manual

GLI

5612. In the Northern Hemisphere, if a sailplane is accelerated or decelerated, the magnetic compass will normally indicate

A—correctly, only when on a north or south heading.
B—a turn toward south while accelerating on a west heading.
C—a turn toward north while decelerating on an east heading.

When on a north or south heading, there is no acceleration or deceleration error. (H314) — AC 61-23C, Chapter 3

GLI

5613. When flying on a heading of west from one thermal to the next, the airspeed is increased to the "speed-to-fly" with the wings level. What will the conventional magnetic compass indicate while the airspeed is increasing?

A—A turn toward the south.
B—A turn toward the north.
C—Straight flight on a heading of 270°.

When on an east or west heading, an acceleration will cause the compass to indicate a turn to the north. (H314) — AC 61-23C, Chapter 3

GLI

5792. Select the true statement concerning oxygen systems that are often installed in sailplanes.

A—Most civilian aircraft oxygen systems use low-pressure cylinders for oxygen storage.
B—When aviation breathing oxygen is not available, hospital or welder's oxygen serves as a good substitute.
C—In case of a malfunction of the main oxygen system a bailout bottle may serve as an emergency oxygen supply.

A bail-out bottle is an emergency oxygen supply and can be used in the event of a malfunction in the main system. (N28) — Soaring Flight Manual

GLI

5794. Which is true regarding the assembly of a glider for flight?

A—It may be accomplished by the pilot.
B—It is not required by regulations for a glider pilot to know this.
C—It must be accomplished under the supervision of an FAA maintenance inspector.

The removal or installation of glider wings and tail surfaces may be accomplished by a certificated (private or better) pilot. (N29) — Soaring Flight Manual

GLI

5769. What corrective action should be taken during a landing if the glider pilot makes the roundout too soon while using spoilers?

A—Leave the spoilers extended and lower the nose slightly.
B—Retract the spoilers and leave them retracted until after touchdown.
C—Retract the spoilers until the glider begins to settle again, then extend the spoilers.

During the round-out, if spoilers are being used and it becomes apparent that the round-out was made too soon or too late, the spoilers should be retracted. Once retracted under either of these conditions, they should not be extended again until the wheel is on the ground, because opening the spoilers close to the ground may cause the sailplane to drop to the runway. (N32) — Soaring Flight Manual

---

Answers

| 5297 | [C] | 5612 | [A] | 5613 | [B] | 5792 | [C] | 5794 | [A] | 5769 | [B] |

GLI
5770. What consideration should be given in the choice of a towplane for use in aerotows?

A—L/D ratio of the glider to be towed.
B—Gross weight of the glider to be towed.
C—Towplane's low-wing loading and low-power loading.

A good towplane has low wing loading and low power loading, with sufficient excess power to get off the ground in much less than runway length and give a reasonable and safe rate and angle of climb, considering the local terrain and gross weight of the sailplane. (N30) — Soaring Flight Manual

GLI
5771. Looseness in a glider's flight control linkage or attachments could result in

A—increased stalling speed.
B—loss of control during an aerotow in turbulence.
C—flutter while flying at near maximum speed in turbulence.

Flutter may be caused by looseness in control cables, linkages, hinges, or play in the wing or empennage attachments. (N20) — Soaring Flight Manual

GLI
5772. A left side slip is used to counteract a crosswind drift during the final approach for landing. An over-the-top spin would most likely occur if the controls were used in which of the following ways? Holding the stick

A—too far back and applying full right rudder.
B—in the neutral position and applying full right rudder.
C—too far to the left and applying full left rudder.

The spin will be in the direction of the applied rudder. Just before the nose drops, the control stick is brought to the aft stop and full rudder is applied in the desired spin direction. Rotation begins immediately and continues as long as the controls are held in this position. (N20) — Soaring Flight Manual

GLI
5791. When flying on a heading of east from one thermal to the next, the airspeed is increased to the speed-to-fly with wings level. What will the conventional magnetic compass indicate while the airspeed is increasing?

A—A turn toward the south.
B—A turn toward the north.
C—Straight flight on a heading of 090°.

When on an east or west heading, an acceleration will cause the compass to indicate a turn to the north. (N22) — Soaring Flight Manual

GLI
5793. The spoilers should be in what position when operating in a strong wind?

A—Extended during both a landing roll or ground operation.
B—Retracted during both a landing roll or ground operation.
C—Extended during a landing roll, but retracted during a ground operation.

The spoilers should remain open during ground operations in strong wind conditions. (N29) — Soaring Flight Manual

GLI
5795. The primary cause of towline slack during aerotows is

A—acceleration.
B—poor coordination.
C—positioning the glider too high.

Slack in the towline can only occur when the glider has a faster speed than the towplane. The most common time to have towline slack is during, and especially after turns. The glider is flying a greater distance, and therefore will have to fly a faster airspeed. (N29) — Soaring Flight Manual

Answers

5770 [B]   5771 [C]   5772 [A]   5791 [B]   5793 [A]   5795 [A]

Chapter 2 Aircraft Systems

GLI
5796. To signal the glider pilot during an aerotow to release immediately, the tow pilot will

A—fishtail the towplane.
B—rock the towplane's wings.
C—alternately raise and lower the towplane's pitch attitude.

The mandatory release signal is the rocking of the towplane's wings. (N30) — Soaring Flight Manual

GLI
5797. During an aerotow, moving from the inside to the outside of the towplane's flightpath during a turn will cause the

A—towline to slacken.
B—glider's airspeed to increase, resulting in a tendency to climb.
C—glider's airspeed to decrease, resulting in a tendency to descend.

When a glider turns on a radius outside that of the towplane, it not only goes faster but also farther. Any increase in speed will cause a climb, assuming there is no change in pitch. (N30) — Soaring Flight Manual

GLI
5798. During an aerotow, is it good operating practice to release from a low-tow position?

A—No. The tow ring may strike and damage the glider after release.
B—No. The towline may snap forward and strike the towplane after release.
C—Yes. Low-tow position is the correct position for releasing from the towplane.

In the low-tow position, upon release, the tow ring may snap back and strike the glider. (N30) — Soaring Flight Manual

GLI
5799. During an aerotow, if slack develops in the towline, the glider pilot should correct this situation by

A—making a shallow-banked coordinated turn to either side.
B—increasing the glider's pitch attitude until the towline becomes taut.
C—yawing the glider's nose to one side with rudder while keeping the wings level with the ailerons.

Yawing should be used to remove slack from the towline. (N30) — Soaring Flight Manual

GLI
5800. During aerotow takeoffs in crosswind conditions, the glider starts drifting downwind after becoming airborne and before the towplane lifts off. The glider pilot should

A—not correct for a crosswind during this part of the takeoff.
B—crab into the wind to remain in the flightpath of the towplane.
C—hold upwind rudder in order to crab into the wind and remain in the flightpath of the towplane.

A crabbing heading should be held to make it easier for the towplane to stay lined up with the runway. (N30) — Soaring Flight Manual

GLI
5801. When should the wing runner raise the glider's wing to the level position in preparation for takeoff?

A—When the towplane pilot fans the towplane's rudder.
B—When the glider pilot is seated and has fastened the safety belt.
C—After the glider pilot gives a thumbs-up signal to take up towline slack.

This indicates the pilot is fully ready to take control of the glider for takeoff. (N30) — Soaring Flight Manual

Answers

5796 [B]   5797 [B]   5798 [A]   5799 [C]   5800 [B]   5801 [C]

GLI
5802. During an aerotow, the sailplane moves to one side of the towplane's flightpath. This was most likely caused by

A—variations in the heading of the towplane.
B—entering wingtip vortices created by the towplane.
C—flying the sailplane in a wing-low attitude or holding unnecessary rudder pressure.

Flying with a wing low, or with unnecessary rudder pressure, can cause a sailplane to move off centerline. (N30) — Soaring Flight Manual

GLI
5803. In which manner should the sailplane be flown while turning during an aerotow? By

A—flying inside the towplane's flightpath.
B—flying outside the towplane's flightpath.
C—banking at the same point in space where the towplane banked and using the same degree of bank and rate of roll.

The bank angle of the glider should be the same as that of the towplane. (N30) — Soaring Flight Manual

GLI
5804. What corrective action should the sailplane pilot take during takeoff if the towplane is still on the ground and the sailplane is airborne and drifting to the left?

A—Crab into the wind to maintain a position directly behind the towplane.
B—Establish a right wing-low drift correction to remain in the flightpath of the towplane.
C—Wait until the towplane becomes airborne before attempting to establish a drift correction.

A crabbing heading should be held to make it easier for the towplane to stay lined up with the runway. (N30) — Soaring Flight Manual

GLI
5805. At what point during an autotow should the glider pilot establish the maximum pitch attitude for the climb?

A—Immediately after takeoff.
B—100 feet above the ground.
C—200 feet above the ground.

The pitch angle should not exceed 15° at 50 feet (indicated altitude), 30° at 100 feet, and 45° at 200 feet. (N31) — Soaring Flight Manual

GLI
5806. When preparing for an autotow with a strong crosswind, where should the glider and towrope be placed?

A—Straight behind the tow car.
B—Obliquely to the line of takeoff on the upwind side of the tow car.
C—Obliquely to the line of takeoff on the downwind side of the tow car.

The tow rope should be laid obliquely to the line of takeoff, to preclude the possibility of the glider overrunning the rope during takeoff. (N31) — Soaring Flight Manual

GLI
5807. Which is true regarding the use of glider tow hooks?

A—The use of a CG hook for auto or winch tows allows the sailplane greater altitude for a given line length.
B—The use of a CG hook for aerotows allows better directional control at the start of the launch than the use of a nose hook.
C—The use of a nose hook for an auto or winch launch reduces structural loading on the tail assembly compared to the use of a CG hook.

There is a tendency to pitchup when using a belly hook on a winch tow. The CG hook allows the sailplane to gain a greater altitude with a given length of line. (N31) — Soaring Flight Manual

Answers

5802 [C]   5803 [C]   5804 [A]   5805 [C]   5806 [C]   5807 [A]

Chapter 2 Aircraft Systems

GLI
5808. GIVEN:

Glider's max auto/winch tow speed	66 MPH
Surface wind (direct headwind)	5 MPH
Wind gradient	4 MPH

When the glider reaches an altitude of 200 feet the auto/winch speed should be

A—42 MPH.
B—46 MPH.
C—56 MPH.

The tow speed for straight auto tow may be calculated as follows:

1. *Subtract the surface wind from the placard speed.*
2. *Subtract an additional 5 miles per hour as a safety factor.*
3. *After the sailplane has climbed to an altitude between 100 and 200 feet, the tow speed should be reduced an additional 10 MPH.*
4. *Subtract the surface wind again to accommodate wind gradient increases.*

(N31) — Soaring Flight Manual

GLI
5809. The towrope breaks when at the steepest segment of the climb during a winch launch. To recover to a normal gliding attitude, the pilot should

A—relax the back stick pressure to avoid excessive loss of altitude.
B—apply forward pressure until the buffeting sound and vibration disappear.
C—move the stick fully forward immediately and hold it there until the nose crosses the horizon.

If the power fails or the towline breaks below 200 feet, the pilot should quickly and smoothly lower the nose until it crosses the horizon, pull the release handle, and land straight ahead making turns only to avoid objects on the ground. (N31) — Soaring Flight Manual

GLI
5810. Which would cause pitch oscillations or porpoising during a winch launch?

A—Excessive winch speed.
B—Insufficient winch speed.
C—Excessive slack in the towline.

Porpoising or a rapid pitch oscillation may occur as the sailplane approaches the top of the climb. This phenomenon occurs as a result of the horizontal stabilizer stalling and unstalling in combination with the downward pull of the tow cable. Launching into rough air, jerky movements of the elevator, or too fast a tow can lead to or aggravate porpoising. (N31) — Soaring Flight Manual

GLI
5811. During a ground tow, the pitch angle of the glider should not exceed

A—10° at 50 feet, 20° at 100 feet, and 45° at 200 feet.
B—15° at 50 feet, 30° at 100 feet, and 45° at 200 feet.
C—15° at 50 feet, 20° at 100 feet, and 40° at 200 feet.

The pitch angle should not exceed 15° at 50 feet (indicated altitude), 30° at 100 feet, and 45° (maximum) at 200 feet. (N31) — Soaring Flight Manual

GLI
5812. To stop pitch oscillation during a winch launch, the pilot should

A—increase the back pressure on the control stick and steepen the angle of climb.
B—relax the back pressure on the control stick and shallow the angle of climb.
C—extend and retract the spoilers several times until the oscillations subside.

With full-up elevator (stick back) the angle of attack of the horizontal tail surface becomes so great that it stalls, thus reducing the downward force and allowing the sailplane nose to pitch down. This eventually results in a decrease in the angle of attack of the horizontal tail surface, reinstating the downward forces which pitches the sailplane nose-up. This develops into a cycling situation and the pilot finds the sailplane is porpoising. The recommended corrective procedure is to release a portion of the back pressure to reduce the angle of climb until the oscillation dampens; then add only part of the back pressure which was released, and thus climb less rapidly without porpoising. (N31) — Soaring Flight Manual

Answers

| 5808 [A] | 5809 [C] | 5810 [A] | 5811 [B] | 5812 [B] |

GLI
5813. What should be expected when making a downwind landing? The likelihood of

A—undershooting the intended landing spot and a faster airspeed at touchdown.
B—overshooting the intended landing spot and a faster groundspeed at touchdown.
C—undershooting the intended landing spot and a faster groundspeed at touchdown.

In this situation, the pilot should aim at the near end of the runway, because of the tailwind increasing the glider's ground speed. (N32) — Soaring Flight Manual

GLI
5814. What corrective action should be taken, if while thermalling at minimum sink speed in turbulent air, the left wing drops while turning to the left?

A—Apply right rudder pressure to slow the rate of turn.
B—Lower the nose before applying right aileron pressure.
C—Apply right aileron pressure to counteract the overbanking tendency.

The left wing is starting to stall, so airspeed should be increased before starting any rolling maneuvers. (N32) — Soaring Flight Manual

GLI
5815. A rule of thumb for flying a final approach is to maintain a speed that is

A—twice the glider's stall speed, regardless of windspeed.
B—twice the glider's stall speed plus half the estimated windspeed.
C—50 percent above the glider's stall speed plus half the estimated windspeed.

Pattern airspeed should be a minimum of stalling speed plus one-half stalling speed plus one-half estimated wind velocity. (N32) — Soaring Flight Manual

GLI
5816. To stop a ground loop to the left after landing a glider, it would be best to lower the

A—right wing in order to shift the CG.
B—left wing to compensate for crosswind.
C—nose skid to the ground and apply wheel brake.

The nose skid is helpful until the glider starts to swerve, when it may dig in and pivot the glider into a ground loop. Once the swerve is underway, locking the wheel will help—use the brakes hard. (N32) — Soaring Flight Manual

GLI
5817. In which situation is a hazardous stall more likely to occur if inadequate airspeed allowance is made for wind velocity gradient?

A—During the approach to a landing.
B—While thermalling at high altitudes.
C—During takeoff and climb while on aerotow.

An unintentional stall is most likely to occur during thermalling or in the landing pattern. Because of the glider's proximity to the ground in the pattern, insufficient altitude may remain for recovery. (N32) — Soaring Flight Manual

GLI
5818. With regard to two or more gliders flying in the same thermal, which is true?

A—All turns should be to the right.
B—Turns should be in the same direction as the highest glider.
C—Turns should be made in the same direction as the first glider to enter the thermal.

If more than one sailplane is circling in the same direction, the first sailplane in the thermal establishes the direction of turn. Each pilot must constantly be alert for other traffic while thermalling. (N33) — Soaring Flight Manual

GLI
5819. Which is true regarding the direction in which turns should be made during slope soaring?

A—All reversing turns should be made to the left.
B—All reversing turns should be made into the wind away from the slope.
C—The upwind turn should be made to the left; the downwind turn should be made to the right.

A downwind turn toward the slope may force the glider into the hillside. Make all turns away from the ridge into the wind. (N33) — Soaring Flight Manual

Answers

| 5813 | [B] | 5814 | [B] | 5815 | [C] | 5816 | [C] | 5817 | [A] | 5818 | [C] |
| 5819 | [B] |

Chapter 2 Aircraft Systems

GLI
5820. Which airspeed should be used when circling within a thermal?

A—Best L/D speed.
B—Maneuvering speed.
C—Minimum sink speed for the angle of bank.

Minimum sink speed allows for the least rate of descent, which in turn allows for maximum climb in lift. The minimum sink speed for the bank angle being flown should be utilized. (N33) — Soaring Flight Manual

GLI
5821. Which is a recommended procedure for an off-field landing?

A—A recommended landing site would be a pasture.
B—Always land into the wind even if you have to land downhill on a sloping field.
C—If the field slopes, it is usually best to land uphill, even with a tailwind.

The beneficial effects of an uphill landing make such a procedure preferable even in a tail wind. (N34) — Soaring Flight Manual

GLI
5822. What would be a proper action or procedure to use if you are getting too low on a cross-country flight in a sailplane?

A—Fly directly into the wind and make a straight-in approach at the end of the glide.
B—Have a suitable landing area selected upon reaching 2,000 feet AGL, and a specific field chosen upon reaching 1,500 feet AGL.
C—Continue on course until descending to 500 feet, then select a field and confine the search for lift to an area within gliding range of a downwind leg for the field you have chosen.

Always have a suitable landing area in sight and specific field at low altitude. The choices should be narrowed down at 2,000 feet AGL and a specific field chosen upon reaching 1,500 feet AGL. (N34) — Soaring Flight Manual

GLI
5823. What is the proper speed to fly when passing through lift with no intention to work the lift?

A—Best L/D speed.
B—Maximum safe speed.
C—Minimum sink speed.

Even if not stopping to work a thermal, it is advisable to slow to the minimum sink airspeed when passing through the area of lift. (N34) — Soaring Flight Manual

GLI
5824. What is the proper airspeed to use when flying between thermals on a cross-country flight against a headwind?

A—The best L/D speed increased by one-half the estimated wind velocity.
B—The best L/D speed decreased by one-half the estimated wind velocity.
C—The minimum sink speed increased by one-half the estimated wind velocity.

Fly at best L/D or faster. When flying cross-country, it is good practice to increase speed-to-fly by one-half of the estimated wind velocity. (N34) — Soaring Flight Manual

Answers

5820 [C]   5821 [C]   5822 [B]   5823 [C]   5824 [A]

Balloon Operations

LTA
5825. What should a pilot do if a small hole is seen in the fabric during inflation?

A— Continue the inflation and make a mental note of the location of the hole for later repair.
B— Instruct a ground crewmember to inspect the hole, and if under 5 inches in length, continue the inflation.
C— Consult the flight manual to determine if the hole is within acceptable damage limits established for the balloon being flown.

Any hole in the fabric is dangerous because it is a weak point, and any stress on the fabric will allow a larger hole to tear in the fabric. Preventive maintenance in the form of mending rips and tears in the bag may be done by the owners, but major repairs must be done by a certificated airframe and powerplant mechanic, who also performs the annual checkup. Local unfamiliarity with the equipment often makes it advisable to contact the manufacturer for maintenance assistance. Consult the flight manual to determine if the hole is within acceptable damage limits established for the balloon being flown. (O277) — How to Fly a Balloon

LTA
5826. Propane is preferred over butane for fuel in hot air balloons because

A— it has a higher boiling point.
B— it has a lower boiling point.
C— butane is very explosive under pressure.

Propane is preferred over butane and other hydrocarbons in balloon design because propane has a lower boiling point (propane -44°F, butane 32°F) and, therefore, a consistently higher vapor pressure for a given temperature. (O171) — Balloon Digest

LTA
5827. On a balloon equipped with a blast valve, the blast valve is used for

A— climbs only.
B— emergencies only.
C— control of altitude.

The blast valve allows control of the fuel used to heat the air. (O220) — Balloon Ground School

LTA
5828. It may be possible to make changes in the direction of flight in a hot air balloon by

A— using the maneuvering vent.
B— operating at different flight altitudes.
C— flying a constant atmospheric pressure gradient.

The pilot might accomplish a change of direction in flight by changing altitude. (O263) — How to Fly a Balloon

LTA
5829. Regarding lift as developed by a hot air balloon, which is true?

A— The higher the temperature of the ambient air, the greater the lift for any given envelope temperature.
B— The greater the difference between the temperature of the ambient air and the envelope air, the greater the lift.
C— The smaller the difference between the temperature of the ambient air and the envelope air, the greater the lift.

The primary lifting force is brought about because of a temperature differential, and therefore a density differential, between the outside air and the air inside the envelope. (O252) — How to Fly a Balloon

LTA
5830. What causes false lift which sometimes occurs during launch procedures?

A— Closing the maneuvering vent too rapidly.
B— Excessive temperature within the envelope.
C— Venturi effect of the wind on the envelope.

False lift is created when there is a wind blowing across the top of an inflated envelope, causing a venturi effect. This lowers the pressure above the balloon, creating a dynamic lifting force. As the balloon is accelerated by the wind, this force decreases. (O220) — Balloon Ground School

Answers

| 5825 | [C] | 5826 | [B] | 5827 | [C] | 5828 | [B] | 5829 | [B] | 5830 | [C] |

Chapter 2 Aircraft Systems

LTA

5831. The lifting forces which act on a hot air balloon are primarily the result of the interior air

A—pressure being greater than ambient pressure.
B—temperature being less than ambient temperature.
C—temperature being greater than ambient temperature.

The theory of balloon flight is basically the theory of lift as applied to an envelope which traps a gas that is lighter, or less dense, than the ambient atmosphere. Hot air is less dense than cool air. (O220) — Balloon Ground School

LTA

5832. While in flight, ice begins forming on the outside of the fuel tank in use. This would most likely be caused by

A—water in the fuel.
B—a leak in the fuel line.
C—vaporized fuel instead of liquid fuel being drawn from the tank into the main burner.

Vaporized fuel being drawn off reduces the tank pressure, allowing liquid fuel in the tank to boil off, reducing the temperature of the tank. (This is a good procedure for reducing tank pressure on a hot day.) (O257) — How to Fly a Balloon

LTA

5833. If ample fuel is available, within which temperature range will propane fuel vaporize sufficiently to provide enough fuel pressure for burner operation during flight?

A—0°F to 30°F.
B—10°F to 30°F.
C—30°F to 90°F.

When ample liquid propane is available, propane will vaporize sufficiently to provide proper operation between 30° and 90°F. (O270) — How to Fly a Balloon

LTA

5834. When landing a balloon, what should the occupant(s) do to minimize landing shock?

A—Be seated on the floor of the basket.
B—Stand back-to-back and hold onto the load ring.
C—Stand with knees slightly bent facing the direction of movement.

By facing forward, the body is best balanced against tipping, and the bent legs will absorb the landing shock. (O265) — How to Fly a Balloon

LTA

5835. One means of vertical control on a gas balloon is

A—by using the rip panel rope.
B—valving gas or releasing ballast.
C—opening and closing the appendix.

Altitude in a gas balloon is controlled by valving gas or releasing ballast. (O261) — How to Fly a Balloon

LTA

5836. To perform a normal descent in a gas balloon, it is necessary to release

A—air.
B—gas.
C—ballast.

By releasing gas, a balloon will have reduced lift and will descend. (O150) — Balloon Digest

LTA

5837. What would cause a gas balloon to start a descent if a cold air mass is encountered and the envelope becomes cooled?

A—The expansion of the gas.
B—The contraction of the gas.
C—A barometric pressure differential.

As gas cools, it contracts, reducing its lift capacity. (O150) — Balloon Digest

LTA

5838. If a balloon inadvertently descends into stratus clouds and is shielded from the Sun, and if no corrections are made, one can expect to descend

A—more slowly.
B—more rapidly.
C—at an unchanged rate.

In order for clouds to form, water vapor condenses to liquid. During this process, the latent heat of condensation of water is about 600 calories per gram. This heat released during condensation is an important source of

Answers

| 5831 | [C] | 5832 | [C] | 5833 | [C] | 5834 | [C] | 5835 | [B] | 5836 | [B] |
| 5837 | [B] | 5838 | [B] | | | | | | | | | |

Chapter 2 Aircraft Systems

energy for the maintenance of thunderstorms, etc. While descending into a cloud layer, a temperature rise can be expected, reducing the difference in temperature between the balloon and the atmosphere increasing the descent. The rate of descent will also increase because of the loss of solar heating. (O220) — Balloon Ground School

LTA
5839. What action is most appropriate when an envelope overtemperature condition occurs?

A—Turn the main burner OFF.
B—Land as soon as practical.
C—Throw all unnecessary equipment overboard.

Overtemping the envelope is dangerous because it weakens the fabric; land as soon as possible if this condition occurs. (O220) — Balloon Ground School

LTA
5840. Which is the proper way to detect a fuel leak?

A—Sight.
B—Use of smell and sound.
C—Check fuel pressure gauge.

Propane under pressure will cause a hissing sound when leaking and it has an artificial odor added to aid in detection. (O220) — Balloon Ground School

LTA
5841. What is the weight of propane?

A—4.2 pounds per gallon.
B—6.0 pounds per gallon.
C—7.5 pounds per gallon.

Propane weighs 4.2 pounds per gallon. (O220) — Balloon Ground School

LTA
5842. What effect, if any, does ambient temperature have on propane tank pressure?

A—It has no effect.
B—As temperature decreases, propane tank pressure decreases.
C—As temperature decreases, propane tank pressure increases.

Propane boils at -40°F. The colder the ambient air, the colder the tank and the less pressure the tank will have. The greater the pressure, the greater the volume of fuel to the burner. (O220) — Balloon Ground School

LTA
5843. Why is it considered a good practice to blast the burner after changing fuel tanks?

A—To check for fuel line leaks.
B—It creates an immediate source of lift.
C—To ensure the new tank is functioning properly.

This check will ensure that a full, properly functioning tank has been selected before its use becomes critical. (O220) — Balloon Ground School

LTA
5844. For what reason is methanol added to the propane fuel of hot air balloons?

A—As a fire retardant.
B—As an anti-icing additive.
C—To reduce the temperature.

Methanol mixes with water and acts as an antifreeze. (O220) — Balloon Ground School

LTA
5845. To respond to a small leak around the stem of a Rego blast valve in a single-burner system balloon, one should

A—turn off the fuel system and make an immediate landing.
B—continue operating the blast valve making very small quick blasts until a good landing field appears.
C—continue operating the blast valve, making long infrequent blasts and opening the handle slightly to reduce leakage until a good landing field appears.

If a propane leak develops anywhere in the fuel system during flight, the pilot should, as soon as possible, close the main tank valves, set open all of the other control valves in the fuel system, and pilot the balloon to a more or less normal landing by opening and closing the main tank valve to control the heat output of the burners. (O170) — Balloon Digest

---

Answers

| 5839 | [B] | 5840 | [B] | 5841 | [A] | 5842 | [B] | 5843 | [C] | 5844 | [B] |
| 5845 | [A] | | | | | | | | | | |

Chapter 2 Aircraft Systems

LTA

5846. Which action would be appropriate if a small leak develops around the stem of the tank valve, and no other tanks have sufficient fuel to reach a suitable landing field?

A—Warm the tank valve leak with your bare hand.
B—Turn the leaking tank handle to the full-open position.
C—Turn off the tank, then slowly reopen to reseat the seal.

Tank valves have a stem seal built in, when the valve is wide open. (O170) — Balloon Digest

LTA

5847. Why should propane lines be bled after use?

A—Fire may result from spontaneous combustion.
B—The propane may expand and rupture the lines.
C—If the temperature is below freezing, the propane may freeze.

The fuel in the lines is in a liquid state. Any heating would cause the fuel to expand and possibly rupture the line. (O220) — Balloon Ground School

LTA

5848. The purpose of the preheating coil as used in hot air balloons is to

A—prevent ice from forming in the fuel lines.
B—warm the fuel tanks for more efficient fuel flow.
C—vaporize the fuel for more efficient burner operation.

Hot air balloon burners perform the following three functions:

1. *Vaporize the liquid propane supplied to them;*
2. *Mix the propane vapor with air to form a combustible mixture; and*
3. *Burn the resulting mixture to form an essentially directional flow of very hot gases.*

All burners commonly in use on hot air balloons have preheat coils surrounding the base of the flame. The liquid propane flows through these coils on its way through the burner. Since the coils are heated directly by the flame, they are hot enough to vaporize the liquid propane flowing through them. If the propane is not vaporized, it does not mix well with the air, and burns in a long, yellow flame which radiates a great amount of heat. Properly vaporized propane burns with a mostly blue flame. (O155) — Balloon Digest

LTA

5849. The best way to determine burner BTU availability is the

A—burner sound.
B—tank quantity.
C—fuel pressure gauge.

BTUs (British Thermal Units) are a measure of heat per a given time. The greater the pressure the greater the volume of fuel and therefore, the greater the heat. (O170) — Balloon Digest

LTA

5850. The practice of allowing the ground crew to lift the balloon into the air is

A—a safe way to reduce stress on the envelope.
B—unsafe because it can lead to a sudden landing at an inopportune site just after lift-off.
C—considered to be a good operating practice when obstacles must be cleared shortly after lift-off.

The practice of allowing ground crew to lift the balloon in an attempt to shove it up into the air is unsafe, for it can lead to a sudden landing at an inopportune site just after lift off. (O30) — Powerline Excerpts

LTA

5851. Why is false lift dangerous?

A—Pilots are not aware of its effect until the burner sound changes.
B—To commence a descent, the venting of air will nearly collapse the envelope.
C—When the balloon's horizontal speed reaches the windspeed, the balloon could descend into obstructions downwind.

Sometimes false lift occurs, caused by the venturi effect produced by the wind blowing across an inflated, but stationary, envelope. This is known as aerodynamic lift, created by relative air movement. When the balloon is released, the relative wind decreases as the balloon accelerates to the wind's speed, and false lift decreases. (O30) — Powerline Excerpts

Answers

| 5846 [B] | 5847 [B] | 5848 [C] | 5849 [C] | 5850 [B] | 5851 [C] |

LTA
5852. If you are over a heavily-wooded area with no open fields in the vicinity and have only about 10 minutes of fuel remaining, you should

A—stay low and keep flying in hope that you will find an open field.
B—climb as high as possible to see where the nearest landing field is.
C—land in the trees while you have sufficient fuel for a controlled landing.

A controlled landing is always best, to minimize damage and injury. (O30) — Powerline Excerpts

LTA
5853. Which precaution should be exercised if confronted with the necessity of having to land when the air is turbulent?

A—Land in the center of the largest available field.
B—Throw propane equipment overboard immediately prior to touchdown.
C—Land in the trees to absorb shock forces, thus cushioning the landing.

Turbulent air can and will suddenly change direction. If the air is turbulent, land in the center of the largest available field. (O30) — Powerline Excerpts

LTA
5854. False lift occurs whenever a balloon

A—ascends rapidly.
B—ascends due to solar assistance.
C—ascends into air moving faster than the air below.

Sometimes false lift occurs, caused by the venturi effect produced by the wind blowing across an inflated, but stationary, envelope. This is known as aerodynamic lift, created by relative air movement. When the balloon is released, the relative wind decreases as the balloon accelerates to the wind's speed, and false lift decreases. (O30) — Powerline Excerpts

LTA
5855. What is the relationship of false lift to the wind? False lift

A—exists only if the surface winds are calm.
B—increases if the vertical velocity of the balloon increases.
C—decreases as the wind accelerates the balloon to the same speed as the wind.

Sometimes false lift occurs, caused by the venturi effect produced by the wind blowing across an inflated, but stationary, envelope. This is known as aerodynamic lift, created by relative air movement. When the balloon is released, the relative wind decreases as the balloon accelerates to the wind's speed, and false lift decreases. (O30) — Powerline Excerpts

LTA
5856. The weigh-off procedure is useful because the

A—pilot can adjust the altimeter to the correct setting.
B—ground crew can assure that downwind obstacles are cleared.
C—pilot will learn what the equilibrium conditions are prior to being committed to fly.

The weigh-off procedure allows a pilot to determine the equilibrium of the balloon before lift-off. (O30) — Powerline Excerpts

LTA
5857. One characteristic of nylon rope is that it

A—is flexible.
B—does not stretch.
C—splinters easily.

Nylon is light and flexible and very strong for its weight. (O30) — Powerline Excerpts

LTA
5858. Why is nylon rope good for tethering a balloon?

A—It does not stretch under tension.
B—It is not flexible and therefore can withstand greater tension without breaking.
C—It stretches under tension, but recovers to normal size when tension is removed, giving it excellent shock absorbing qualities.

Nylon has great elasticity; therefore it is a good shock absorber. (O30) — Powerline Excerpts

Answers

| 5852 | [C] | 5853 | [A] | 5854 | [C] | 5855 | [C] | 5856 | [C] | 5857 | [A] |
| 5858 | [C] | | | | | | | | | | |

Chapter 2 Aircraft Systems

LTA

5859. One advantage nylon rope has over manila rope is that it

A—will not stretch.
B—is nearly three times as strong.
C—does not tend to snap back if it breaks.

Nylon is a synthetic material and is not as affected by sunlight, and is nearly 3 times as strong as manila rope. (O30) — Powerline Excerpts

LTA

5860. A pilot should be aware that drag ropes constructed of hemp or nylon

A—should be a maximum of 100 feet long and used only in gas balloons.
B—can be considered safe because they will not conduct electricity.
C—can conduct electricity when contacting powerlines carrying 600 volts or more current if they are not clean and dry.

Any material will conduct electricity through the dirt and water on it. (O30) — Powerline Excerpts

LTA

5861. If powerlines become a factor during a balloon flight, a pilot should know that

A—it is safer to contact the lines than chance ripping.
B—contact with powerlines creates no great hazard for a balloon.
C—it is better to chance ripping at 25 feet above the ground than contacting the lines.

Powerlines are the most hazardous obstacle to ballooning. (O30) — Powerline Excerpts

LTA

5862. The windspeed is such that it is necessary to deflate the envelope as rapidly as possible during a landing. When should the deflation port (rip panel) be opened?

A—Prior to ground contact.
B—The instant the gondola contacts the surface.
C—As the balloon skips off the surface the first time and the last of the ballast has been discharged.

In a high-wind landing, ripping out at one foot above the ground for each mile per hour of speed will minimize ground slide and potential damage to the balloon. (O30) — Powerline Excerpts

LTA

5863. The term "to weigh off" as used in ballooning means to determine the

A—standard weight and balance of the balloon.
B—static equilibrium of the balloon as loaded for flight.
C—amount of gas required for an ascent to a preselected altitude.

The weigh-off procedure allows a pilot to determine the equilibrium of the balloon before lift-off. (O30) — Powerline Excerpts

LTA

5864. Superheat is a term used to describe the condition which exists

A—when the surrounding air is at least 10° warmer than the gas in the envelope.
B—when the Sun heats the envelope surface to a temperature at least 10° greater than the surrounding air.
C—relative to the difference in temperature between the gas in the envelope and the surrounding air caused by the Sun.

Superheat is the term used to describe the difference in temperature between the gas in the envelope and the surrounding air. (P01) — Goodyear Airship Operations Manual

LTA

5865. How does the pilot know when pressure height has been reached? Liquid in the gas

A—and air manometers will fall below the normal level.
B—manometer will fall and the liquid in the air manometer will rise above normal levels.
C—manometer will rise and the liquid in the air manometer will fall below normal levels.

The pilot will know when pressure height has been reached when liquid in the gas manometer rises and the liquid in the air manometer falls below normal levels. (P01) — Goodyear Airship Operations Manual

Answers

| 5859 | [B] | 5860 | [C] | 5861 | [C] | 5862 | [A] | 5863 | [B] | 5864 | [C] |
| 5865 | [C] | | | | | | | | | | |

Airship Operations

LTA
5866. The ballonet volume of an airship envelope with respect to the total gas volume is approximately

A—15 percent.
B—25 percent.
C—30 percent.

Ballonet volume is about 25% of total gas volume. (P01) — Goodyear Airship Operations Manual

LTA
5867. The pressure height with any airship is that height at which

A—both ballonets are empty.
B—both ballonets are inflated.
C—gas pressure is 3 inches of water.

Pressure height is reached when the ballonets are deflated. (P01) — Goodyear Airship Operations Manual

LTA
5868. If both engines fail while en route, an airship should be

A—brought to a condition of equilibrium as soon as possible and free-ballooned.
B—trimmed nose-heavy to use the airship's negative dynamic lift to fly the airship down to the landing site.
C—trimmed nose-light to use the airship's positive dynamic lift to control the angle and rate of descent to the landing site.

An airship without engine power must be flown as a free balloon. (P03) — Goodyear Airship Operations Manual

LTA
5869. If an airship in flight is either light or heavy, the unbalanced condition must be overcome

A—by valving air.
B—aerodynamically.
C—by releasing ballast.

A light or heavy condition must be overcome aerodynamically. (P04) — Goodyear Airship Operations Manual

LTA
5870. Maximum headway in an airship is possible only under which condition?

A—Slightly nosedown.
B—Slightly tail down.
C—Flying in equilibrium.

Flying in equilibrium will produce the smallest frontal area and least drag. (P04) — Goodyear Airship Operations Manual

LTA
5871. To accomplish maximum headway, the airship must be kept

A—at equilibrium.
B—heavy and flown dynamically positive.
C—heavy by the bow and light by the stern.

Flying in equilibrium will produce the smallest frontal area and least drag. (P04) — Goodyear Airship Operations Manual

LTA
5872. Damper valves should normally be kept closed during a maximum rate climb to altitude because any air forced into the system would

A—decrease the volume of gas within the envelope.
B—decrease the purity of the gas within the envelope.
C—increase the amount of air to be exhausted, resulting in a lower rate of ascent.

Air entering through the damper valves would have to be exhausted as well as the air in the ballonets. (P05) — Goodyear Airship Operations Manual

LTA
5873. When checking gas pressure (pressure height) of an airship during a climb, the air damper valves should be

A—opened.
B—closed.
C—opened aft and closed forward.

Any ram pressure will keep the ballonet pressure too high and prevent deflation at pressure height. Therefore, the air damper valves should be closed when checking gas pressure during a climb. (P05) — Goodyear Airship Operations Manual

Answers

| 5866 | [B] | 5867 | [A] | 5868 | [A] | 5869 | [B] | 5870 | [C] | 5871 | [A] |
| 5872 | [C] | 5873 | [B] | | | | | | | | | | |

Chapter 2 **Aircraft Systems**

LTA
5874. Which take-off procedure is considered to be most hazardous?

A—Failing to apply full engine power properly on all takeoffs, regardless of wind.
B—Maintaining only 50 percent of the maximum permissible positive angle of inclination.
C—Maintaining a negative angle of inclination during takeoff after elevator response is adequate for controllability.

The most hazardous takeoff condition would be maintaining a negative angle of inclination during takeoff, after elevator response is adequate for stability. (P11) — Goodyear Airship Operations Manual

LTA
5875. The purpose of a ground weigh-off is to determine the

A—useful lift of the airship.
B—gross weight of the airship.
C—static condition of the airship and the condition of trim.

The purpose of the ground weigh-off is to determine the static condition of the airship and the condition of trim. (P11) — Goodyear Airship Operations Manual

LTA
5876. When operating an airship with the ballonet air valve in the automatic forward position, the aft valve locks should not be engaged with either after-damper open because

A—ballonet overinflation and rupture may occur.
B—the aircraft will enter an excessive bow-high attitude.
C—the aircraft will enter an excessive stern-high attitude.

Under the conditions described, the aft ballonet could inflate, causing a bow-high condition. (P11) — Goodyear Airship Operations Manual

LTA
5877. Which action is necessary to perform a normal descent in an airship?

A—Valve gas.
B—Valve air.
C—Take air into the aft ballonets.

An airship is normally flown heavy and so a decrease in power will result in a descent. Valving gas will also cause a descent. (P11) — Goodyear Airship Operations Manual

LTA
5878. To land an airship that is 250 pounds heavy when the wind is calm, the best landing can usually be made if the airship is

A—in trim.
B—nose-heavy approximately 20°.
C—tail-heavy approximately 20°.

A heavy airship should be landed tail heavy. (P11) — Goodyear Airship Operations Manual

LTA
5879. A heavy airship flying dynamically with air ballasted forward to overcome a climbing tendency and slowed down for a weigh-off in the air prior to landing, will be very bow heavy. This condition must be corrected prior to landing by

A—ballasting air aft.
B—discharging forward ballast.
C—dumping fuel from the forward tanks.

Air must be ballasted aft to overcome the bow-heavy condition. (P11) — Goodyear Airship Operations Manual

LTA
5880. If an airship should experience failure of both engines during flight and neither engine can be restarted, what initial immediate action must the pilot take?

A—Immediate preparations to operate the airship as a free balloon are necessary.
B—The airship must be driven down to a landing before control and envelope shape are lost.
C—The emergency auxiliary power unit must be started for electrical power to the airscoop blowers so that ballonet inflation can be maintained.

An airship without engine power must be flown as a free balloon. (P11) — Goodyear Airship Operations Manual

Answers

| 5874 | [C] | 5875 | [C] | 5876 | [B] | 5877 | [A] | 5878 | [C] | 5879 | [A] |
| 5880 | [A] | | | | | | | | | | |

LTA
5881. Critical factors affecting the flight characteristics and controllability of an airship are

A—airspeed and power.
B—static and dynamic trim.
C—temperature and atmospheric density.

Critical factors affecting the flight characteristics and controllability of an airship are airspeed and power. (P12) — Goodyear Airship Operations Manual

Airship IFR Operations

LTA
5555. For airship IFR operations off established airways, ROUTE OF FLIGHT portion of an IFR flight plan should list VOR navigational aids which are no more than

A—40 miles apart.
B—70 miles apart.
C—80 miles apart.

Operation off established airways below 18,000 feet MSL use aids not more than 80 NM apart. These aids are depicted on Enroute Low-Altitude Charts. (J15) — AIM ¶5-1-7

LTA
5562. When operating an airship under IFR with a VFR-on-top clearance, what altitude should be maintained?

A—The last IFR altitude assigned by ATC.
B—An IFR cruising altitude appropriate to the magnetic course being flown.
C—A VFR cruising altitude appropriate to the magnetic course being flown and as restricted by ATC.

When operating in VFR conditions with an ATC authorization to "Maintain VFR-On-Top/maintain VFR conditions," pilots on IFR flight plans must fly at the appropriate VFR altitude as prescribed in 14 CFR §91.159. When operating below 18,000 feet MSL and:

1. *On a magnetic course of 0° through 179°, any odd thousand-foot MSL altitude plus 500 feet.*
2. *On a magnetic course of 180° through 359°, any even thousand-foot MSL altitude plus 500 feet.*

(J14) — 14 CFR §91.179 and §91.159

LTA
5563. Does the ATC term, "cruise 3000", apply to airship IFR operations?

A—No, this term applies to airplane IFR operations only.
B—Yes, it means that any assigned altitude can be vacated without notifying ATC.
C—Yes, in part, it authorizes the pilot to commence the approach at the destination airport at the pilot's discretion.

The term "cruise" may be used instead of "maintain" to assign a block of airspace to a pilot, from the minimum IFR altitude up to and including the altitude specified in the cruise clearance. The pilot may level off at any intermediate altitude within this block of airspace. Climb/descent within the block is to be made at the discretion of the pilot. However, once the pilot starts descent and verbally reports leaving an altitude in the block, he may not return to that altitude without additional ATC clearance. Also, it is approval for the pilot to proceed to and make an approach to the destination airport. (J33) — AIM ¶4-4-3

LTA
5603. You are flying an airship under an IFR flight plan and experience two-way communications radio failure while in VFR conditions. In this situation, you should continue your flight under

A—VFR and land as soon as practicable.
B—VFR and proceed to your flight-plan destination.
C—IFR and maintain the last assigned route and altitude to your flight-plan destination.

A radio failure in VFR conditions requires that the aircraft remain VFR and land as soon as practicable. (B08) — 14 CFR §91.185(b)

Answers

5881 [A]   5555 [C]   5562 [C]   5563 [C]   5603 [A]

Chapter 3
Flight Instruments

Airspeed Indicator *3–3*

Altitude Definitions *3–6*

Magnetic Compass *3–6*

Gyroscopic Instruments and Systems *3–7*

Attitude Instrument Flying *3–8*

Airspeed Indicator

The airspeed indicator in a light airplane shows some of the airspeed limitations of the aircraft by means of colored arcs. On aircraft manufactured prior to 1978, these arcs are calibrated airspeed. The arcs on later aircraft are indicated airspeed.

The white arc is the flap operating range. The low-speed end of the white arc is V_{S0}, which is the stalling speed, or the minimum steady-flight speed in the landing configuration. The high-speed end of the white arc is the maximum flap extended speed. Flight at airspeeds greater than V_{FE} with the flaps extended can impose excessive loads on the flaps and wing structure.

The green arc is the normal operating range. The low-speed end is V_{S1}, which is the stalling speed or the minimum steady-flight speed in a specified configuration. At the high-speed end of the green arc is V_{NO} (maximum structural cruising speed).

The yellow arc begins at V_{NO} and continues to the red line, V_{NE} (never exceed speed). Operations may be conducted only in smooth air and with caution.

Other speed limitations which are not color-coded on the airspeed indicator include:

V_S—Stalling speed or minimum steady flight speed at which the airplane is controllable.

V_F—Design flap speed.

V_{LE}—Maximum landing gear extended speed.

V_A—Design maneuvering speed. If severe turbulence (for example, significant clear air turbulence) is encountered during flight, the pilot should reduce the airspeed to the design maneuvering speed. In addition to setting the power and trimming to obtain an airspeed at or below maneuvering speed, the wings should be kept level, and allow slight variations of airspeed and altitude. This technique will help minimize the wing load factor in severe turbulence. Maneuvering speed is also the maximum speed at which full or abrupt control movements may be made. Maneuvering speed decreases as gross weight decreases. *See* Figure 3-1.

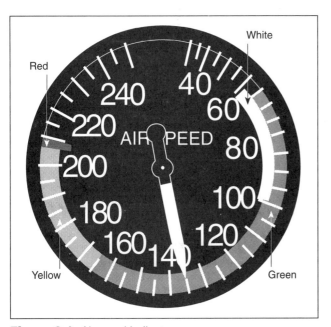

Figure 3-1. Airspeed indicator

AIR, GLI

5233. (Refer to Figure 5.) The vertical line from point D to point G is represented on the airspeed indicator by the maximum speed limit of the

A—green arc.
B—yellow arc.
C—white arc.

The high speed limit of the green arc is the maximum speed for normal operation. This is designated as maximum structural cruising speed (V_{NO}), which is line D to G. (H66) — AC 61-23C, Chapter 3

Answer (B) is incorrect because the high speed limit of the yellow arc is the never exceed speed (V_{NE}), which is line E to F. Answer (C) is incorrect because the high speed limit of the white arc is the maximum flap extended speed (V_{FE}), which is not indicated on the chart.

Answers

5233 [A]

Chapter 3 **Flight Instruments**

AIR, GLI, MIL
5013. Which is the correct symbol for the stalling speed or the minimum steady flight speed in a specified configuration?

A—V_S.
B—V_{S1}.
C—V_{S0}.

V_{S1} is the stalling speed or the minimum steady flight speed obtained in a specific configuration. (A02) — 14 CFR §1.2

Answer (A) is incorrect because V_S is the stalling speed or the minimum steady flight speed at which the airplane is controllable. Answer (C) is incorrect because V_{S0} is the stalling speed or the minimum steady flight speed in the landing configuration.

AIR, GLI, MIL
5014. Which is the correct symbol for the stalling speed or the minimum steady flight speed at which the airplane is controllable?

A—V_S.
B—V_{S1}.
C—V_{S0}.

V_S is the stalling speed or the minimum steady flight speed at which the airplane is controllable. (A02) — 14 CFR §1.2

Answer (B) is incorrect because V_{S1} is the stalling speed or the minimum steady flight speed obtained in a specified configuration. Answer (C) is incorrect because V_{S0} is the stalling speed or the minimum steady flight speed in the landing configuration.

AIR, GLI, MIL
5015. 14 CFR Part 1 defines V_F as

A—design flap speed.
B—flap operating speed.
C—maximum flap extended speed.

V_F is the design flap speed. (A02) — 14 CFR §1.2

Answer (B) is incorrect because the flap operating range is indicated by the white arc on the airspeed indicator. Answer (C) is incorrect because V_{FE} is the maximum flap extended speed.

AIR, MIL
5016-1. 14 CFR Part 1 defines V_{LE} as

A—maximum landing gear extended speed.
B—maximum landing gear operating speed.
C—maximum leading edge flaps extended speed.

V_{LE} is the maximum landing gear extended speed. (A02) — 14 CFR §1.2

Answer (B) is incorrect because V_{LO} is the maximum landing gear operating speed. Answer (C) is incorrect because maximum leading edge flaps extended speed is not defined in 14 CFR Part 1.

AIR, GLI, MIL
5016-2. 14 CFR Part 1 defines V_{NE} as

A—maximum nose wheel extend speed.
B—never-exceed speed.
C—maximum landing gear extended speed.

V_{NE} is the never-exceed speed. (A02) — 14 CFR §1.2

Answer (A) is incorrect because this speed is not defined in 14 CFR Part 1. Answer (C) is incorrect because this is V_{LE}.

AIR, GLI, MIL
5016-3. 14 CFR Part 1 defines V_Y as

A—speed for best rate of descent.
B—speed for best angle of climb.
C—speed for best rate of climb.

V_Y is the speed for best rate of climb. (A02) — 14 CFR §1.2

Answer (A) is incorrect because this is not defined in 14 CFR Part 1. Answer (B) is incorrect because this is V_X.

AIR, GLI, MIL
5177. Which airspeed would a pilot be unable to identify by the color coding of an airspeed indicator?

A—The never-exceed speed.
B—The power-off stall speed.
C—The maneuvering speed.

Maneuvering speed is not color-coded on airspeed indicators. (H312) — AC 61-23C, Chapter 3

Answer (A) is incorrect because the never-exceed speed is identified by the red line at the high-speed end of the yellow arc. Answer (B) is incorrect because the power-off stall speed is identified by the low-speed end of the white arc for landing configuration (V_{S0}), and the low-speed end of the green arc for clean configuration (V_{S1}).

ALL
5604. Why should flight speeds above V_{NE} be avoided?

A—Excessive induced drag will result in structural failure.
B—Design limit load factors may be exceeded, if gusts are encountered.
C—Control effectiveness is so impaired that the aircraft becomes uncontrollable.

Answers

| 5013 | [B] | 5014 | [A] | 5015 | [A] | 5016-1 | [A] | 5016-2 | [B] | 5016-3 | [C] |
| 5177 | [C] | 5604 | [B] | | | | | | | | |

Any speed above V_{NE} can cause damage; therefore, flight above this speed should be avoided even in smooth air. (H312) — AC 61-23C, Chapter 3

Answer (A) is incorrect because induced drag decreases with increased airspeed. Answer (C) is incorrect because control effectiveness increases with increased airspeed.

AIR
5605. Maximum structural cruising speed is the maximum speed at which an airplane can be operated during

A—abrupt maneuvers.
B—normal operations.
C—flight in smooth air.

The maximum structural cruising speed (V_{NO}) is the speed at which exceeding the load limit factor may cause permanent deformation of the airplane structure. This is the maximum speed for normal operation. (H312) — AC 61-23C, Chapter 3

Answer (A) is incorrect because design maneuvering speed (V_A) is the maximum speed for abrupt maneuvers. Answer (C) is incorrect because the yellow arc identifies the range where flight is only recommended in smooth air.

AIR
5669. A pilot is entering an area where significant clear air turbulence has been reported. Which action is appropriate upon encountering the first ripple?

A—Maintain altitude and airspeed.
B—Adjust airspeed to that recommended for rough air.
C—Enter a shallow climb or descent at maneuvering speed.

In an area where significant clear air turbulence (CAT) has been reported or is forecast, it is suggested that the pilot adjust the speed to fly at the recommended rough air speed on encountering the first ripple, since the intensity of such turbulence may build up rapidly. (H66) — AC 61-23C, Chapter 3

Answers (A) and (C) are incorrect because the appropriate action is to adjust airspeed to design maneuvering speed (V_A).

AIR
5670. If severe turbulence is encountered during flight, the pilot should reduce the airspeed to

A—minimum control speed.
B—design-maneuvering speed.
C—maximum structural cruising speed.

Design maneuvering speed (V_A) is the maximum speed at which the maximum load limit can be imposed (either by gust or full deflection of the control surfaces) without causing structural damage. (H66) — AC 61-23C, Chapter 3

Answer (A) is incorrect because in turbulence, minimum control speed would result in the airplane stalling or significant control problems. Answer (C) is incorrect because the maximum structural cruising speed is faster than V_A.

AIR
5741. Which is the best technique for minimizing the wing-load factor when flying in severe turbulence?

A—Change power settings, as necessary, to maintain constant airspeed.
B—Control airspeed with power, maintain wings level, and accept variations of altitude.
C—Set power and trim to obtain an airspeed at or below maneuvering speed, maintain wings level, and accept variations of airspeed and altitude.

In severe turbulence, set power and trim to obtain an airspeed at or below maneuvering speed; this helps avoid exceeding the aircraft's maximum load factor. Attempt to maintain constant attitude, and accept airspeed and altitude variations caused by gusts. (I30) — AC 00-6A, Chapter 11

Answers (A) and (B) are incorrect because it is not possible to maintain a constant airspeed in severe turbulence.

ALL
5601. Calibrated airspeed is best described as indicated airspeed corrected for

A—installation and instrument error.
B—instrument error.
C—non-standard temperature.

Calibrated airspeed is the indicated airspeed corrected for position (or installation), and instrument errors. (H312) — AC 61-23C, Chapter 3

ALL
5602. True airspeed is best described as calibrated airspeed corrected for

A—installation or instrument error.
B—non-standard temperature.
C—altitude and non-standard temperature.

True airspeed is indicated airspeed after it has been corrected for nonstandard temperature and pressure altitude. (H312) — AC 61-23C, Chapter 3

Answers

5605 [B] 5669 [B] 5670 [B] 5741 [C] 5601 [A] 5602 [C]

Chapter 3 **Flight Instruments**

Altitude Definitions

Indicated altitude—the altitude indicated on an altimeter set to the current local altimeter setting.

Pressure altitude—the altitude indicated on an altimeter when it is set to the standard sea level pressure of 29.92 inches of mercury (29.92" Hg). Above 18,000 feet MSL, flight levels, which are pressure altitudes, are flown.

Density altitude—pressure altitude corrected for a non-standard temperature. The performance tables of an aircraft are based on density altitude.

True altitude—the exact height above mean sea level. Calculation of true altitude does not always yield a correct figure. Atmospheric conditions may deviate from the standard temperature and pressure lapse rates used in the computation of true altitude.

ALL
5740. To determine pressure altitude prior to takeoff, the altimeter should be set to

A—the current altimeter setting.
B—29.92" Hg and the altimeter indication noted.
C—the field elevation and the pressure reading in the altimeter setting window noted.

Pressure altitude can be determined by setting the altimeter to 29.92" Hg and reading the altimeter indication. (I04) — AC 61-27C, Chapter 4

Answers (A) and (C) are incorrect because these would indicate field elevation, or true altitude.

ALL, MIL
5114. What altimeter setting is required when operating an aircraft at 18,000 feet MSL?

A—Current reported altimeter setting of a station along the route.
B—29.92" Hg.
C—Altimeter setting at the departure or destination airport.

Each person operating an aircraft shall maintain the cruising altitude or flight level of that aircraft, as the case may be, by reference to an altimeter that is set, when operating at or above 18,000 feet MSL, to 29.92" Hg. (B08) — 14 CFR §91.121

Magnetic Compass

The Magnetic Compass is the only self-contained directional instrument in the aircraft. It is affected by deviation error. Magnetic disturbances (magnetic fields) within an aircraft deflect the compass needles from alignment with magnetic north. Each aircraft will affect a magnetic compass differently, and the direction and magnitude of the error varies with heading and with the electrical systems in use. Compensating magnets are used to minimize this type of error as much as possible. Any remaining error is noted on the compass correction card.

ALL
5178. Which statement is true about magnetic deviation of a compass? Deviation

A—varies over time as the agonic line shifts.
B—varies for different headings of the same aircraft.
C—is the same for all aircraft in the same locality.

Deviation depends, in part, on the heading of the aircraft. The difference between the direction indicated by a magnetic compass not installed in an airplane, and one that is installed in an airplane, is deviation. (H314) — AC 61-23C, Chapter 3

Answer (A) is incorrect because variation varies over time as the agonic line shifts. Answer (C) is incorrect because deviation varies from aircraft to aircraft.

Answers
5740 [B]   5114 [B]   5178 [B]

Chapter 3 **Flight Instruments**

Gyroscopic Instruments and Systems

Gyroscopes ("gyros") exhibit two important principles—rigidity in space and precession. Of the seven basic flight instruments, three are controlled by gyroscopes:

- Attitude indicator
- Turn coordinator/turn-and-slip indicator
- Heading indicator

The turn coordinator/turn-and-slip indicator is the only one addressed on the test. The turn coordinator is designed to show roll rate, rate of turn, and quality of turn. See Figure 3-2. The turn-and-slip indicators are gyroscopically-operated instruments designed to show the rate of turn and quality of turn. The turn-and-slip indicator does not show roll rate. See Figure 3-3.

A single needle-width deflection on the 2-minute indicator means that the aircraft is turning at 3° per second, or standard rate (2 minutes for a 360° turn). On the 4-minute indicator, a single needle-width deflection shows when the aircraft is turning at 1-1/2° per second, or half-standard rate (4 minutes for a 360° turn).

Before starting the engine, the turn needle should be centered and the race full of fluid. During a taxiing turn, the needle will indicate a turn in the proper direction and the ball will show a skid. An electric turn-and-slip, or turn coordinator, acts as a backup system in case of a failure of the vacuum-powered gyros.

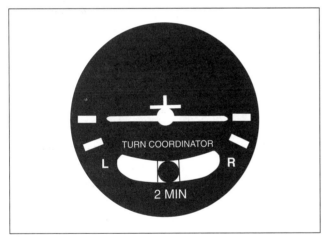

Figure 3-2. Turn coordinator

Figure 3-3. Turn-and-slip indicator

AIR
5268. What is an operational difference between the turn coordinator and the turn-and-slip indicator? The turn coordinator

A—is always electric; the turn-and-slip indicator is always vacuum-driven.
B—indicates bank angle only; the turn-and-slip indicator indicates rate of turn and coordination.
C—indicates roll rate, rate of turn, and coordination; the turn-and-slip indicator indicates rate of turn and coordination.

The turn coordinator indicates roll rate in addition to rate of turn and coordination. The turn-and-slip indicator only indicates rate of turn and coordination. (I04) — AC 61-27C, Chapter 4

Answer (A) is incorrect because both these instruments are usually electrically driven. Answer (B) is incorrect because a turn coordinator does not indicate bank angle.

Answers

5268 [C]

Chapter 3 **Flight Instruments**

AIR, RTC, LTA

5269. What is an advantage of an electric turn coordinator if the airplane has a vacuum system for other gyroscopic instruments?

A—It is a backup in case of vacuum system failure.
B—It is more reliable than the vacuum-driven indicators.
C—It will not tumble as will vacuum-driven turn indicators.

An electric turn coordinator provides a backup in case the vacuum system fails. (I04) — AC 61-27C, Chapter 4

Answers (B) and (C) are incorrect because both the vacuum-driven and electrically-driven indicators are reliable, and both can tumble.

AIR, RTC, LTA

5270. If a standard rate turn is maintained, how long would it take to turn 360°?

A—1 minute.
B—2 minutes.
C—3 minutes.

A standard rate turn means the aircraft is turning at a rate of 3° per second. 360° divided by 3° per second is equal to 120 seconds, or 2 minutes. (I05) — AC 61-27C, Chapter 5

Attitude Instrument Flying

The four flight fundamentals involved in maneuvering an aircraft are: straight-and-level flight, turns, climbs, and descents.

AIR

5191. Name the four fundamentals involved in maneuvering an aircraft.

A—Power, pitch, bank, and trim.
B—Thrust, lift, turns, and glides.
C—Straight-and-level flight, turns, climbs, and descents.

Maneuvering the airplane is generally divided into four flight fundamentals:

1. *Straight-and-level*
2. *Turns*
3. *Climbs*
4. *Descents*

(H55) — FAA-H-8083-3, Chapter 4

Answers

5269 [A]   5270 [B]   5191 [C]

Chapter 4
Regulations

Pilot Certificate Types and Privileges   *4–3*

Medical Certificates   *4–5*

Pilot Logbooks   *4–6*

High-Performance, Complex and Tailwheel Airplanes   *4–7*

Recent Flight Experience: Pilot-In-Command   *4–9*

Change of Address   *4–10*

Towing   *4–10*

Responsibility and Authority of the Pilot-In-Command   *4–11*

Preflight Action   *4–14*

Seatbelts   *4–15*

Portable Electronic Devices   *4–17*

Fuel Requirements   *4–17*

Transponder Requirements   *4–18*

Supplemental Oxygen   *4–19*

Instrument and Equipment Requirements   *4–19*

Restricted, Limited and Experimental Aircraft: Operating Limitations   *4–20*

Emergency Locator Transmitter (ELT)   *4–21*

Truth in Leasing   *4–22*

Operating Near Other Aircraft and Right-of-Way Rules   *4–22*

Speed Limits   *4–25*

Aerobatic Flight   *4–26*

Aircraft Lights   *4–26*

Minimum Altitudes   *4–28*

Maintenance Responsibility   *4–29*

Aircraft Inspections   *4–30*

Continued

Chapter 4 **Regulations**

Maintenance Records   *4–31*

Maintenance, Preventative Maintenance, Rebuilding and Alteration   *4–33*

NTSB Part 830   *4–34*

Rotorcraft Regulations   *4–36*

Glider Regulations   *4–37*

Lighter-Than-Air Regulations   *4–38*

LTA Fundamentals of Instructing   *4–41*

Chapter 4 **Regulations**

Pilot Certificate Types and Privileges

Although "FAR" is used as the acronym for "Federal Aviation Regulations," and found throughout the regulations themselves and hundreds of other publications, the FAA is now actively discouraging its use. "FAR" also means "Federal Acquisition Regulations." To eliminate any possible confusion, the FAA cites the federal aviation regulations with reference to Title 14 of the Code of Federal Regulations. For example, "FAR Part 91.3" is referenced as "14 CFR Part 91 Section 3."

The holder of a **Student Pilot Certificate** is limited to flight with an instructor pilot until certain requirements are met, after which solo flight may be authorized. A student pilot may neither carry passengers nor fly for compensation or hire.

A **Private Pilot Certificate** grants almost unlimited solo, passenger carrying, and cargo transport privileges, as long as the flying is not done for compensation or hire. A private pilot may share the operating expenses of a flight with his/her passengers. He/she may also act as pilot-in-command in connection with business or employment if the flight is only incidental to that business.

The holder of a **Commercial Pilot Certificate** may act as pilot-in-command of an aircraft carrying persons or property for compensation or hire. To carry passengers for hire on cross-country flight of more than 50 nautical miles, or at night, a commercial airplane pilot must hold an appropriate instrument rating.

An **Airline Transport Pilot** has the privileges of a commercial pilot with an instrument rating. He/she may also instruct other pilots in air transportation service in aircraft of the category, class, and type for which he/she is rated. Part 121 and Part 135 regulations address the situations in which an ATP rating is required, such as when acting as pilot-in-command of a multi-engine commuter flight.

Aircraft category and class ratings in which a pilot is qualified are placed on the pilot's certificate. A type rating is also required to act as pilot-in-command of a large aircraft (in excess of 12,500 pounds maximum gross takeoff weight) or of a turbojet-powered aircraft of any weight. Private or commercial pilots wishing to fly under instrument flight rules must also have an instrument rating placed on their certificates.

Additional ratings may be granted when a pilot has achieved the required level of skill and knowledge and has successfully completed an in-flight test. In the case of an instrument rating, the pilot must also pass an FAA Knowledge Exam.

All pilot certificates (except student) are valid indefinitely unless surrendered, suspended, or revoked. There are no expiration dates.

A pilot must have in his/her possession, or readily accessible in the aircraft, the following documents when operating an aircraft:

1. A pilot's certificate; and
2. A current medical certificate (except for glider and free balloon pilots).

AIR, RTC, LTA, MIL
5019. Which of the following are considered aircraft class ratings?

A—Transport, normal, utility, and acrobatic.
B—Airplane, rotorcraft, glider, and lighter-than-air.
C—Single-engine land, multiengine land, single-engine sea, and multiengine sea.

Aircraft class ratings, with respect to airmen, include single-engine land, multi-engine land, single-engine sea, and multi-engine sea, helicopter, gyroplane, airship, and free balloon. (A20) — 14 CFR §1.1, §61.5(b)

Answer (A) is incorrect because these are aircraft categories with respect to the certification of aircraft. Answer (B) is incorrect because these are categories of aircraft with respect to airmen.

Answers
5019 [C]

Chapter 4 Regulations

ALL, MIL
5020. Does a commercial pilot certificate have a specific expiration date?

A—No, it is issued without an expiration date.
B—Yes, it expires at the end of the 24th month after the month in which it was issued.
C—No, but commercial privileges expire if a flight review is not satisfactorily completed each 12 months.

Any pilot certificate, other than a Student Pilot Certificate, is issued without a specific expiration date. (A20) — 14 CFR §61.19(c)

AIR, RTC, LTA, MIL
5022. When is the pilot in command required to hold a category and class rating appropriate to the aircraft being flown?

A—All solo flights.
B—On practical tests given by an examiner or FAA Inspector.
C—On flights when carrying another person.

Unless a person holds a category, class, and type rating (if a class and type rating is required) that applies to the aircraft, that person may not act as pilot-in-command of an aircraft that is carrying another person, or is operated for compensation or hire. (A20) — 14 CFR §61.31

Answer (A) is incorrect because a pilot can solo an aircraft with an authorized instructor's logbook endorsement. Answer (B) is incorrect because a pilot may act as pilot-in-command when testing for a category and class rating with a duly authorized examiner.

AIR, RTC, MIL
5023. Unless otherwise authorized, the pilot in command is required to hold a type rating when operating any

A—aircraft that is certificated for more than one pilot.
B—aircraft of more than 12,500 pounds maximum certificated takeoff weight.
C—multiengine aircraft having a gross weight of more than 12,000 pounds.

A person may not act as pilot-in-command of any of the following aircraft unless he/she holds a type rating for that aircraft:

1. *A large aircraft (except lighter-than-air), more than 12,500 pounds maximum certificated takeoff weight.*
2. *A helicopter, for operations requiring an airline transport pilot certificate.*
3. *A turbojet-powered airplane.*
4. *Other aircraft specified by the Administrator through aircraft type certificate procedures.*

(A20) — 14 CFR §1.1, §61.31(a)

AIR, MIL
5039. What limitation is imposed on a newly certificated commercial pilot–airplane, if that person does not hold an instrument rating? The carriage of passengers

A—for hire on cross-country flights is limited to 50 NM for night flights, but not limited for day flights.
B—or property for hire on cross-country flights at night is limited to a radius of 50 NM.
C—for hire on cross-country flights in excess of 50 NM or for hire at night is prohibited.

A person who applies for a commercial pilot certificate with an airplane category and does not hold an instrument rating in the same category and class will be issued a commercial pilot certificate that contains the limitation, "The carriage of passengers for hire in (airplanes) on cross-country flights in excess of 50 NM or at night is prohibited." (A24) — 14 CFR §61.133

ALL, MIL
5018. Commercial pilots are required to have a valid and appropriate pilot certificate in their personal possession when

A—piloting for hire only.
B—carrying passengers only.
C—acting as pilot in command.

No person may act as pilot-in-command or in any other capacity as a required pilot flight crew member of a civil aircraft unless a current pilot certificate and, except for glider and free balloon pilots, an appropriate current medical certificate is in his/her possession. (A20) — 14 CFR §61.3

Answers (A) and (B) are incorrect because having the appropriate pilot and medical certificates is required regardless of the type of operation.

Answers

5020 [A]   5022 [C]   5023 [B]   5039 [C]   5018 [C]

ALL, MIL
5111. No person may operate an aircraft in simulated instrument flight conditions unless the

A—other control seat is occupied by at least an appropriately rated commercial pilot.
B—pilot has filed an IFR flight plan and received an IFR clearance.
C—other control seat is occupied by a safety pilot, who holds at least a private pilot certificate and is appropriately rated.

No person may operate a civil aircraft in simulated instrument flight unless the other control seat is occupied by a safety pilot who possesses at least a private pilot certificate with category and class ratings appropriate to the aircraft being flown. (B08) — 14 CFR §91.109

ALL, MIL
5126-1. A person with a Commercial Pilot certificate may act as pilot in command of an aircraft for compensation or hire, if that person

A—is qualified in accordance with 14 CFR part 61 and with the applicable parts that apply to the operation.
B—is qualified in accordance with 14 CFR part 61 and has passed a pilot competency check given by an authorized check pilot.
C—holds appropriate category, class ratings, and meets the recent flight experience requirements of 14 CFR part 61.

Not only must the pilot be qualified in accordance with Part 61, the pilot must also comply with Part 91, the regulations which govern the operation of the flight. (A24) — 14 CFR §61.31

ALL, MIL
5126-2. A person with a commercial pilot certificate may act as pilot in command of an aircraft carrying persons or property for compensation or hire, if that person

A—holds appropriate category, class ratings, and meets the recent flight experience requirements of 14 CFR part 61.
B—is qualified in accordance with 14 CFR part 61 and with the applicable parts that apply to the operation.
C—is qualified in accordance with 14 CFR part 61 and has passed a pilot competency check given by an authorized check pilot.

Not only must the pilot be qualified in accordance with Part 61, the pilot must also comply with Part 91, the regulations which govern the operation of the flight. (A24) — 14 CFR §61.31

Medical Certificates

Student pilot, recreational pilot, and private pilot operations, other than glider and balloon pilots, require a Third-Class Medical Certificate. A Third-Class Medical Certificate issued before September 16, 1996, expires at the end of the 24th month after the month of the date of examination shown on the certificate. Certificates issued on or after September 16, 1996, expire at the end of:

1. The 36th month after the month of the date of the examination shown on the certificate if the person has not reached his or her 40th birthday on or before the date of examination; or

2. The 24th month after the month of the date of examination shown on the certificate if the person has reached his or her 40th birthday on or before the date of the examination.

Continued

Answers

5111 [C]   5126-1 [A]   5126-2 [B]

Chapter 4 **Regulations**

The holder of a Second-Class Medical Certificate may exercise commercial privileges during the first 12 calendar months, but the certificate is valid only for private pilot privileges during the following (12 or 24) calendar months, depending on the applicant's age.

The holder of a First-Class Medical Certificate may exercise Airline Transport Pilot privileges during the first 6 calendar months, commercial privileges during the following 6 calendar months, and private pilot privileges during the following (12 or 24) calendar months, depending on the applicant's age.

Each type of medical certificate is valid through the last day of the month (of the month it expires), regardless of the day the physical examination was given.

AIR, RTC, LTA, MIL
5021. A second-class medical certificate issued to a commercial pilot on April 10, this year, permits the pilot to exercise which of the following privileges?
A—Commercial pilot privileges through April 30, next year.
B—Commercial pilot privileges through April 10, 2 years later.
C—Private pilot privileges through, but not after, March 31, next year.

A Second-Class Medical Certificate expires at the end of the last day of:
1. *The 12th month after the month of the date of examination shown on the certificate, for operations requiring a Commercial Pilot Certificate or an air traffic control tower operator certificate, and*
2. *The 24th or 36th (depending on the applicant's age) month after the month of the date of examination shown on the certificate, for operations requiring a Private, Recreational, or Student Pilot Certificate.*

(A20) — 14 CFR §61.23(b)

Answer (B) is incorrect because a Second-Class medical is valid for commercial operations for 12 months, and expires on the last day of the month. Answer (C) is incorrect because a Second-Class medical is valid for private pilot operations for 24 or 36 months (depending on the applicant's age).

Pilot Logbooks

A pilot must log:

- The time required for an added certificate or rating; or
- The time necessary for meeting his/her recent flight experience requirements.

The logging of other flight time is not required.

A pilot may log as **pilot-in-command (PIC)** only that flight time during which he/she is the sole manipulator of the controls, or during the time he/she acts as PIC on an aircraft which requires more than one pilot due to the aircraft's certification or the regulations under which the flight is conducted.

A pilot may log as **second-in-command (SIC)** time all flight time during which he/she acts as SIC of an aircraft in which more than one pilot is required due to the aircraft's certification or the regulations under which the flight is conducted.

An Airline Transport Pilot may log as pilot-in-command time, all the time during which he/she acts as pilot-in-command.

Answers
5021 [A]

ALL, MIL
5026. What flight time must be documented and recorded, by a pilot exercising the privileges of a commercial certificate?

A— Flight time showing aeronautical training and experience to meet requirements for a certificate, rating, or flight review.
B— All flight time flown for compensation or hire.
C— Only flight time for compensation or hire with passengers aboard which is necessary to meet the recent flight experience requirements.

The aeronautical training and experience needed to meet the requirements for a certificate or rating, and/or the recent flight experience requirements, must be documented and recorded. The logging of other flight time is not required. (A20) — 14 CFR §61.51

AIR, RTC, MIL
5025. What flight time may a pilot log as second in command?

A— All flight time while acting as second in command in aircraft configured for more than one pilot.
B— All flight time when qualified and occupying a crewmember station in an aircraft that requires more than one pilot.
C— Only that flight time during which the second in command is the sole manipulator of the controls.

A person may log second-in-command time only for that flight time during which that person is qualified in accordance with the second-in-command requirements of this part, and occupies a crewmember station in an aircraft that requires more than one pilot by the aircraft's type certificate. (A20) — 14 CFR §61.51

High-Performance, Complex and Tailwheel Airplanes

To act as pilot-in-command of an airplane that has more than 200 horsepower, or is equipped with retractable landing gear, flaps and a controllable propeller, a person must receive flight instruction and obtain a logbook endorsement of competency from a certified flight instructor. This endorsement is required only one time.

No person may serve as second-in-command (SIC) of a large airplane, or a turbojet-powered airplane type certificated for more than one required flight crew member unless he/she holds:

1. At least a current Private Pilot Certificate with appropriate category and class ratings; and
2. An appropriate instrument rating, in the case of flight under IFR.

AIR, MIL
5024. To act as pilot in command of an airplane that is equipped with retractable landing gear, flaps, and controllable-pitch propeller, a person is required to

A— make at least six takeoffs and landings in such an airplane within the preceding 6 months.
B— receive and log ground and flight training in such an airplane, and obtain a logbook endorsement certifying proficiency.
C— hold a multiengine airplane class rating.

No person may act as PIC of a complex airplane (an airplane that has a retractable landing gear, flaps, and a controllable pitch propeller), unless the person has received and logged ground and flight training from an authorized instructor in a complex airplane, or in a flight simulator or flight training device that is representative of a complex airplane, and has been found proficient in the operation and systems of the airplane; and received a one-time endorsement in the pilot's logbook from an authorized instructor who certifies the person is proficient to operate a complex airplane. (A20) — 14 CFR §61.31

Answers

5026 [A]   5025 [B]   5024 [B]

AIR, MIL
5106. To act as pilot-in-command of an airplane with more than 200 horsepower, a person is required to

A—receive and log ground and flight training from a qualified pilot in such an airplane.
B—obtain an endorsement from a qualified pilot stating that the person is proficient to operate such an airplane.
C—receive and log ground and flight training from an authorized instructor in such an airplane.

To act as pilot-in-command of a high-performance airplane (an airplane with an engine of more than 200 horsepower), the person must receive and log ground and flight training from an authorized instructor in a high-performance airplane and has been found proficient in the operation and systems of the airplane. In addition, the pilot must receive a one-time endorsement in the pilot's logbook from an authorized instructor who certifies that the person is proficient to operate a high-performance airplane. The training and endorsement is not required if the person has logged flight time as pilot-in-command of a high-performance airplane prior to August 4, 1997. (A20) — 14 CFR §61.31

AIR, MIL
5107. To serve as pilot in command of an airplane that is certified for more than one pilot crewmember, and operated under Part 91, a person must

A—complete a flight review within the preceding 24 calendar months.
B—receive and log ground and flight training from an authorized flight instructor.
C—complete a pilot-in-command proficiency check within the preceding 12 calendar months in an airplane that is type certificated for more than one pilot.

To serve as PIC of an aircraft that is type certificated for more than one required pilot flight crewmember, a person must complete a PIC proficiency check in the aircraft that is type certificated for more than one required pilot flight crewmember within the preceding 12 calendar months. Additionally, the pilot must complete a PIC proficiency check in the particular type of aircraft in which that person will serve as PIC within the preceding 24 calendar months. (A20) — 14 CFR §61.58

AIR, MIL
5108. To serve as second in command of an airplane that is certificated for more than one pilot crewmember, and operated under Part 91, a person must

A—receive and log flight training from an authorized flight instructor in the type of airplane for which privileges are requested.
B—hold at least a commercial pilot certificate with an airplane category rating.
C—within the last 12 months become familiar with the required information, and perform and log pilot time in the type of airplane for which privileges are requested.

Under Part 91 no person may serve as a second-in-command of an aircraft type certificated for more than one required pilot flight crewmember unless that person has within the previous 12 calendar months become familiar with information for the specific type aircraft for which second-in-command privileges are requested and performed and logged pilot time in the type of aircraft or in a flight simulator that represents the type of aircraft for which second-in-command privileges are requested. Also, no person may serve as a second-in-command of an aircraft type certificated for more than one required pilot flight crewmember unless that person holds at least a current private pilot certificate with the appropriate category and class rating. (A20) — 14 CFR §61.55

AIR, MIL
5128. To act as pilot in command of a tailwheel airplane, without prior experience, a pilot must

A—log ground and flight training from an authorized instructor.
B—pass a competency check and receive an endorsement from an authorized instructor.
C—receive and log flight training from an authorized instructor.

No person may act as pilot-in-command of a tailwheel airplane unless that person has received and logged flight training from an authorized instructor in a tailwheel airplane and received an endorsement in the person's logbook from an authorized instructor who found the person proficient in the operation of a tailwheel airplane. (A20) — 14 CFR §61.31

Answers

5106 [C]   5107 [C]   5108 [C]   5128 [C]

Recent Flight Experience: Pilot-In-Command

No person may act as PIC of an aircraft unless within the preceding 24 calendar months he/she has accomplished a **flight review**. This review is given in an aircraft for which he/she is rated by an appropriately certificated instructor or other person designated by the FAA. Satisfactory accomplishment of this flight review will be endorsed in his/her logbook. If the pilot takes a proficiency check (for a certificate or a new rating), it counts for the flight review. A biennial flight review would then be required at the end of the 24th calendar month following that proficiency check. The review must include a minimum of 1 hour flight instruction and 1 hour ground instruction.

No person may act as PIC of an aircraft carrying passengers unless, within the preceding 90 days, he/she has made three takeoffs and landings (touch and go is allowed if the aircraft is not a tail-wheel airplane) as the sole manipulator of the controls in an aircraft of the same category, class and type (if required). In order to carry passengers in a tailwheel (conventional gear) airplane, the takeoffs and landings must be made to a full stop, and they must be in a tailwheel airplane.

No person may act as PIC of an aircraft carrying passengers during the period from 1 hour after sunset to 1 hour before sunrise unless, within the preceding 90 days, he/she has made at least three takeoffs and three landings to a full stop during that period in the same category and class of aircraft to be used.

ALL, MIL
5027. If a pilot does not meet the recency of experience requirements for night flight and official sunset is 1900 CST, the latest time passengers should be carried is

A—1959 CST.
B—1900 CST.
C—1800 CST.

No person may act as PIC of an aircraft carrying passengers during the period beginning 1 hour after sunset and ending 1 hour before sunrise, unless within the preceding 90 days that person has made at least three takeoffs and three landings to a full stop during the period beginning 1 hour after sunset and ending 1 hour before sunrise. (A20) — 14 CFR §61.57

ALL, MIL
5028. Prior to carrying passengers at night, the pilot in command must have accomplished the required takeoffs and landings in

A—any category aircraft.
B—the same category and class of aircraft to be used.
C—the same category, class, and type of aircraft (if a type rating is required).

No person may act as pilot-in-command of an aircraft carrying passengers during the period beginning 1 hour after sunset and ending 1 hour before sunrise unless within the preceding 90 days, he/she has made at least three takeoffs and three landings to a full stop during that period in the same category, class, and type (if a type rating is required) of aircraft to be used. (A20) — 14 CFR §61.57(b)

ALL, MIL
5031. To act as pilot in command of an aircraft under 14 CFR Part 91, a commercial pilot must have satisfactorily accomplished a flight review or completed a proficiency check within the preceding

A—6 calendar months.
B—12 calendar months.
C—24 calendar months.

No person may act as pilot-in-command of an aircraft unless, within the preceding 24 calendar months, he/she has accomplished a flight review, completed a pilot proficiency check, or has completed one or more phases of an FAA-sponsored pilot proficiency award program. (A20) — 14 CFR §61.56

Answers
5027 [A]   5028 [C]   5031 [C]

Chapter 4 **Regulations**

Change of Address

If a pilot changes his/her permanent address without notifying the FAA Airman Certification Branch in writing within 30 days, he/she may not exercise the privileges of his/her certificate. For notification of address change, write to:

Department of Transportation
Federal Aviation Administration
Airman Certification Branch
Box 25082
Oklahoma City, OK 73125

ALL, MIL
5032. Pilots who change their permanent mailing address and fail to notify the FAA Airmen Certification Branch of this change, are entitled to exercise the privileges of their pilot certificate for a period of

A—30 days.
B—60 days.
C—90 days.

The holder of a pilot or flight instructor certificate who has made a change in his/her permanent mailing address may not after 30 days from the date he/she moved, exercise the privileges of his/her certificate unless he/she has notified the FAA in writing. (A20) — 14 CFR §61.60

Towing

A certificated pilot may not act as pilot-in-command of an aircraft towing a glider unless there is a minimum of 100 hours of pilot flight time in powered aircraft entered in the pilot's logbook, and within the preceding 12 months, he/she has made at least three actual or simulated glider tows while accompanied by a qualified pilot.

No pilot may tow anything other than a glider with an aircraft, except in accordance with the terms of a certificate of waiver issued by the Administrator.

AIR, GLI, MIL
5033. To act as pilot in command of an airplane towing a glider, the tow pilot is required to have

A—a logbook endorsement from an authorized glider instructor certifying receipt of ground and flight training in gliders, and be proficient with techniques and procedures for safe towing of gliders.
B—at least a private pilot certificate with a category rating for powered aircraft, and made and logged at least three flights as pilot or observer in a glider being towed by an airplane.
C—a logbook record of having made at least three flights as sole manipulator of the controls of a glider being towed by an airplane.

To act as pilot-in-command of an aircraft towing a glider, a person must have a logbook endorsement from an authorized instructor who certifies that the person has received ground and flight training in gliders and is proficient in the techniques and procedures essential to the safe towing of gliders, including airspeed limitations. (A21) — 14 CFR §61.69

Answer (B) is incorrect because the PIC must have logged at least three flights as the sole manipulator of the controls, not as an observer, of an aircraft towing a glider or simulating glider-towing flight procedures while accompanied by an authorized pilot. Answer (C) is incorrect because any person who, before May 17, 1967, has made and logged 10 or more flights as PIC of an aircraft towing a glider in accordance with a certificate of waiver need not comply with this rule.

Answers

5032 [A]   5033 [A]

AIR, MIL
5034. To act as pilot in command of an airplane towing a glider, a pilot must have accomplished, within the preceding 12 months, at least

A—three actual glider tows under the supervision of a qualified tow pilot.
B—three actual or simulated glider tows while accompanied by a qualified tow pilot.
C—ten flights as pilot in command of an aircraft while towing a glider.

To act as PIC, the pilot must make at least three actual or simulated glider tows while accompanied by a qualified pilot or make at least three flights as pilot-in-command of a glider towed by an aircraft. (A21) — 14 CFR §61.69

AIR, MIL
5055. Which is required to operate an aircraft towing an advertising banner?

A—Approval from ATC to operate in Class E airspace.
B—A certificate of waiver issued by the Administrator.
C—A safety link at each end of the towline which has a breaking strength not less than 80 percent of the aircraft's gross weight.

A certificate of waiver is required for towing objects other than gliders. (B12) — 14 CFR §91.311

Answer (A) is incorrect because approval from ATC to operate in Class E airspace is only required in IFR conditions. Answer (C) is incorrect because the towline breaking strength requirement is for towing gliders.

GLI
5127. To act as pilot in command of a glider, using self-launch procedures, that person must hold a pilot certificate with a glider rating and have accomplished

A—a competency flight check given by an authorized flight instructor.
B—ground and flight training in self-launch procedures and operations, and possess a logbook endorsement from a flight instructor certifying such proficiency.
C—appropriate flight training and meet recent experience in self-launch operations.

No person may act as pilot-in-command of a glider using self-launch procedures, unless that person has satisfactorily accomplished ground and flight training on self-launch procedures and operations and has received an endorsement from an authorized instructor who certifies in that pilot's logbook that the pilot has been found proficient in self-launch procedures and operations. (A20) — 14 CFR §61.31

Responsibility and Authority of the Pilot-In-Command

The pilot-in-command is directly responsible for the safety of his/her aircraft. The PIC is the final authority as to the operation of the aircraft. In the interest of safety, the PIC can deviate from any of these regulations to the extent necessary to meet an emergency. Each pilot-in-command who deviates from a regulation must, if requested, send a written report of that deviation to the Administrator.

No pilot-in-command may allow any object to be dropped from an aircraft in flight that creates a hazard to persons or property. However, this does not prohibit the dropping of objects if reasonable precautions are taken to avoid injury or damage to persons or property.

Operate, with respect to aircraft, means use, cause to use or authorize to use aircraft, for the purpose of air navigation including the piloting of aircraft, with or without the right of legal control (as owner, lessee, or otherwise).

Operational control, with respect to a flight, means the exercise of authority over initiating, conducting or terminating a flight.

Commercial operator means a person who, for compensation or hire, engages in the carriage by aircraft in air commerce of persons or property, other than as an air carrier or foreign air carrier. Where it is doubtful that an operation is for "compensation or hire," the test applied is whether the carriage by air is merely incidental to the person's other business or is, in itself, a major enterprise for profit.

Answers
5034 [B]   5055 [B]   5127 [B]

Chapter 4 **Regulations**

ALL, MIL
5044. What action must be taken when a pilot in command (PIC) deviates from any rule in 14 CFR Part 91?

A—Upon landing, report the deviation to the FAA Flight Standards District Office.
B—Advise ATC of the pilot-in-command's intentions.
C—Upon the request of the Administrator, send a written report of that deviation to the Administrator.

Each pilot-in-command who deviates from a rule in an emergency shall, upon request, send a written report of that deviation to the Administrator. (B07) — 14 CFR §91.3(c)

ALL, MIL
5047. A pilot in command (PIC) of a civil aircraft may not allow any object to be dropped from that aircraft in flight

A—if it creates a hazard to persons and property.
B—unless the PIC has permission to drop any object over private property.
C—unless reasonable precautions are taken to avoid injury to property.

No pilot-in-command of a civil aircraft may allow any object to be dropped from an aircraft in flight that creates a hazard to persons or property. However, this does not prohibit the dropping of objects if reasonable precautions are taken to avoid injury or damage to persons or property. (B07) — 14 CFR §91.15

ALL, MIL
5011. Regulations which refer to operate relate to that person who

A—acts as pilot in command of the aircraft.
B—is the sole manipulator of the aircraft controls.
C—causes the aircraft to be used or authorizes its use.

"Operate," with respect to aircraft, means use, cause to use, or authorize to use aircraft. (A01) — 14 CFR §1.1

Answer (A) is incorrect because the pilot-in-command may not necessarily be the operator of the aircraft. Answer (B) is incorrect because the sole manipulator may not necessarily be the operator of the aircraft.

ALL, MIL
5012. Regulations which refer to the operational control of a flight are in relation to

A—the specific duties of any required crewmember.
B—acting as the sole manipulator of the aircraft controls.
C—exercising authority over initiating, conducting, or terminating a flight.

"Operational control," with respect to a flight, means the exercise of authority over initiating, conducting, or terminating a flight. (A01) — 14 CFR §1.1

Answer (A) is incorrect because assigning specific duties to a crew member is only a small part of exercising operational control. Answer (B) is incorrect because acting as sole manipulator of an aircraft is a flight crew responsibility.

ALL, MIL
5010. Regulations which refer to commercial operators relate to that person who

A—is the owner of a small scheduled airline.
B—for compensation or hire, engages in the carriage by aircraft in air commerce of persons or property, as an air carrier.
C—for compensation or hire, engages in the carriage by aircraft in air commerce of persons or property, other than as an air carrier.

"Commercial operator" means a person who, for compensation or hire, engages in the carriage by aircraft in air commerce of persons or property, other than as an air carrier. (A01) — 14 CFR §1.1

ALL, MIL
5109. What person is directly responsible for the final authority as to the operation of the airplane?

A—Certificate holder.
B—Pilot in command.
C—Airplane owner/operator.

The pilot-in-command of an aircraft is directly responsible for, and is the final authority as to, the operation of that aircraft. (B07) — 14 CFR §91.3

Answers

| 5044 | [C] | 5047 | [A] | 5011 | [C] | 5012 | [C] | 5010 | [C] | 5109 | [B] |

Chapter 4 **Regulations**

ALL, MIL
5141. A pilot convicted of operating a motor vehicle while either intoxicated by, impaired by, or under the influence of alcohol or a drug is required to provide a

A—written report to the FAA Civil Aeromedical Institute (CAMI) within 60 days after the motor vehicle action.
B—written report to the FAA Civil Aviation Security Division (AMC-700) not later than 60 days after the conviction.
C—notification of the conviction to an FAA Aviation Medical Examiner (AME) not later than 60 days after the motor vehicle action.

Each person holding a pilot certificate shall provide a written report of each motor vehicle action to the FAA, Civil Aviation Security Division (AMC-700), not later than 60 days after the motor vehicle action. (A20) — 14 CFR §61.15

ALL, MIL
5142. A pilot convicted of a motor vehicle offense involving alcohol or drugs is required to provide a written report to the

A—nearest FAA Flight Standards District Office (FSDO) within 60 days after such action.
B—FAA Civil Aeronautical Institute (CAMI) within 60 days after the conviction.
C—FAA Civil Aviation Security Division (AMC-700) within 60 days after such action.

Each person holding a pilot certificate shall provide a written report of each motor vehicle action to the FAA, Civil Aviation Security Division (AMC-700), not later than 60 days after the motor vehicle action. (A20) — 14 CFR §61.15

ALL, MIL
5143. A pilot convicted for the violation of any Federal or State statute relating to the process, manufacture, transportation, distribution, or sale of narcotic drugs is grounds for

A—a written report to be filed with the FAA Civil Aviation Security Division (AMC-700) not later than 60 days after the conviction.
B—notification of this conviction to the FAA Civil Aeromedical Institute (CAMI) within 60 days after the conviction.
C—suspension or revocation of any certificate, rating, or authorization issued under 14 CFR Part 61.

A conviction for the violation of any Federal or State statute relating to the growing, processing, manufacture, sale, disposition, possession, transportation, or importation of narcotic drugs, marijuana, or depressant or stimulant drugs or substances is grounds for: (1) Denial of an application for any certificate, rating, or authorization for a period of up to 1 year after the date of final conviction; or (2) Suspension or revocation of any certificate, rating, or authorization. (A20) — 14 CFR §61.15

ALL, MIL
5144. A pilot convicted of operating an aircraft as a crewmember under the influence of alcohol, or using drugs that affect the person's faculties, is grounds for a

A—written report to be filed with the FAA Civil Aviation Security Division (AMC-700) not later than 60 day after the conviction.
B—written notification to the FAA Civil Aeromedical Institute (CAMI) within 60 days after the conviction.
C—denial of an application for an FAA certificate or rating.

No person may act or attempt to act as a crewmember of a civil aircraft within 8 hours after the consumption of any alcoholic beverage, while under the influence of alcohol, while using any drug that affects the person's faculties in any way contrary to safety; or while having 0.04 percent by weight or more alcohol in the blood. Committing any of these acts is grounds for: (1) Denial of an application for a certificate, rating, or authorization for a period of up to 1 year after the date of that act; or (2) Suspension or revocation of any certificate, rating, or authorization. (A20) — 14 CFR §61.15

Answers

5141 [B]   5142 [C]   5143 [C]   5144 [C]

Chapter 4 **Regulations**

Preflight Action

Each pilot-in-command, before starting a flight, shall familiarize him/herself with all available information concerning that flight. This information must include runway lengths and takeoff and landing distance information for airports of intended use. If the flight will not be in the vicinity of an airport or will be under IFR, the PIC must also check:

- Available weather reports and forecasts
- Fuel requirements
- Alternatives available if the flight cannot be completed as planned
- Any known traffic delays as advised by ATC

No person may operate an aircraft unless it has within it:

1. An appropriate and current Airworthiness Certificate, displayed at the cabin or cockpit entrance so that it is legible to passengers or crew.
2. A Registration Certificate issued to its owner.
3. An approved flight manual, manual material, markings and placards or any combination of these, showing the operating limitations of the aircraft.

ALL, MIL
5045. Who is responsible for determining if an aircraft is in condition for safe flight?

A—A certificated aircraft mechanic.
B—The pilot in command.
C—The owner or operator.

The pilot-in-command of an aircraft is responsible for determining whether that aircraft is in condition for safe flight. The pilot shall discontinue the flight when unairworthy mechanical, electrical, or structural conditions occur. (B07) — 14 CFR §91.7(b)

ALL, MIL
5046. When operating a U.S.-registered civil aircraft, which document is required by regulation to be available in the aircraft?

A—A manufacturer's Operations Manual.
B—A current, approved Airplane Flight Manual.
C—An Owner's Manual.

No person may operate an aircraft unless it has within it:

1. *An appropriate and current Airworthiness Certificate, displayed at the cabin or cockpit entrance so that it is legible to passengers or crew.*
2. *A Registration Certificate issued to its owner.*
3. *An approved flight manual, manual material, markings and placards or any combination of these, showing the operating limitations of the aircraft.*

(B07) — 14 CFR §91.203

ALL, MIL
5049-1. When is preflight action required, relative to alternatives available, if the planned flight cannot be completed?

A—IFR flights only.
B—any flight not in the vicinity of an airport.
C—any flight conducted for hire or compensation.

For a flight under IFR or flight not in the vicinity of an airport, the pilot must be familiar with weather reports and forecasts, fuel requirements, alternatives available if the planned flight cannot be completed, and any known traffic delays of which he/she has been advised by ATC. (B08) — 14 CFR §91.103(a)

ALL, MIL
5049-2. The required preflight action relative to weather reports and fuel requirements is applicable to

A—any flight conducted for compensation or hire.
B—any flight not in the vicinity of an airport.
C—IFR flights only.

Before beginning a flight under IFR or a flight not in the vicinity of an airport, each PIC shall become familiar with all available information concerning that flight, including weather reports and forecasts, fuel requirements, alternatives available if the planned flight cannot be completed, and any known traffic delays of which the pilot-in-command has been advised by ATC. (B08) — 14 CFR §91.103

Answers

5045 [B] 5046 [B] 5049-1 [B] 5049-2 [B]

ALL, MIL
5050-1. Before beginning any flight under IFR, the pilot in command must become familiar with all available information concerning that flight. In addition, the pilot must

A—be familiar with all instrument approaches at the destination airport.
B—list an alternate airport on the flight plan and confirm adequate takeoff and landing performance at the destination airport.
C—be familiar with the runway lengths at airports of intended use, and the alternatives available if the flight cannot be completed.

For a flight under IFR or flight not in the vicinity of an airport, the pilot must be familiar with weather reports and forecasts, fuel requirements, alternatives available if the planned flight cannot be completed, and any known traffic delays of which he/she has been advised by ATC. For any flight, the pilot must be familiar with runway lengths at airports of intended use, and the following takeoff and landing distance information: For civil aircraft for which an approved airplane or rotorcraft flight manual containing takeoff and landing distance data is required, the takeoff and landing distance data contained therein. (B08) — 14 CFR §91.103

Answer (A) is incorrect because only pilots filing an IFR flight must be familiar with the instrument approaches at their destination airport and possible alternates. Answer (B) is incorrect because an alternate is not required if weather is VFR at the destination.

ALL, MIL
5050-2. Before beginning any flight under IFR, the pilot in command must become familiar with all available information concerning that flight. In addition, the pilot must

A—list an alternate airport on the flight plan, and confirm adequate takeoff and landing performance at the destination airport.
B—be familiar with all instrument approaches at the destination airport.
C—be familiar with the runway lengths at airports of intended use, weather reports, fuel requirements, and alternatives available, if the planned flight cannot be completed.

Each PIC shall, before beginning a flight, become familiar with all available information concerning that flight. This information must include, for a flight under IFR or a flight not in the vicinity of an airport, weather reports and forecasts, fuel requirements, alternatives available if the planned flight cannot be completed, and any known traffic delays of which the pilot-in-command has been advised by ATC. For any flight, runway lengths at airports of intended use, and takeoff and landing distance information must be determined. (B08) — 14 CFR §91.103

Answers (A) and (B) are incorrect because an alternate airport need only be listed if it is warranted by the forecast weather conditions, and the pilot must be familiar with any instrument approach procedure prior to flying that procedure.

Seatbelts

During takeoff and landing, and while en route, each required flight crewmember shall be at his/her station with seatbelt fastened, unless the crewmember has to leave in connection with the operation of the aircraft or physiological needs. Also, each required flight crewmember must keep the shoulder harness fastened during takeoff and landing, unless the crewmember would be unable to perform his/her duties with the shoulder harness fastened.

No pilot may takeoff or land a civil aircraft unless the PIC ensures that each person on board has been notified to fasten his/her safety belt and ensures that each person on board knows how to operate the safely belt. Each person on board must occupy a seat with a seatbelt properly secured during takeoffs and landings. (A person who has not reached his/her second birthday may be held by an adult.)

ALL, MIL
5051-1. Required flight crewmembers' safety belts must be fastened

A—only during takeoff and landing.
B—while the crewmembers are at their stations.
C—only during takeoff and landing when passengers are aboard the aircraft.

Required flight crewmembers must keep their safety belts fastened while at their stations. (B08) — 14 CFR §91.105(a)(2)

Answers

5050-1 [C]   5050-2 [C]   5051-1 [B]

Chapter 4 **Regulations**

ALL, MIL

5051-2. Each required flight crewmember is required to keep his or her shoulder harness fastened

A—during takeoff and landing only when passengers are aboard the aircraft.
B—while the crewmembers are at their stations, unless he or she is unable to perform required duties.
C—during takeoff and landing, unless he or she is unable to perform required duties.

Each required flight crewmember of a U.S.-registered civil aircraft shall, during takeoff and landing, keep his or her shoulder harness fastened while at his or her assigned duty station. This rule does not apply if the seat at the crewmember's station is not equipped with a shoulder harness, or the crewmember would be unable to perform required duties with the shoulder harness fastened. (B08) — 14 CFR §91.105

Answer (A) is incorrect because the shoulder harness must be used on takeoff and landing regardless of the passenger complement. Answer (B) is incorrect because the regulation specifies shoulder harness use during takeoff and landing.

AIR, MIL

5052. With U.S.-registered civil airplanes, the use of safety belts is required during movement on the surface, takeoffs, and landings for

A—safe operating practice, but not required by regulations.
B—each person over 2 years of age on board.
C—commercial passenger operations only.

No pilot may cause to be moved on the surface, takeoff, or land a U.S.-registered civil aircraft unless the PIC of that aircraft ensures that each person on board has been notified to fasten his or her safety belt and, if installed, his or her shoulder harness. A person who has not reached his/her second birthday may be held by an adult. (B08) — 14 CFR §91.107

AIR, MIL

5110-1. Operating regulations for U.S.-registered civil airplanes require that during movement on the surface, takeoffs, and landings, a seat belt and shoulder harness (if installed) must be properly secured about each

A—flight crewmember only.
B—person on board.
C—flight and cabin crewmembers.

No pilot may cause to be moved on the surface, takeoff, or land a U.S.-registered civil aircraft unless the pilot-in-command of that aircraft ensures that each person on board (over two years of age) has been notified to fasten his or her safety belt and, if installed, his or her shoulder harness. (B08) — 14 CFR §91.107

RTC, MIL

5110-2. Operating regulations for U.S.-registered civil helicopters require that during movement on the surface, takeoffs, and landings, a seat belt and shoulder harness (if installed) must be properly secured about each

A—person on board.
B—flight and cabin crewmembers.
C—flight crew member only.

No pilot may cause to be moved on the surface, takeoff, or land a U.S.-registered civil aircraft unless the pilot-in-command of that aircraft ensures that each person on board (over two years of age) has been notified to fasten his or her safety belt and, if installed, his or her shoulder harness. (B08) — 14 CFR §91.107

Answers

5051-2 [C]   5052 [B]   5110-1 [B]   5110-2 [A]

Portable Electronic Devices

No person may operate or allow the operation of any portable electronic device on board an aircraft:

1. Operated by an air carrier or commercial operator; or
2. While it is operated under IFR.

Exceptions to this regulation are:

- Voice recorders
- Hearing aids
- Heart pacemakers
- Electric shavers
- Anything else the PIC has determined will not cause interference with navigation or communications systems.

ALL, MIL

5056-1. Portable electronic devices which may cause interference with the navigation or communication system may not be operated on a U.S.-registered civil aircraft being flown

A—along Federal airways.
B—within the U.S.
C—in air carrier operations.

Portable electronic devices may not be operated on aircraft operated by a holder of an air carrier operating certificate. (B07) — 14 CFR §91.21(a)(1)

Answers (A) and (B) are incorrect because this rule pertains only to IFR and commercial flights.

ALL, MIL

5056-2. Portable electronic devices which may cause interference with the navigation or communication system may not be operated on U.S.-registered civil aircraft being operated

A—under IFR.
B—in passenger carrying operations.
C—along Federal airways.

Any portable electronic device that the operator of the aircraft has determined could cause interference with the navigation or communication systems may not be used on any aircraft operated under IFR. (B07) — 14 CFR §91.21

Fuel Requirements

No person may begin a flight in an airplane under VFR unless there is enough fuel to get to the first point of intended landing and, assuming normal cruise speed,

- During the day, to fly after that for at least 30 minutes
- At night, to fly after that for at least 45 minutes.

No person may operate an aircraft in IFR conditions unless it carries enough fuel (considering available weather reports and forecasts) to:

1. Fly to the first airport of intended landing;
2. Fly from that airport to the alternate, if required; and
3. Fly thereafter for 45 minutes at normal cruising speed;
4. If a standard instrument approach is prescribed for the first airport of intended landing; and
5. For at least 1 hour before to 1 hour after the ETA at the airport, weather reports and forecasts indicate:
   a. The ceiling will be at least 2,000 feet above the airport elevation; and
   b. Visibility will be at least 3 miles.

Answers

5056-1 [C]   5056-2 [A]

Chapter 4 **Regulations**

AIR, LTA, MIL
5059. If weather conditions are such that it is required to designate an alternate airport on your IFR flight plan, you should plan to carry enough fuel to arrive at the first airport of intended landing, fly from that airport to the alternate airport, and fly thereafter for

A—30 minutes at slow cruising speed.
B—45 minutes at normal cruising speed.
C—1 hour at normal cruising speed.

Civil aircraft in IFR conditions must have sufficient fuel (considering weather reports, forecasts, and conditions) to fly to the first airport of intended landing, then to an alternate, then to fly for 45 minutes at normal cruising speed. (B10) — 14 CFR §91.167(a)(3)

Transponder Requirements

A coded transponder with altitude reporting capability is required for flight in all airspace of the 48 contiguous states and the District of Columbia at and above 10,000 feet MSL and below the floor of a Class A airspace, excluding the airspace at and below 2,500 feet AGL.

ATC may authorize deviations on a continuing basis, or for individual flights, for operations of aircraft without a transponder, in which case the request for a deviation must be submitted to the ATC facility having jurisdiction over the airspace concerned at least 1 hour before the proposed operation.

ALL, MIL
5060. A coded transponder equipped with altitude reporting equipment is required for

A—Class A, Class B, and Class C airspace areas.
B—all airspace of the 48 contiguous U.S. and the District of Columbia at and above 10,000 feet MSL (including airspace at and below 2,500 feet above the surface).
C—both answer A and B.

Mode C (encoding) transponders are required in Class A, B, and C airspace. (B11) — 14 CFR §91.215(b)(1)

Answers (B) and (C) are incorrect because a transponder is required in all airspace of the 48 contiguous U.S. and the District of Columbia at and above 10,000 feet MSL, excluding airspace at and below 2,500 feet AGL.

AIR, RTC, MIL
5061. In the contiguous U.S., excluding the airspace at and below 2,500 feet AGL, an operable coded transponder equipped with Mode C capability is required in all airspace above

A—10,000 feet MSL.
B—12,500 feet MSL.
C—14,500 feet MSL.

With some balloon, glider, and no-electrical-system exceptions, Mode C (encoding) transponders are required in all airspace above 10,000 feet MSL excluding airspace at or below 2,500 feet AGL. (B11) — 14 CFR §91.215(b)(5)(i)

ALL, MIL
5072-2. What transponder equipment is required for airplane operations within Class B airspace? A transponder

A—with 4096 code or Mode S, and Mode C capability.
B—with 4096 code capability is required except when operating at or below 1,000 feet AGL under the terms of a letter of agreement.
C—is required for airplane operations when visibility is less than 3 miles.

Unless otherwise authorized or directed by ATC, no person may operate an aircraft in a Class B airspace area unless the aircraft is equipped with an operating transponder and automatic altitude reporting equipment that has an operable coded radar beacon transponder having either Mode 3/A 4096 code capability, or a Mode S capability, and that aircraft is equipped with automatic pressure altitude reporting equipment having a Mode C capability. (B11) — 14 CFR §91.131 and §91.215

Answers
5059 [B]   5060 [A]   5061 [A]   5072-2 [A]

Supplemental Oxygen

No person may operate a civil aircraft:

1. At cabin pressure altitudes above 12,500 feet MSL up to and including 14,000 feet MSL, unless the required minimum flight crew uses supplemental oxygen for that part of the flight that is more than 30 minutes duration;

2. At cabin pressure altitudes above 14,000 feet MSL, unless the required minimum flight crew uses supplemental oxygen during the entire flight at those altitudes; and

3. At cabin pressure altitudes above 15,000 feet MSL, unless each occupant of the aircraft is provided with oxygen.

ALL, MIL
5063. In accordance with 14 CFR Part 91, supplemental oxygen must be used by the required minimum flight-crew for that time exceeding 30 minutes while at cabin pressure altitudes of

A— 10,500 feet MSL up to and including 12,500 feet MSL.
B— 12,000 feet MSL up to and including 18,000 feet MSL.
C— 12,500 feet MSL up to and including 14,000 feet MSL.

No person may operate a civil aircraft of U.S. registry:

1. *At cabin pressure altitudes above 12,500 feet MSL up to and including 14,000 feet MSL, unless the required minimum flight crew is provided with and uses supplemental oxygen for that part of the flight at those altitudes that is of more than 30 minutes duration;*

2. *At cabin pressure altitudes above 14,000 feet MSL, unless the required minimum flight crew is provided with and uses supplemental oxygen during the entire flight time at those altitudes; and*

3. *At cabin pressure altitudes above 15,000 feet MSL, unless each occupant of the aircraft is provided with supplemental oxygen.*

(B11) — 14 CFR §91.211(a)

ALL, MIL
5064. What are the oxygen requirements when operating at cabin pressure altitudes above 15,000 feet MSL?

A— Oxygen must be available for the flightcrew.
B— Oxygen is not required at any altitude in a free balloon.
C— The flightcrew must use and passengers must be provided oxygen.

No person may operate a civil aircraft of U.S. registry:

1. *At cabin pressure altitudes above 12,500 feet MSL up to and including 14,000 feet MSL, unless the required minimum flight crew is provided with and uses supplemental oxygen for that part of the flight at those altitudes that is of more than 30 minutes duration;*

2. *At cabin pressure altitudes above 14,000 feet MSL, unless the required minimum flight crew is provided with and uses supplemental oxygen during the entire flight time at those altitudes; and*

3. *At cabin pressure altitudes above 15,000 feet MSL, unless each occupant of the aircraft is provided with supplemental oxygen.*

(B11) — 14 CFR §91.211(a)

Answer (A) is incorrect because the flight crew must use oxygen above 14,000 feet MSL. Answer (B) is incorrect because oxygen requirements apply to all aircraft.

Instrument and Equipment Requirements

If a flight is being conducted for hire over water and beyond power-off gliding distance from shore, approved flotation gear readily available to each occupant and at least one pyrotechnic signaling device are required. Also, if the flight is conducted for hire at night, one electric landing light is needed. An operating anti-collision system is required for all night flights. This equipment is in addition to that required for non-commercial operations.

Answers
5063 [C]   5064 [C]

AIR, RTC, MIL

5066. Which is required equipment for powered aircraft during VFR night flights?

A—Flashlight with red lens if the flight is for hire.
B—An electric landing light if the flight is for hire.
C—Sensitive altimeter adjustable for barometric pressure.

An electric landing light is required only if the night flight is for hire. (B11) — 14 CFR §91.205(c)(4)

Answer (A) is incorrect because there is no specific requirement for flashlights and the color of the lens. Answer (C) is incorrect because sensitive altimeters are only required for IFR flight.

AIR, MIL

5067-1. Approved flotation gear, readily available to each occupant, is required on each airplane if it is being flown for hire over water,

A—in amphibious aircraft beyond 50 NM from shore.
B—beyond power-off gliding distance from shore.
C—more than 50 statute miles from shore.

Approved flotation gear readily available to each occupant is required on each aircraft if it is being flown for hire over water beyond power-off gliding distance from shore. "Shore" is defined as the area of the land adjacent to the water that is above the high water mark, excluding land areas which are intermittently under water. (B11) — 14 CFR §91.205(b)(12)

Answer (A) is incorrect because the flotation gear requirement applies to all aircraft operated for hire when flying beyond power-off gliding distance from shore. Answer (C) is incorrect because flotation gear is not required if the aircraft remains within power-off gliding distance from shore.

RTC, MIL

5067-2. Approved flotation gear, readily available to each occupant, is required on each helicopter if it is being flown for hire over water,

A—more than 50 statute miles from shore.
B—beyond power-off gliding distance from shore.
C—in amphibious aircraft beyond 50 NM from shore.

If the aircraft is operated for hire over water and beyond power-off gliding distance from shore, approved flotation gear must be readily available to each occupant. (B11) — 14 CFR §91.205

Restricted, Limited and Experimental Aircraft: Operating Limitations

No person may operate a restricted, limited or experimentally certificated civil aircraft carrying passengers or property for compensation or hire. In addition, no person may operate a restricted category aircraft:

- Over a densely populated area
- In a congested airway
- Near a busy airport where passenger transport operations are conducted.

ALL, MIL

5069. The carriage of passengers for hire by a commercial pilot is

A—not authorized in "utility" category aircraft.
B—not authorized in "limited" category aircraft.
C—authorized in "restricted" category aircraft.

Operations for compensation or hire are not authorized in limited category aircraft, experimental aircraft, and restricted category aircraft. Operations using utility category aircraft (such as Cessna 152) are authorized. (B12) — 14 CFR §91.315

Answer (A) is incorrect because the carriage of passengers for hire is permitted in normal and utility category aircraft. Answer (C) is incorrect because the carriage of passengers for hire is not permitted in restricted category aircraft.

ALL, MIL

5129. No person may operate an aircraft that has an experimental airworthiness certificate

A—under instrument flight rules (IFR).
B—when carrying property for hire.
C—when carrying persons or property for hire.

No person may operate an aircraft that has an experimental certificate while carrying persons or property for compensation or hire. (B12) — 14 CFR §91.319

Answers

| 5066 | [B] | 5067-1 | [B] | 5067-2 | [B] | 5069 | [B] | 5129 | [C] |

AIR, MIL
5068-2. Which is true with respect to operating limitations of a "restricted" category airplane?

A—A pilot of a "restricted" category airplane is required to hold a commercial pilot certificate.
B—A "restricted" category airplane is limited to an operating radius of 25 miles from its home base.
C—No person may operate a "restricted" category airplane carrying passengers or property for compensation or hire.

No person may operate a restricted category civil aircraft carrying persons or property for compensation or hire. (B12) — 14 CFR §91.313

AIR, MIL
5068-3. Which is true with respect to operating limitations of a "primary" category airplane?

A—A "primary" category airplane is limited to a specified operating radius from its home base.
B—No person may operate a "primary" category airplane carrying passengers or property for compensation or hire.
C—A pilot of a "primary" category airplane must hold a commercial pilot certificate when carrying passengers for compensation or hire.

No person may operate a primary category aircraft carrying persons or property for compensation or hire. (B12) — 14 CFR §91.325

Emergency Locator Transmitter (ELT)

Except as listed below, all aircraft must have on board an ELT that:

1. Is attached to the airplane in such a manner as to minimize the possibility of damage in a crash;
2. Transmits on 121.5 and 243.0 MHz;
3. Has batteries which must be replaced (or recharged, if the battery is rechargeable) after 1 hour of cumulative use, or when 50% of their useful shelf life has expired (or in the case or rechargeable batteries, when 50% of the useful life of the charge has expired). This date must be stamped on the outside of the battery case and entered in the aircraft logbook.

Aircraft and operations that do not need ELTs are:

1. Those ferrying an aircraft for an ELT installation or repair;
2. Training flights within a 50-mile radius from the airport;
3. Turbojet-powered aircraft; or
4. Agricultural operations.

Testing of ELTs should be carried out only during the first 5 minutes of any hour for no more than three sweeps, unless coordinated with ATC.

AIR, RTC, MIL
5070. The maximum cumulative time that an emergency locator transmitter may be operated before the rechargeable battery must be recharged is

A—30 minutes.
B—45 minutes.
C—60 minutes.

ELT batteries must be replaced or recharged when the transmitter has been in use for more than one cumulative hour. (B11) — 14 CFR §91.207(c)(1)

Answers

5068-2 [C]   5068-3 [B]   5070 [C]

Chapter 4 **Regulations**

Truth in Leasing

To operate a large civil U.S. aircraft that is leased, the lessee must mail a copy of the lease to the Aircraft Registry Technical Section, Box 25724, Oklahoma City, Oklahoma, 73125, within 24 hours of its execution.

AIR, MIL
5071. No person may operate a large civil U.S. aircraft which is subject to a lease, unless the lessee has mailed a copy of the lease to the FAA Aircraft Registration Branch, Technical Section, Oklahoma City, OK within how many hours of its execution?

A—24.
B—48.
C—72.

The lessee must mail a copy of the lease to the Aircraft Registry Technical Section, Box 25724, Oklahoma City, Oklahoma, 73125, within 24 hours of its execution. (B07) — 14 CFR §91.23(c)(1)

Operating Near Other Aircraft and Right-of-Way Rules

No person may operate an aircraft so close to another aircraft as to create a collision hazard. Aircraft carrying passengers for hire may not be flown in formation. **Formation flying** on flights not carrying passengers for hire is allowed, if the pilots of all the aircraft involved are in agreement.

When weather conditions permit, it is each pilot's responsibility to see and avoid other traffic, regardless of whether the flight is being conducted under Visual Flight Rules or Instrument Flight Rules. An aircraft in distress has the right-of-way over all others.

Aircraft on final approach to land, or while landing, have the right-of-way over other aircraft in flight or on the surface. If two aircraft are approaching the airport for the purpose of landing, the lower aircraft has the right-of-way, but the pilot shall not take advantage of this rule to cut in front of or to overtake the other aircraft.

When aircraft of the same category are converging, the aircraft to the other's right has the right-of-way. If the aircraft are of different categories, the order of right-of-way is:

1. Balloon
2. Glider
3. Airship
4. Airplane or rotorcraft

If two aircraft are approaching head-on, each pilot shall alter course to the right. *See* Figure 4-1. If two aircraft of the same category are converging because one is overtaking the other, the one being overtaken has the right-of-way and the overtaking aircraft must pass well clear to the right. *See* Figure 4-2.

Any aircraft towing another aircraft or refueling in flight has the right-of-way over all other engine-driven aircraft.

Answers
5071 [A]

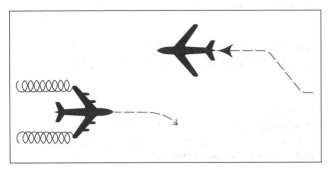

Figure 4-1. Aircraft approaching head-on

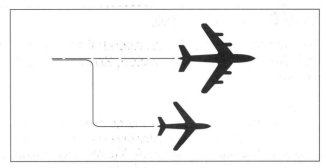

Figure 4-2. One aircraft overtaking another

ALL, MIL
5073-1. Which is true with respect to formation flights? Formation flights are

A—authorized when carrying passengers for hire with prior arrangement with the pilot in command of each aircraft in the formation.
B—not authorized when visibilities are less than 3 SM.
C—not authorized when carrying passengers for hire.

No person may operate an aircraft carrying passengers for hire in formation flight. (B08) — 14 CFR §91.111(c)

Answer (A) is incorrect because when carrying passengers for hire, formation flights are prohibited. Answer (B) is incorrect because formation flights are authorized regardless of visibility.

ALL, MIL
5073-2. Which is true with respect to operating near other aircraft in flight? They are

A—not authorized, when operated so close to another aircraft they can create a collision hazard.
B—not authorized, unless the pilot in command of each aircraft is trained and found competent in formation.
C—authorized when carrying passengers for hire, with prior arrangement with the pilot in command of each aircraft in the formation.

No person may operate an aircraft so close to another aircraft as to create a collision hazard. No person may operate an aircraft in formation flight except by arrangement with the pilot-in-command of each aircraft in the formation. No person may operate an aircraft, carrying passengers for hire, in formation flight. (B08) — 14 CFR §91.111

ALL, MIL
5073-3. Which is true with respect to formation flights? Formation flights are

A—not authorized, except by arrangement with the pilot in command of each aircraft.
B—not authorized, unless the pilot in command of each aircraft is trained and found competent in formation.
C—authorized when carrying passengers for hire, with prior arrangement with the pilot in command of each aircraft in the formation.

No person may operate an aircraft so close to another aircraft as to create a collision hazard. No person may operate an aircraft in formation flight except by arrangement with the pilot-in-command of each aircraft in the formation. No person may operate an aircraft, carrying passengers for hire, in formation flight. (B08) — 14 CFR §91.111

AIR, RTC, MIL
5074. While in flight a helicopter and an airplane are converging at a 90° angle, and the helicopter is located to the right of the airplane. Which aircraft has the right-of-way, and why?

A—The helicopter, because it is to the right of the airplane.
B—The helicopter, because helicopters have the right-of-way over airplanes.
C—The airplane, because airplanes have the right-of-way over helicopters.

When aircraft of the same category are converging at approximately the same altitude (except head-on, or nearly so) the aircraft to the other's right has the right-of-way. Aircraft and rotorcraft are treated as equally maneuverable, so neither aircraft has right-of-way over the other. (B08) — 14 CFR §91.113(d)

Answers

5073-1 [C]   5073-2 [A]   5073-3 [A]   5074 [A]

Chapter 4 Regulations

ALL, MIL
5075. Two aircraft of the same category are approaching an airport for the purpose of landing. The right-of-way belongs to the aircraft

A—at the higher altitude.
B—at the lower altitude, but the pilot shall not take advantage of this rule to cut in front of or to overtake the other aircraft.
C—that is more maneuverable, and that aircraft may, with caution, move in front of or overtake the other aircraft.

When two or more aircraft are approaching an airport for the purpose of landing, the lower aircraft has the right-of-way. However, a pilot shall not take advantage of this rule to overtake or cut in front of another aircraft that is on final approach to land. (B08) — 14 CFR §91.113(g)

AIR, RTC, MIL
5076-1. Airplane A is overtaking airplane B. Which airplane has the right-of-way?

A—Airplane A; the pilot should alter course to the right to pass.
B—Airplane B; the pilot should expect to be passed on the right.
C—Airplane B; the pilot should expect to be passed on the left.

Each aircraft that is being overtaken has the right-of-way, and each pilot of an overtaking aircraft shall alter course to the right to pass well clear. (B08) — 14 CFR §91.113(f)

AIR, RTC, MIL
5076-2. An airplane is overtaking a helicopter. Which aircraft has the right-of-way?

A—Helicopter; the pilot should expect to be passed on the right.
B—Airplane; the airplane pilot should alter course to the left to pass.
C—Helicopter; the pilot should expect to be passed on the left.

When weather conditions permit, regardless of whether an operation is conducted under instrument flight rules or visual flight rules, vigilance shall be maintained by each person operating an aircraft so as to see and avoid other aircraft. Each aircraft that is being overtaken has the right-of-way and each pilot of an overtaking aircraft shall alter course to the right to pass well clear. (B08) — 14 CFR §91.113

AIR, RTC, MIL
5076-3. During a night operation, the pilot of aircraft #1 sees only the green light of aircraft #2. If the aircraft are converging, which pilot has the right-of-way? The pilot of aircraft

A—#2; aircraft #2 is to the left of aircraft #1.
B—#2; aircraft #2 is to the right of aircraft #1
C—#1; aircraft #1 is to the right of aircraft #2.

When aircraft of the same category are converging at approximately the same altitude (except head-on, or nearly so), the aircraft to the other's right has the right-of-way. The green light indicates aircraft #1 is looking at the right wing of aircraft #2, which means aircraft #1 is to the right of aircraft #2. (B08) — 14 CFR §91.113

AIR, RTC, MIL
5076-4. A pilot flying a single-engine airplane observes a multiengine airplane approaching from the left. Which pilot should give way?

A—The pilot of the multiengine airplane should give way; the single-engine airplane is to its right.
B—The pilot of the single-engine airplane should give way; the other airplane is to the left.
C—Each pilot should alter course to the right.

When aircraft of the same category are converging at approximately the same altitude (except head-on, or nearly so), the aircraft to the other's right has the right-of-way. (B08) — 14 CFR §91.113

Answers

5075 [B]   5076-1 [B]   5076-2 [A]   5076-3 [C]   5076-4 [A]

Chapter 4 **Regulations**

Speed Limits

If ATC assigned an airspeed, it must be maintained within plus or minus 10 knots. All aircraft must observe the speed limits (all speeds are shown in knots and are indicated airspeed) as illustrated in Figure 4-3.

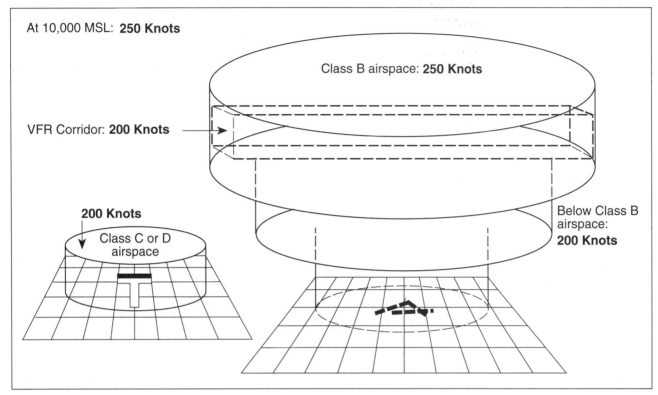

Figure 4-3. Maximum speed limits

AIR, RTC, MIL
5077. What is the maximum indicated airspeed allowed in the airspace underlying Class B airspace?

A—156 knots.
B—200 knots.
C—230 knots.

No person may operate an aircraft in the airspace underlying Class B airspace, or in a VFR corridor designated through Class B airspace, at an indicated airspeed of more than 200 knots (230 MPH). (B08) — 14 CFR §91.117(c)

AIR, RTC, MIL
5078. Unless otherwise authorized or required by ATC, the maximum indicated airspeed permitted when at or below 2,500 feet AGL within 4 NM of the primary airport of a Class C, or D airspace is

A—180 knots.
B—200 knots.
C—230 knots.

Unless otherwise authorized or required by ATC, no person may operate an aircraft within 4 NM of the primary airport of Class C or D airspace at an indicated airspeed of more than 200 knots (230 MPH). (B08) — 14 CFR §91.117(b)

Answers

5077 [B]   5078 [B]

Chapter 4 **Regulations**

ALL, MIL
5112. If the minimum safe speed for any particular operation is greater than the maximum speed prescribed in 14 CFR Part 91, the

A—operator must have a Memorandum of Agreement (MOA) with the controlling agency.
B—aircraft may be operated at that speed.
C—operator must have a Letter of Agreement with ATC.

If the minimum safe airspeed for any particular operation is greater than the maximum speed prescribed in Part 91, the aircraft may be operated at that minimum speed. (B08) — 14 CFR §91.117

Aerobatic Flight

Aerobatic (acrobatic) flight means an intentional maneuver involving an abrupt change in an aircraft's altitude, an abnormal attitude, or abnormal acceleration unnecessary for normal flight.

The pilot must take extra caution during aerobatic flight that there is no danger of collision with other aircraft. In addition, a pilot may not conduct aerobatic flight:

- Over any congested area of a city, town or settlement
- Over an open air assembly of persons
- Within the lateral boundaries of Class B, C, D, or E airspace designated for an airport, or a federal airway
- At less than 1,500 feet AGL
- When the flight visibility is less than 3 miles.

AIR, GLI, MIL
5079. What is the minimum altitude and flight visibility required for acrobatic flight?

A—1,500 feet AGL and 3 miles.
B—2,000 feet MSL and 2 miles.
C—3,000 feet AGL and 1 mile.

The minimum altitude for aerobatic flight is 1,500 feet AGL. The minimum visibility required for aerobatic flight is 3 miles. (B08) — 14 CFR §91.303(e), (f)

Aircraft Lights

No person may, during the period from sunset to sunrise, operate an aircraft unless it has lighted position lights. The right wing-tip position light is green, the left is red and the tail white. Each aircraft must also have an approved anti-collision light system.

AIR, RTC, MIL
5065. Which is required equipment for powered aircraft during VFR night flights?

A—Anticollision light system.
B—Gyroscopic direction indicator.
C—Gyroscopic bank-and-pitch indicator.

An approved anticollision light system is required for flight under VFR between sunset and sunrise. (B11) — 14 CFR §91.205(c)(3)

Answers (B) and (C) are incorrect because a gyroscopic direction indicator and gyroscopic bank-and-pitch indicator is required for IFR flights.

Answers
5112 [B]   5079 [A]   5065 [A]

ALL, MIL

5080-1. If not equipped with required position lights, an aircraft must terminate flight

A—at sunset.
B—30 minutes after sunset.
C—1 hour after sunset.

No person may, during the period from sunset to sunrise (or, in Alaska, during the period a prominent unlighted object cannot be seen from a distance of 3 statute miles or the sun is more than 6° below the horizon) operate an aircraft unless it has lighted position lights. (B11) — 14 CFR §91.209(a)

ALL, MIL

5080-2. If an aircraft is not equipped with an electrical or anticollision light system, no person may operate that aircraft

A—after sunset to sunrise.
B—after dark.
C—1 hour after sunset.

No person may during the period from sunset to sunrise operate an aircraft that is equipped with an anticollision light system, unless it has lighted anticollision lights. However, the anticollision lights need not be lighted when the pilot-in-command determines that, because of operating conditions, it would be in the interest of safety to turn the lights off. (B11) — 14 CFR §91.209

LTA

5080-3. Operation of a balloon, during the period of sunset to sunrise, requires that it be equipped and lighted with

A—red and green position lights.
B—approved aviation red and white lights.
C—a steady aviation white position light and a red or white anticollision light.

No person may, during the period from sunset to sunrise, operate an aircraft unless it has lighted position lights. Forward position lights must consist of a red and a green light spaced laterally as far apart as practicable with the red light on the left side, and the green light on the right side. The rear position light must be a white light mounted as far aft as practicable. (B11) — 14 CFR §91.209 and §29.1385

LTA

5080-4. Operation of a lighter-than-air airship, during the period of sunset to sunrise, requires it be equipped and lighted with

A—position lights and aviation red or white anticollision light system.
B—approved aviation red and white lights.
C—position lights.

No person may, during the period from sunset to sunrise, operate an aircraft unless it has lighted position lights. The airplane must have an approved anticollision light system and each anticollision light must be either aviation red or aviation white. (B11) — 14 CFR §91.209 and §23.1401

ALL

5666. What is the general direction of movement of the other aircraft if during a night flight you observe a steady white light and a rotating red light ahead and at your altitude? The other aircraft is

A—headed away from you.
B—crossing to your left.
C—approaching you head-on.

The pilot is seeing the rotating beacon and the white light on the tail. The other aircraft is headed away. (H63) — FAA-H-8083-3, Chapter 10

Answer (B) is incorrect because you would see a steady red light if the other aircraft was crossing to your left. Answer (C) is incorrect because you would see both the red and green wing-tip position lights if the other aircraft were approaching you head-on.

Answers

5080-1   [A]   5080-2   [A]   5080-3   [A]   5080-4   [A]   5666   [A]

Chapter 4 **Regulations**

Minimum Altitudes

Except when necessary for takeoff or landing, or unless otherwise authorized by the Administrator, the minimum altitude for IFR flight is:

1. 2,000 feet above the highest obstacle within a horizontal distance of 4 NM from the course to be flown over an area designated as a mountainous area; or
2. 1,000 feet above the highest obstacle within a horizontal distance of 4 NM from the course to be flown over terrain in other areas.

ALL, MIL
5092. Except when necessary for takeoff or landing or unless otherwise authorized by the Administrator, the minimum altitude for IFR flight is

A—3,000 feet over all terrain.
B—3,000 feet over designated mountainous terrain; 2,000 feet over terrain elsewhere.
C—2,000 feet above the highest obstacle over designated mountainous terrain; 1,000 feet above the highest obstacle over terrain elsewhere.

At a minimum, IFR flight must remain at 2,000 feet AGL in mountainous areas and 1,000 feet AGL elsewhere. (B10) — 14 CFR §91.177(a)(2)

RTC, MIL
5113-1. Minimum safe altitude rules require that helicopter pilots

A—not fly lower than 500 feet, except when necessary for takeoff or landing.
B—comply with routes and altitudes prescribed by the FAA.
C—not fly closer than 500 feet to any person, vessel, vehicle, or structure.

Helicopters may be operated at less than minimum altitudes if the operation is conducted without hazard to persons or property on the surface. In addition, each person operating a helicopter shall comply with any routes or altitudes specifically prescribed for helicopters by the Administrator. (B08) — 14 CFR §91.119

RTC, MIL
5113-2. Minimum safe altitude rules authorize helicopter pilots to

A—fly at less than 500 feet.
B—fly at less than 500 feet if they do not create a hazard to persons or property on the surface.
C—fly closer than 500 feet to any person, vehicle, vessel, or structure on the surface.

Helicopters may be operated at less than minimum altitudes if the operation is conducted without hazard to persons or property on the surface. In addition, each person operating a helicopter shall comply with any routes or altitudes specifically prescribed for helicopters by the Administrator. (B08) — 14 CFR §91.119

Answers
5092 [C]　　　5113-1 [B]　　　5113-2 [B]

Maintenance Responsibility

The owner or operator of an aircraft holds primary responsibility for:

1. Maintaining that aircraft in an airworthy condition;
2. Having required inspections performed; and
3. Ensuring that maintenance personnel make the required entries in the aircraft maintenance records indicating that the aircraft has been approved for return to service.

Operation of an aircraft after maintenance, rebuilding or alteration is prohibited unless:

- The aircraft has been approved for return to service by authorized maintenance personnel
- The required maintenance record entry has been made.

In addition, after any major alteration or repair that may have substantially affected an aircraft's flight characteristics or flight operation, it must be test flown before passengers may be carried. The test pilot must be appropriately rated and must hold at least a Private Pilot Certificate. The aircraft's documents must indicate that it was test flown and approved for return to service by an appropriately-rated pilot.

ALL, MIL
5093. Who is primarily responsible for maintaining an aircraft in an airworthy condition?

A—The lead mechanic responsible for that aircraft.
B—Pilot in command or operator.
C—Operator or owner of the aircraft.

The owner or operator of an aircraft is primarily responsible for maintaining that aircraft in an airworthy condition, including compliance with Airworthiness Directives. (B13) — 14 CFR §91.403(a)

Answer (A) is incorrect because mechanics work under the advisement of the owner/operator. Answer (B) is incorrect because the pilot is only responsible for determining airworthiness before flight.

ALL, MIL
5097. If an aircraft's operation in flight was substantially affected by an alteration or repair, the aircraft documents must show that it was test flown and approved for return to service by an appropriately-rated pilot prior to being operated

A—under VFR or IFR rules.
B—with passengers aboard.
C—for compensation or hire.

No person may carry any person (meaning passengers) other than crewmembers in an aircraft that has been maintained, rebuilt, or altered in a manner that may have appreciably changed its flight characteristics or substantially affected its operation in flight, until an appropriately-rated pilot with at least a Private Pilot Certificate flies the aircraft, makes an operational check of the maintenance performed or alteration made, and logs the flight in the aircraft records. (B13) — 14 CFR §91.407(b)

Answers

5093 [C] 5097 [B]

Chapter 4 **Regulations**

Aircraft Inspections

Each aircraft must have had an **annual inspection** performed within the preceding 12 calendar months. That inspection must be recorded in the aircraft and engine logbooks, and is valid through the last day of the month, regardless of the issuance date.

In addition to an annual inspection, if an aircraft is used to carry passengers for hire or is used for flight instruction, it must have an inspection every 100 hours. The **100-hour limitation** may be exceeded by not more than 10 hours, if necessary, to reach a location where the inspection can be done. The excess time, however, must be included in calculating the following 100 hours of time in service. An annual inspection may be substituted for 100-hour inspection.

A **progressive maintenance program** requires that an aircraft be maintained and inspected at specified intervals (for example, every 50 hours) and according to a specific sequence (for example, four operations per cycle). A typically-approved progressive maintenance program meets the requirements of both annual and 100-hour inspections when the scheduled operations are completed within 12 months.

In order to be approved for operation at all, each transponder must be inspected and tested within 24 calendar months. This inspection must be noted in the appropriate logbooks.

The validity of an airworthiness Certificate is maintained by an appropriate return-to-service notation in the aircraft maintenance records at the completion of any required inspections and maintenance.

ALL, MIL
5096. A standard airworthiness certificate remains in effect as long as the aircraft receives

A—required maintenance and inspections.
B—an annual inspection.
C—an annual inspection and a 100-hour inspection prior to their expiration dates.

No person may operate an aircraft unless, within the preceding 12 calendar months, it has had an annual inspection and has been approved for return to service by an authorized person or an inspection for the issuance of an airworthiness certificate. No person may operate an aircraft carrying passengers for hire or give flight instruction for hire in an aircraft which that person provides, unless within the preceding 100 hours of time in service the aircraft has received an annual or 100-hour inspection. An aircraft inspected in accordance with an approved aircraft progressive inspection program may be authorized as compliant. (B13) — 14 CFR §91.409

ALL, MIL
5099. An aircraft carrying passengers for hire has been on a schedule of inspection every 100 hours of time in service. Under which condition, if any, may that aircraft be operated beyond 100 hours without a new inspection?

A—The aircraft may be flown for any flight as long as the time in service has not exceeded 110 hours.
B—The aircraft may be dispatched for a flight of any duration as long as 100 hours has not been exceeded at the time it departs.
C—The 100-hour limitation may be exceeded by not more than 10 hours if necessary to reach a place at which the inspection can be done.

The 100-hour limitation may be exceeded by not more than 10 hours if necessary to reach a place the inspection can be done. The excess time, however, is included in computing the next 100 hours of time in service. (B13) — 14 CFR §91.409(b)

Answers (A) and (B) are incorrect because the 10-hour grace period only applies to ferrying an aircraft to a maintenance facility, for the 100-hour inspection.

Answers
5096 [A]   5099 [C]

ALL, MIL
5100. Which is true concerning required maintenance inspections?

A—A 100-hour inspection may be substituted for an annual inspection.
B—An annual inspection may be substituted for a 100-hour inspection.
C—An annual inspection is required even if a progressive inspection system has been approved.

The annual is considered the more rigorous inspection and can be substituted for the 100-hour, not vice versa. (B13) — 14 CFR §91.409(a), (b), (c)

Answer (A) is incorrect because an annual inspection may be substituted for a 100-hour inspection. Answer (C) is incorrect because a progressive inspection can be done in place of an annual inspection.

ALL, MIL
5101. An ATC transponder is not to be used unless it has been tested, inspected, and found to comply with regulations within the preceding

A—30 days.
B—12 calendar months.
C—24 calendar months.

No person may use an ATC transponder unless, within the preceding 24 calendar months, that ATC transponder has been tested, inspected, and found to comply with regulations. (B13) — 14 CFR §91.413(c)

ALL, MIL
5105. If an ATC transponder installed in an aircraft has not been tested, inspected, and found to comply with regulations within a specified period, what is the limitation on its use?

A—Its use is not permitted.
B—It may be used when in Class G airspace.
C—It may be used for VFR flight only.

No person may use an ATC transponder unless, within the preceding 24 calendar months, that ATC transponder has been tested, inspected and found to comply with regulations. (B13) — 14 CFR §91.413(a)

Maintenance Records

With the exception of work performed under 14 CFR §91.171, records of the following work must be kept until the work is repeated or superseded by other work, or for one year after the work is performed:

- Records of maintenance or alterations
- Records of 100-hour, annual, progressive and other required or approved inspections.

The records of the following must be retained and transferred with the aircraft at the time the aircraft is sold:

- Total time in service of the airframe
- Current status of life-limited parts of each airframe, engine, propeller, rotor, and appliance
- Current status of inspections and airworthiness directives
- A list of all current major alterations.

Answers

5100 [B]   5101 [C]   5105 [A]

Chapter 4 **Regulations**

A recording tachometer cannot be substituted for required aircraft maintenance records.

Each manufacturer or agency that grants zero time to a rebuilt engine shall enter, in a new record:

- A signed statement of the date the engine was rebuilt
- Each change made as required by airworthiness directives
- Each change made in compliance with manufacturer service bulletins.

Old maintenance records may be discarded when an engine is rebuilt.

ALL, MIL
5095. After an annual inspection has been completed and the aircraft has been returned to service, an appropriate notation should be made

A—on the airworthiness certificate.
B—in the aircraft maintenance records.
C—in the FAA-approved flight manual.

Each owner or operator of an aircraft shall have that aircraft inspected as prescribed by regulations and shall, between required inspections, have discrepancies repaired as prescribed in 14 CFR Part 43. In addition, each owner or operator shall ensure that maintenance personnel make appropriate entries in the aircraft maintenance records indicating that the aircraft has been approved for return to service. (B13) — 14 CFR §91.405(b)

ALL, MIL
5102. Aircraft maintenance records must include the current status of the

A—applicable airworthiness certificate.
B—life-limited parts of only the engine and airframe.
C—life-limited parts of each airframe, engine, propeller, rotor, and appliance.

Each registered owner or operator shall keep records containing the following information: the current status of life-limited parts of each airframe, engine, propeller, rotor, and appliance. (B13) — 14 CFR §91.417(a)(2)(ii)

AIR, RTC, MIL
5104. A new maintenance record being used for an aircraft engine rebuilt by the manufacturer must include previous

A—operating hours of the engine.
B—annual inspections performed on the engine.
C—changes as required by Airworthiness Directives.

Each manufacturer or agency that grants zero time to an engine rebuilt by it shall enter, in the new record, each change made as required by Airworthiness Directives. (B13) — 14 CFR §91.421(b)(2)

Answer (A) is incorrect because a rebuilt engine is considered to start with zero time. Answer (B) is incorrect because a record of previous maintenance is not required for a rebuilt engine.

Answers
5095 [B]   5102 [C]   5104 [C]

Chapter 4 **Regulations**

Maintenance, Preventative Maintenance, Rebuilding and Alteration

The holder of a pilot certificate issued under Part 61 may perform **preventative maintenance** (in accordance with 14 CFR Part 43, Appendix A) on any aircraft owned or operated by him/her that is not in air carrier service. Two of the conditions that apply are as follows:

1. Preventive maintenance performed by a certificated pilot must be logged, and kept in the aircraft maintenance records; and
2. Work considered as preventive maintenance is generally that which does nothing to alter the weight, CG, or flight controls, and does not require tampering with key components (prop, struts, etc.).

14 CFR Part 39 prescribes **Airworthiness Directives (ADs)** that apply to an aircraft and its parts. When aircraft operational or mechanical problems are discovered, manufacturers may issue service bulletins recommending corrective measures. However, when an unsafe condition exists and that condition is likely to exist or develop in other products of the same or similar type or design, the FAA will issue an Airworthiness Directive. Airworthiness Directives are considered to be amendments to regulations. Compliance is mandatory and is the responsibility of the owner or operator of that aircraft. Noncompliance with ADs renders an aircraft unairworthy. No person may operate an aircraft to which an AD applies, except in accordance with the requirements of that AD. Compliance with the provisions of each AD must be recorded in the aircraft maintenance records.

ALL, MIL
5098. Which is correct concerning preventive maintenance, when accomplished by a pilot?

A—A record of preventive maintenance is not required.
B—A record of preventive maintenance must be entered in the maintenance records.
C—Records of preventive maintenance must be entered in the FAA-approved flight manual.

A person holding at least a Private Pilot Certificate may approve an aircraft for return to service after performing preventative maintenance under the provisions of Part 43.3(g) and shall make an entry in the maintenance record of that maintenance. (B13) — 14 CFR §43.5, §43.9

ALL, MIL
5094. Assuring compliance with an Airworthiness Directive is the responsibility of the

A—pilot in command and the FAA certificated mechanic assigned to that aircraft.
B—pilot in command of that aircraft.
C—owner or operator of that aircraft.

The owner or operator of an aircraft is primarily responsible for maintaining that aircraft in an airworthy condition, including compliance with Part 39 (Airworthiness Directives). (B13) — 14 CFR §91.403(a)

Answers (A) and (B) are incorrect because the pilots are responsible for determining airworthiness before flight, and the mechanics work under the advisement of the owner/operator.

ALL, MIL
5103. Which is true relating to Airworthiness Directives (ADs)?

A—ADs are advisory in nature and are, generally, not addressed immediately.
B—Noncompliance with ADs renders an aircraft unairworthy.
C—Compliance with ADs is the responsibility of maintenance personnel.

No person may operate a product to which an airworthiness directive applies except in accordance with the requirements of that airworthiness directive. (B13) — 14 CFR §39.3

Answer (A) is incorrect because ADs are mandatory. Answer (C) is incorrect because compliance with ADs, along with other maintenance regulations, is the responsibility of the owner/operator.

Answers

5098 [B]   5094 [C]   5103 [B]

Chapter 4 **Regulations**

NTSB Part 830

Part 830 deals with the reporting of aircraft accidents and incidents. An operator is responsible to the NTSB, not the FAA, for all rules pertaining to this part. Part 830 also deals with the preservation of aircraft wreckage, mail, cargo, and records. An operator will notify the nearest NTSB office immediately if any of the following occur:

1. An aircraft accident, meaning—

   a. A fatality or serious injury; or

   b. Any substantial damage. This means any damage which adversely affects the structural strength, performance, or flight characteristics of the aircraft; (damage to the landing gear, wheels, tires, flaps, engine accessories, brakes, wing tips or small puncture holes in the skin or fabric are not considered substantial damage);

2. An aircraft overdue and believed to have been involved in an accident; or

3. Any of the following incidents:

   a. An inflight fire;

   b. Aircraft collision in flight;

   c. Inability of a flight crew member to perform his/her duties due to illness or injury; or

   d. Flight control system malfunction or failure.

In addition to the immediate notification, the pilot or operator will file a written report:

- In the case of an accident, within 10 days
- In the case of an overdue aircraft, within 7 days if the aircraft is still missing
- In the case of an incident, upon request.

ALL, MIL
5001. Notification to the NTSB is required when there has been substantial damage

A—which requires repairs to landing gear.
B—to an engine caused by engine failure in flight.
C—which adversely affects structural strength or flight characteristics.

The operator of an aircraft shall immediately, and by the most expeditious means available, notify the nearest NTSB field office when an aircraft accident occurs. An aircraft accident is an occurrence associated with the operation of an aircraft which takes place between the time any person boards the aircraft with the intention of flight and all such persons have disembarked, and in which any person suffers death or serious injury, or in which the aircraft receives substantial damage which adversely affects the structural strength, performance, or flight characteristics of the aircraft. (G10) — 49 CFR §830.2, §830.5

ALL, MIL
5002. NTSB Part 830 requires an immediate notification as a result of which incident?

A—Engine failure for any reason during flight.
B—Damage to the landing gear as a result of a hard landing.
C—Any required flight crewmember being unable to perform flight duties because of illness.

The operator of an aircraft shall immediately, and by the most expeditious means available, notify the nearest NTSB field office of the inability of any required flight crew member to perform normal flight duties as a result of injury or illness. (G11) — 49 CFR §830.2, §830.5

Answers (A) and (B) are incorrect because these are not considered "substantial damage" requiring immediate notification.

Answers
5001 [C]   5002 [C]

Chapter 4 **Regulations**

ALL, MIL
5003-1. Which incident would require that the nearest NTSB field office be notified immediately?

A—In-flight fire.
B—Ground fire resulting in fire equipment dispatch.
C—Fire of the primary aircraft while in a hangar which results in damage to other property of more than $25,000.

The operator of an aircraft shall immediately and by the most expeditious means available notify the nearest NTSB field office of any in-flight fire. (G11) — 49 CFR §830.5(a)(4)

Answers (B) and (C) are incorrect because the regulation specifies "in-flight" fires only.

ALL, MIL
5003-2. Which airborne incident would require that the nearest NTSB field office be notified immediately?

A—Cargo compartment door malfunction or failure.
B—Cabin door opened in-flight.
C—Flight control system malfunction or failure.

Immediate notification is required by the operator of any civil aircraft when a flight control system malfunction or failure occurs. (G11) — 49 CFR §830.5

Answers (A) and (B) are incorrect because neither a cargo compartment door malfunction nor a cabin door opening would require immediate notification to the NTSB, unless substantial damage affecting the structural integrity of the aircraft resulted.

ALL, MIL
5004-1. While taxiing for takeoff, a small fire burned the insulation from a transceiver wire. What action would be required to comply with NTSB Part 830?

A—No notification or report is required.
B—A report must be filed with the avionics inspector at the nearest FAA field office within 48 hours.
C—An immediate notification must be filed by the operator of the aircraft with the nearest NTSB field office.

The operator of an aircraft shall immediately and by the most expeditious means available notify the nearest NTSB field office of any in-flight fire. This rule specifies "in-flight" fires only. (G11) — 49 CFR §830.5

Answers (B) and (C) are incorrect because an immediate report is only required if certain items occur, such as an in-flight fire, and reports are made to the NTSB, not the FAA.

ALL, MIL
5004-2. While taxiing on the parking ramp, the landing gear, wheel, and tire are damaged by striking ground equipment. What action would be required to comply with NTSB Part 830?

A—An immediate notification must be filed by the operator of the aircraft with the nearest NTSB field office.
B—A report must be filed with the nearest FAA field office within 7 days.
C—No notification or report is required.

An accident is defined as the occurrence of a fatality or serious injury or substantial damage to the aircraft. Substantial damage specifically excludes damage to the landing gear, wheels and tires. (G11) — 49 CFR §830.2

Answer (A) is incorrect because no notification is required due to damaged landing gear alone. Answer (B) is incorrect because notification of an overdue aircraft requires a report within 7 days if the aircraft is still missing, not damage to the landing gear.

ALL, MIL
5005. During flight a fire which was extinguished burned the insulation from a transceiver wire. What action is required by regulations?

A—No notification or report is required.
B—A report must be filed with the avionics inspector at the nearest FAA Flight Standards District Office within 48 hours.
C—An immediate notification by the operator of the aircraft to the nearest NTSB field office.

The operator of an aircraft shall immediately and by the most expeditious means available notify the nearest NTSB field office of any in-flight fire. (G11) — 49 CFR §830.5

Answer (A) is incorrect because only an in-flight fire requires immediate notification. Answer (B) is incorrect because no report to the avionics inspector is required.

ALL, MIL
5006. When should notification of an aircraft accident be made to the NTSB if there was substantial damage and no injuries?

A—Immediately.
B—Within 10 days.
C—Within 30 days.

Answers

5003-1 [A]   5003-2 [C]   5004-1 [A]   5004-2 [C]   5005 [C]   5006 [A]

The operator of an aircraft shall immediately, and by the most expeditious means available, notify the nearest NTSB field office when an aircraft accident occurs. An aircraft accident is an occurrence associated with the operation of an aircraft which takes place between the time any person boards the aircraft with the intention of flight and all such persons have disembarked, and in which any person suffers death or serious injury, or in which the aircraft receives substantial damage which adversely affects the structural strength, performance, or flight characteristics of the aircraft. (G11) — 49 CFR §830.2, §830.5

Answer (B) is incorrect because 10 days is the time specified to file a detailed aircraft accident report with the NTSB. Answer (C) is incorrect because 30 days is not a deadline specified in NTSB Part 830.

ALL, MIL
5007. The operator of an aircraft that has been involved in an incident is required to submit a report to the nearest field office of the NTSB

A—within 7 days.
B—within 10 days.
C—only if requested to do so.

A written report of an incident need be filed only if requested by the NTSB. (G13) — 49 CFR §830.2, §830.15

Answer (A) is incorrect because 7 days is the time limitation for reporting overdue (missing) aircraft. Answer (B) is incorrect because 10 days is the limitation on filing a report for accidents.

ALL, MIL
5008. Within how many days of an accident is an accident report required to be filed with the nearest NTSB field office?

A—2 days.
B—7 days.
C—10 days.

The operator of an aircraft shall file a report within 10 days after an accident. (G13) — 49 CFR §830.2, §830.15

Answer (A) is incorrect because 2 days is not a reporting requirement in NTSB Part 830. Answer (B) is incorrect because 7 days is the limitation with respect to an overdue aircraft that is missing.

Rotorcraft Regulations

RTC, MIL
5029. To act as pilot in command of a gyroplane carrying passengers, what must the pilot do in that gyroplane to meet recent daytime flight experience requirements?

A—Make nine takeoffs and landings within the preceding 30 days.
B—Make three takeoffs and landings to a full stop within the preceding 90 days.
C—Make three takeoffs and landings within the preceding 90 days.

No person may act as pilot-in-command carrying passengers unless within the preceding 90 days he/she has made three takeoffs and three landings in the category and class of aircraft to be used. (A20) — 14 CFR §61.57(a)

RTC, MIL
5058. To begin a flight in a rotorcraft under VFR, there must be enough fuel to fly to the first point of intended landing and, assuming normal cruise speed, to fly thereafter for at least

A—20 minutes.
B—30 minutes.
C—45 minutes.

Twenty minutes of fuel is required for rotorcraft beyond the first point of intended landing, assuming normal cruising speed. (B09) — 14 CFR §91.151(b)

Answers

5007 [C]   5008 [C]   5029 [C]   5058 [A]

RTC, MIL
5068-1. Which is true with respect to operating limitations of a "restricted" category helicopter?

A—A "restricted" category helicopter is limited to an operating radius of 25 miles from its home base.
B—A pilot of a "restricted" category helicopter is required to hold a commercial pilot certificate.
C—No person may operate a "restricted" category helicopter carrying property or passengers for compensation or hire.

No person may operate a restricted category civil aircraft carrying persons or property for compensation or hire. (B12) — 14 CFR §91.313(c)

RTC, MIL
5072-1. What transponder equipment is required for helicopter operations within Class B airspace? A transponder

A—with 4096 code and Mode C capability.
B—is required for helicopter operations when visibility is less than 3 miles.
C—with 4096 code capability is required except when operating at or below 1,000 feet AGL under the terms of a letter of agreement.

Helicopters must use Mode C transponders in Class B airspace. (B11) — 14 CFR §91.215(b)(1)

Answer (B) is incorrect because transponder regulations are not tied to visibility. Answer (C) is incorrect because the Mode C requirement would also require a waiver.

RTC, MIL
5087. Basic VFR weather minimums require at least what visibility for operating a helicopter within Class D airspace?

A—1 mile.
B—2 miles.
C—3 miles.

Visibility in controlled airspace below 10,000 feet is 3 miles. (B08) — 14 CFR §91.155

RTC, MIL
5086. Which minimum flight visibility and distance from clouds is required for a day VFR helicopter flight in Class G airspace at 3,500 feet MSL over terrain with an elevation of 1,900 feet MSL?

A—Visibility–3 miles; distance from clouds–1,000 feet below, 1,000 feet above, and 1 mile horizontally.
B—Visibility–3 miles; distance from clouds–500 feet below, 1,000 feet above, and 2,000 feet horizontally.
C—Visibility–1 mile; distance from clouds–500 feet below, 1,000 feet above, and 2,000 feet horizontally.

The chopper is at 1,600 feet AGL. In Class G airspace, more than 1,200 feet above the surface but less than 10,000 feet MSL, during day VFR operations, helicopters are required to maintain 1 mile visibility, and 500 feet below, 1,000 above, and 2,000 feet horizontal distance from clouds. (B08) — 14 CFR §91.155(a), (b)

Glider Regulations

GLI, LTA
5035-1. What is the minimum age requirement for a person to be issued a commercial pilot certificate for the operation of gliders?

A—17 years.
B—18 years.
C—16 years.

A person must be at least 18 years of age to hold a Commercial Pilot Certificate. (A22) — 14 CFR §61.123

GLI
5038. To exercise the privileges of a commercial pilot certificate with a glider category rating, what medical certification is required?

A—No medical certification is required.
B—At least a second-class medical certificate when carrying passengers for hire.
C—A statement by the pilot certifying he/she has no known physical defects that makes him/her unable to pilot a glider.

A person is not required to hold a medical certificate when exercising the privileges of a pilot certificate with a glider category rating. (A20) — 14 CFR §61.23

Answers

5068-1 [C]   5072-1 [A]   5087 [C]   5086 [C]   5035-1 [B]   5038 [A]

GLI
5084. When flying a glider above 10,000 feet MSL and more than 1,200 feet AGL, what minimum flight visibility is required?

A—3 NM.
B—5 NM.
C—5 SM.

The only area requiring 5 statute miles visibility is 10,000 feet MSL and up, when above 1,200 feet AGL. (B08) — 14 CFR §91.155(a), §61.3(e)(3)

Lighter-Than-Air Regulations

LTA
5035-2. What is the minimum age requirement for a person to be issued a student pilot certificate for the operation of balloons?

A—16 years.
B—15 years.
C—14 years.

A person must be at least 14 years of age to hold a Student Pilot Certificate for the operation of a glider or balloon. (A22) — 14 CFR §61.83

LTA
5030. No pilot may act as pilot in command of an airship under IFR or in weather conditions less than the minimums prescribed for VFR unless that pilot has, within the past 6 months, performed and logged under actual or simulated instrument conditions, at least

A—six instrument approaches, holding procedures, intercepting and tracking courses, or passed an instrument proficiency check in an airship.
B—three instrument approaches and logged 3 hours of instruments.
C—six instrument flights and six approaches.

To act as pilot-in-command under IFR, the pilot must, within the preceding 6 calendar months, have performed and logged under actual or simulated instrument conditions at least six instrument approaches, holding procedures, and intercepting and tracking of courses through the use of navigation systems. (A20) — 14 CFR §61.57

LTA
5036. To operate a balloon in solo flight, a student pilot must have a logbook endorsement by an authorized instructor who gave the flight training within the preceding

A—30 days.
B—60 days.
C—90 days.

A student pilot may not operate an aircraft in solo flight unless his/her Student Pilot Certificate is endorsed, and unless within the preceding 90 days his/her pilot logbook has been endorsed, by an authorized flight instructor. (A22) — 14 CFR §61.87(m)

LTA
5037. To exercise the privileges of a commercial pilot certificate with a lighter-than-air category, balloon class rating, what medical certification is required?

A—Statement by pilot certifying that he/she has no known physical defects that makes him/her unable to act as pilot of a balloon.
B—At least a current second-class medical certificate when carrying passengers for hire.
C—No medical certification is required.

A person is not required to hold a medical certificate when exercising the privileges of a pilot certificate with a balloon class rating. (A20) — 14 CFR §61.23

Answers

5084   [C]   5035-2   [C]   5030   [A]   5036   [C]   5037   [C]

LTA
5131. A person with a commercial pilot certificate with a lighter-than-air, balloon rating may give

A—flight training and conduct practical tests for balloon certification.
B—balloon ground and flight training and endorsements that are required for a flight review, or recency-of-experience requirements.
C—ground and flight training and endorsements that are required for balloon and airship ratings.

A person with a commercial pilot certificate with a lighter-than-air, balloon rating may: give flight and ground training in a balloon for the issuance of a certificate or rating, give an endorsement for a pilot certificate with a balloon rating, endorse a student pilot certificate or logbook for solo operating privileges in a balloon, and give flight and ground training and endorsements that are required for a flight review, an operating privilege, or recency-of-experience requirements. (A24) — 14 CFR §61.133

LTA
5132. A person who makes application for a commercial pilot certificate with a balloon rating, using a balloon with an airborne heater, will be

A—limited to balloon, with an airborne heater.
B—authorized to conduct ground and flight training in a balloon with an airborne heater or gas balloon.
C—authorized both airborne heater or gas balloon.

A person who applies for commercial pilot certificate with a balloon rating, using a balloon with an airborne heater, will be restricted to exercising the privileges of that certificate to a balloon with an airborne heater. (A24) — 14 CFR §61.133

LTA
5040. A commercial pilot who gives flight instruction in lighter-than-air category aircraft must keep a record of such instruction for a period of

A—1 year.
B—2 years.
C—3 years.

The records of instruction given shall be retained by the flight instructor separately or in his/her logbook for at least three years. (A26) — 14 CFR §61.189(b)

LTA
5041. What is the maximum amount of flight instruction an authorized instructor may give in any 24 consecutive hours?

A—4 hours.
B—6 hours.
C—8 hours.

An instructor may not conduct more than eight hours of flight instruction in any period of 24 consecutive hours. (A26) — 14 CFR §61.195(a)

LTA
5042. A student pilot may not operate a balloon in solo flight unless that pilot has

A—made and logged at least 10 balloon flights under the supervision of an authorized instructor.
B—received and logged flight training from an authorized instructor and demonstrated satisfactory proficiency and safety on the required maneuvers and procedures.
C—received a minimum of 5 hours of flight training in a balloon from an authorized instructor.

A student pilot who is receiving training in a balloon must receive and log flight training for the following maneuvers and procedures:

1. *Layout and assembly procedures;*
2. *Proper flight preparation procedures, including preflight planning and preparation, and aircraft systems;*
3. *Ascents and descents;*
4. *Landing and recovery procedures;*
5. *Emergency procedures and equipment malfunctions;*
6. *Operation of hot air or gas source, ballast, valves, vents, and rip panels, as appropriate;*
7. *Use of deflation valves or rip panels for simulating an emergency;*
8. *The effects of wind on climb and approach angles; and*
9. *Obstruction detection and avoidance techniques.*

(A26) — 14 CFR §61.87

Answers
5131 [B]   5132 [A]   5040 [C]   5041 [C]   5042 [B]

Chapter 4 Regulations

LTA

5048. Which person is directly responsible for the prelaunch briefing of passengers for a balloon flight?

A—Crew chief.
B—Safety officer.
C—Pilot in command.

The pilot-in-command of an aircraft is directly responsible for, and is the final authority as to, the operation of that aircraft. As part of crew briefing and preparation, the pilot-in-command briefs crewmembers and occupants on their duties and responsibilities in all areas of the flight including inflation, tether, in-flight, landing, emergency, and recovery procedures. (B08) — 14 CFR §61.189

LTA

5057. The use of certain portable electronic devices is prohibited on airships that are being operated under

A—IFR.
B—VFR.
C—DVFR.

Certain portable electronic devices may not be operated on any aircraft which is operating under IFR. (B07) — 14 CFR §91.21(a)(2)

LTA

5081. If a balloon is not equipped for night flight and official sunset is 1730 EST, the latest a pilot may operate that balloon and not violate regulations is

A—1629 EST.
B—1729 EST.
C—1759 EST.

No person may, during the period from sunset to sunrise (or, in Alaska, during the period a prominent unlighted object cannot be seen from a distance of 3 statute miles or the sun is more than 6° below the horizon) operate an aircraft unless it has lighted position lights. The glow from the burner is not enough; balloons can be equipped for night flight with unique dangling position lights. (B11) — 14 CFR §91.209(a)

LTA

5590. A balloon flight through a restricted area is

A—never permitted.
B—permitted anytime, but caution should be exercised because of high-speed military aircraft.
C—permitted at certain times, but only with prior permission by the appropriate authority.

No person may operate an aircraft within a restricted area contrary to the restrictions imposed, or within a prohibited area, unless he/she has the permission of the using or controlling agency. (J37) — 14 CFR §73.13

LTA

5554. (Refer to Figure 52, point 2.)

GIVEN:

Sacramento (SAC) tower reports wind ... 290 at 10 kts
Highest balloon flight altitude .... 1,200 MSL

If you depart for a 2-hour balloon flight from SAC airport (point 2), which response best describes what ATC requires of you?

A—You will have to contact Sacramento Approach Control.
B—You must communicate with Sacramento Approach Control because you will enter the Alert Area.
C—Your flightpath will require communications with Sacramento Executive (SAC) control tower and not with Sacramento Approach Control.

The floor of the Class C airspace here is 1,600 feet MSL, the top is 4,100 feet MSL. The flight path of the balloon will pass under the floor of Class B airspace; therefore, no communication is required with Sacramento Approach Control. However, communications must be established with SAC tower since the balloon will be flying through the Class D airspace. (J08) — Sectional Chart Legend

Answers

5048 [C]   5057 [A]   5081 [B]   5590 [C]   5554 [C]

LTA Fundamentals of Instructing

LTA
5882. A change in behavior as a result of experience can be defined as

A—learning.
B—knowledge.
C—understanding.

As a result of a learning experience, an individual's way of perceiving, thinking, feeling, and doing may change. Thus, learning can be defined as a change in behavior as a result of experience. (H20) — FAA-H-8083-9, Chapter 1

LTA
5883. In levels of learning, what are the steps of progression?

A—Application, understanding, rote, and correlation.
B—Rote, understanding, application, and correlation.
C—Correlation, rote, understanding, and application.

The lowest level, rote learning, is the ability to repeat back something which one has been taught, without understanding or being able to apply what has been learned. Progressively higher levels of learning are understanding what has been taught, achieving the skill to apply what has been learned to perform correctly, and associating and correlating what has been learned with things previously learned or subsequently encountered. (H20) — FAA-H-8083-9, Chapter 1

LTA
5884. In the learning process, fear or the element of threat will

A—inspire the student to improve.
B—narrow the student's perceptual field.
C—decrease the rate of associative reactions.

Fear adversely affects students' perception by narrowing their perceptual field. Confronted with threat, students tend to limit their attention to the threatening object or condition. (H20) — FAA-H-8083-9, Chapter 1

LTA
5885. What is the basis of all learning?

A—Insight.
B—Perception.
C—Motivation.

Initially, all learning comes from perceptions directed to the brain by one or more of the five senses (sight, hearing, touch, smell, and taste). (H20) — FAA-H-8083-9, Chapter 1

LTA
5886. While material is being taught, students may be learning other things as well. What is the additional learning called?

A—Residual learning.
B—Conceptual learning.
C—Incidental learning.

Learning is multifaceted. While learning the subject at hand, students may be learning other things as well. They may be developing attitudes about aviation—good or bad—depending on what they experience. Under a skillful instructor, for example, they may learn self-reliance. This learning is sometimes called "incidental," but it may have a great impact on the total development of the student. (H20) — FAA-H-8083-9, Chapter 1

LTA
5887. Students learn best when they are willing to learn. This feature of LAWS OF LEARNING is referred to as the law of

A—recency.
B—readiness.
C—willingness.

Individuals learn best when they are ready to learn, and they do not learn much if they see no reason for learning. (H20) — FAA-H-8083-9, Chapter 1

LTA
5888. Perceptions result when a person

A—gives meaning to sensations.
B—groups together bits of information.
C—responds to visual cues first, then aural cues, and relates these cues to ones previously learned.

Perceptions result when a person gives meaning to sensations. (H20) — FAA-H-8083-9, Chapter 1

Answers

| 5882 [A] | 5883 [B] | 5884 [B] | 5885 [B] | 5886 [C] | 5887 [B] |
| 5888 [A] | | | | | |

Chapter 4 Regulations

LTA
5889. Which is true? Motivations

A—should be obvious to be useful.
B—must be tangible to be effective.
C—may be very subtle and difficult to identify.

Motivations may be very subtle and difficult to identify or they may be obvious. (H20) — FAA-H-8083-9, Chapter 1

LTA
5890. To effectively motivate students, an instructor should

A—promise rewards.
B—appeal to their pride and self-esteem.
C—maintain pleasant personal relationships, even if necessary to lower standards.

Positive motivations are provided by the promise or achievement of rewards. (H20) — FAA-H-8083-9, Chapter 1

LTA
5891. Motivations in the form of reproof and threats should be avoided with all but the student who is

A—bored.
B—discouraged.
C—overconfident.

Negative motivations in the form of reproof and threats should be avoided with all but the most overconfident and impulsive students. (H20) — FAA-H-8083-9, Chapter 1

LTA
5892. The level of learning at which a person can repeat something without understanding is called

A—rote learning.
B—basic learning.
C—random learning.

The lowest level, rote learning, is the ability to repeat back something which one has been taught, without understanding or being able to apply what has been learned. (H20) — FAA-H-8083-9, Chapter 1

LTA
5893. The level of learning at which the student becomes able to associate an element which has been learned with other blocks of learning is called the level of

A—application.
B—association.
C—correlation.

The highest level of learning, which should be the objective of all instruction, is that level at which the student becomes able to associate an element which has been learned with other segments or "blocks" of learning or accomplishment. This level is called "correlation." (H20) — FAA-H-8083-9, Chapter 1

LTA
5894. To ensure proper habits and correct techniques during training, an instructor should

A—never repeat subject matter already taught.
B—use the "building-block" technique of instruction.
C—introduce tasks which are difficult and challenging to the student.

It is the instructor's responsibility to insist on correct techniques and procedures from the outset of training to provide proper habit patterns. It is much easier to foster proper habits from the beginning of training than to correct faulty ones later. This is the basic reason for the building block technique of instruction, in which each simple task is performed acceptably and correctly before the next learning task is introduced. (H20) — FAA-H-8083-9, Chapter 1

LTA
5895. Before a student can concentrate on learning, which of these human needs must be satisfied first?

A—Social needs.
B—Safety needs.
C—Physical needs.

At the broadest level are the physical needs. Individuals are first concerned with their need for food, rest, exercise, and protection from the elements. Until these needs are satisfied to a reasonable degree, they cannot concentrate on learning or self-expression. (H21) — FAA-H-8083-9, Chapter 2

Answers

| 5889 | [C] | 5890 | [A] | 5891 | [C] | 5892 | [A] | 5893 | [C] | 5894 | [B] |
| 5895 | [C] | | | | | | | | | | |

LTA
5896. Although defense mechanisms can serve a useful purpose, they can also be a hindrance because they

A—alleviate the cause of problems.
B—can result in delusional behavior.
C—involve self-deception and distortion of reality.

Because they involve some self-deception and distortion of reality, defense mechanisms do not solve problems. (H21) — FAA-H-8083-9, Chapter 2

LTA
5897. When a student asks irrelevant questions or refuses to participate in class activities, it usually is an indication of the defense mechanism known as

A—aggression.
B—resignation.
C—substitution.

An aggressive student may ask irrelevant questions, refuse to participate in the activities of the class, or disrupt activities within their own group. (H21) — FAA-H-8083-9, Chapter 2

LTA
5898. Taking physical or mental flight is a defense mechanism that students use when they

A—want to escape from frustrating situations.
B—become bewildered and lost in the advanced phase of training.
C—attempt to justify actions that otherwise would be unacceptable.

Students often escape from frustrating situations by taking flight, physical or mental. (H21) — FAA-H-8083-9, Chapter 2

LTA
5899. When a student uses excuses to justify inadequate performance, it is an indication of the defense mechanism known as

A—aggression.
B—resignation.
C—rationalization.

If students cannot accept the real reasons for their behavior, they may rationalize. This device permits them to substitute excuses for reasons. Moreover, they can make those excuses plausible and acceptable to themselves. (H21) — FAA-H-8083-9, Chapter 2

LTA
5900. When students become so frustrated they no longer believe it possible to work further, they usually display which defense mechanism?

A—Aggression.
B—Resignation.
C—Rationalization.

Students may become so frustrated that they lose interest and give up. They may no longer believe it profitable or even possible to work further. (H21) — FAA-H-8083-9, Chapter 2

LTA
5901. A student who is daydreaming is engaging in the defense mechanism known as

A—flight.
B—substitution.
C—rationalization.

Students often escape from frustrating situations by taking flight, physical or mental. (H21) — FAA-H-8083-9, Chapter 2

LTA
5902. Which of these instructor actions would more likely result in students becoming frustrated?

A—Presenting a topic or maneuver in great detail.
B—Covering up instructor mistakes or bluffing when the instructor is in doubt.
C—Telling the students that their work is unsatisfactory without explanation.

If a student has made an earnest effort but is told that the work is not satisfactory, with no other explanation, frustration occurs. (H21) — FAA-H-8083-9, Chapter 2

LTA
5903. The effectiveness of communication between the instructor and the student is measured by the degree of

A—motivation manifested by the student.
B—similarity between the idea transmitted and the idea received.
C—attention the student gives to the instructor during a lesson.

Communication's effectiveness is measured by the similarity between the idea transmitted and the idea received. (H22) — FAA-H-8083-9, Chapter 3

Answers

| 5896 | [C] | 5897 | [A] | 5898 | [A] | 5899 | [C] | 5900 | [B] | 5901 | [A] |
| 5902 | [C] | 5903 | [B] | | | | | | | | | | |

LTA
5904. To communicate effectively, instructors must

A—utilize highly organized notes.
B—display an authoritarian attitude.
C—display a positive, confident attitude.

An instructor's attitude must be positive if he/she is to communicate effectively. Communicators must be confident. (H22) — FAA-H-8083-9, Chapter 3

LTA
5905. Probably the greatest single barrier to effective communication is the

A—use of inaccurate statements.
B—use of abstractions by the communicator.
C—lack of a common core of experience between communicator and receiver.

Probably the greatest single barrier to effective communication is the lack of a common core of experience between communicator and receiver. (H22) — FAA-H-8083-9, Chapter 3

LTA
5906. What is the proper sequence in which the instructor should employ the four basic steps in the teaching process?

A—Explanation, demonstration, practice, and evaluation.
B—Explanation, trial and practice, evaluation, and review.
C—Preparation, presentation, application, and review and evaluation.

The teaching of new materials, as reflected in any of the lists, can be broken down into the steps of:

1. *Preparation;*
2. *Presentation;*
3. *Application; and*
4. *Review and evaluation.*

(H23) — FAA-H-8083-9, Chapter 4

LTA
5907. Evaluation of student performance and accomplishment during a lesson should be based on the

A—student's background and past experiences.
B—objectives and goals that were established in the lesson plan.
C—student's actual performance as compared to an arbitrary standard.

The evaluation of student performance and accomplishment during a lesson should be based on the objectives and goals that were established in the instructor's lesson plan. (H23) — FAA-H-8083-9, Chapter 4

LTA
5908. To enhance a student's acceptance of further instruction, the instructor should

A—keep the student informed of his/her progress.
B—continually prod the student to maintain motivational levels.
C—establish performance standards a little above the student's actual ability.

The failure of the instructor to ensure that students are cognizant of their progress, or lack of it, may impose a barrier between them. Though it may be slight, it may make further instruction more difficult. (H23) — FAA-H-8083-9, Chapter 4

LTA
5909. The method of arranging lesson material from the simple to complex, past to present, and known to unknown, is one that

A—the instructor should avoid.
B—creates student thought pattern departures.
C—indicates the relationship of the main points of the lesson.

The instructor must logically organize the material to show the relationships of the main points. The instructor usually shows these primary relationships by developing the main points in one of the following ways: From the past to the present; from the simple to the complex; from the known to the unknown; and from the more frequently used to the least frequently used. (H24) — FAA-H-8083-9, Chapter 5

LTA
5910. When teaching from the KNOWN to the UNKNOWN, an instructor is using the student's

A—anxieties and insecurities.
B—previous experiences and knowledge.
C—previously held opinions, both valid and invalid.

From known to unknown—by using something the student already knows as the point of departure, the instructor can lead into new ideas and concepts. (H24) — FAA-H-8083-9, Chapter 5

Answers

| 5904 | [C] | 5905 | [C] | 5906 | [C] | 5907 | [B] | 5908 | [A] | 5909 | [C] |
| 5910 | [B] | | | | | | | | | | |

Chapter 4 Regulations

LTA
5911. In developing a lesson, the instructor must logically organize explanations and demonstrations to help the student

A—understand the separate items of knowledge.
B—understand the relationships of the main points of the lesson.
C—learn by rote so that performance of the procedure will become automatic.

The instructor must logically organize the material to show the relationships of the main points. (H24) — FAA-H-8083-9, Chapter 5

LTA
5912. Which should be the first step in preparing a lecture?

A—Organizing the material.
B—Researching the subject.
C—Establishing the objective and desired outcome.

The following four steps should be followed in the planning phase of preparation:

1. *Establishing the objective and desired outcomes;*
2. *Researching the subject;*
3. *Organizing the material; and*
4. *Planning productive classroom activities.*

(H24) — FAA-H-8083-9, Chapter 5

LTA
5913. What is one advantage of a lecture?

A—It provides for student participation.
B—Many ideas can be presented in a short time.
C—Maximum attainment in all types of learning outcomes is possible.

In a lecture, the instructor can present many ideas in a relatively short time. (H24) — FAA-H-8083-9, Chapter 5

LTA
5914. In a "guided discussion," lead-off questions should usually begin with

A—"why..."
B—"when..."
C—"where..."

Lead off questions should usually begin with "how" or "why." (H24) — FAA-H-8083-9, Chapter 5

LTA
5915. What are the essential steps in the "demonstration/performance" method of teaching?

A—Demonstration, practice, and evaluation.
B—Demonstration, student performance, and evaluation.
C—Explanation, demonstration, student performance, instructor supervision, and evaluation.

The demonstration—performance method of teaching has five essential phases:

1. *Explanation;*
2. *Demonstration;*
3. *Student performance;*
4. *Instructor supervision; and*
5. *Evaluation.*

(H24) — FAA-H-8083-9, Chapter 5

LTA
5916. Which is true about an instructor's critique of a student's performance?

A—It must be given in written form.
B—It should be subjective rather than objective.
C—It is a step in the learning process, not in the grading process.

A critique is not a step in the grading process. It is a step in the learning process. (H25) — FAA-H-8083-9, Chapter 6

LTA
5917. The purpose of a critique is to

A—identify only the student's faults and weaknesses.
B—give a delayed evaluation of the student's performance.
C—provide direction and guidance to raise the level of the student's performance.

A critique should provide direction and guidance to raise the level of the student's performance. (H25) — FAA-H-8083-9, Chapter 6

LTA
5918. When an instructor critiques a student, it should always be

A—done in private.
B—subjective rather than objective.
C—conducted immediately after the student's performance.

Answers

5911 [B]	5912 [C]	5913 [B]	5914 [A]	5915 [C]	5916 [C]
5917 [C]	5918 [C]				

A critique should come immediately after a student's individual or group performance, while the details of the performance are easy to recall. (H25) — FAA-H-8083-9, Chapter 6

LTA
5919. Proper quizzing by the instructor during a lesson can have which of these results?

A—It identifies points which need emphasis.
B—It encourages rote response from students.
C—It permits the introduction of new material which was not covered previously.

A quiz identifies points which need more emphasis. (H26) — FAA-H-8083-9, Chapter 6

LTA
5920. To be effective in oral quizzing during the conduct of a lesson, a question should

A—center on only one idea.
B—include a combination of where, how, and why.
C—be easy for the student at that particular stage of training.

Effective questions center on only one idea. (H26) — FAA-H-8083-9, Chapter 6

LTA
5921. A written test has validity when it

A—yields consistent results.
B—samples liberally whatever is being measured.
C—actually measures what it is supposed to measure and nothing else.

A measuring instrument, including a written test, is valid when it actually measures what it is supposed to measure and nothing else. (H26) — FAA-H-8083-9, Chapter 6

LTA
5922. A written test which has reliability is one which

A—yields consistent results.
B—measures small differences in the achievement of students.
C—actually measures what it is supposed to measure and nothing else.

A reliable measuring instrument, including a written test, is one which yields consistent results. (H26) — FAA-H-8083-9, Chapter 6

LTA
5923. A written test is said to be comprehensive when it

A—yields consistent results.
B—includes all levels of difficulty.
C—samples liberally whatever is being measured.

To be comprehensive, a measuring instrument, including a written test, must sample liberally whatever is being measured. (H26) — FAA-H-8083-9, Chapter 6

LTA
5924. Which is true concerning the use of visual aids? They

A—should be used to emphasize key points in a lesson.
B—ensure getting and holding the student's attention.
C—should not be used to cover a subject in less time.

The aids should be concentrated on the key points. (H27) — FAA-H-8083-9, Chapter 7

LTA
5925. Instructional aids used in the teaching/learning process should be

A—self-supporting and should require no explanation.
B—compatible with the learning outcomes to be achieved.
C—selected prior to developing and organizing the lesson plan.

Aids should be simple and compatible with the learning outcomes to be achieved. (H27) — FAA-H-8083-9, Chapter 7

LTA
5926. The professional relationship between the instructor and the student should be based upon

A—the need to disregard the student's personal faults, interests, or problems.
B—setting the learning objectives very high so that the student is continually challenged.
C—the mutual acknowledgement that they are important to each other and both are working toward the same objective.

The professional relationship of the instructor with the student should be based on a mutual acknowledgment that both the student and the instructor are important to each other, and that both are working toward the same objective. (H30) — FAA-H-8083-9, Chapter 7

Answers

| 5919 | [A] | 5920 | [A] | 5921 | [C] | 5922 | [A] | 5923 | [C] | 5924 | [A] |
| 5925 | [B] | 5926 | [C] | | | | | | | | |

LTA
5927. Which is true regarding professionalism as an instructor?

A—Anything less than sincere performance destroys the effectiveness of the professional instructor.
B—To achieve professionalism, actions and decisions must be limited to standard patterns and practices.
C—A single definition of professionalism would encompass all of the qualifications and considerations which must be present before true professionalism can exist.

In instructors, anything less than sincere performance is quickly detected, and immediately destroys their effectiveness. (H30) — FAA-H-8083-9, Chapter 8

LTA
5928. An instructor can most effectively maintain a high level of student motivation by

A—making each lesson a pleasurable experience.
B—easing the standards for an apprehensive student.
C—continually challenging the student to meet the highest objectives of training.

By making each lesson a pleasurable experience for the student, the flight instructor can maintain a high-level of student motivation. (H30) — FAA-H-8083-9, Chapter 8

LTA
5929. Faulty performance due to student overconfidence should be corrected by

A—high praise when no errors are made.
B—increasing the standard of performance for each lesson.
C—providing strong, negative evaluation at the end of each lesson.

Apt students can also create problems. Because they make few mistakes, they may assume that the correction of errors is unimportant. Such overconfidence soon results in faulty performance. For such students, a good instructor will constantly raise the standard of performance for each lesson, demanding greater effort. (H30) — FAA-H-8083-9, Chapter 8

LTA
5930. What should an instructor do with a student who assumes that correction of errors is unimportant?

A—Invent student deficiencies.
B—Try to reduce the student's overconfidence.
C—Raise the standards of performance, demanding greater effort.

Apt students can also create problems. Because they make few mistakes, they may assume that the correction of errors is unimportant. Such overconfidence soon results in faulty performance. For such students, a good instructor will constantly raise the standard of performance for each lesson, demanding greater effort. (H30) — FAA-H-8083-9, Chapter 8

LTA
5931. What should an instructor do if a student's slow progress is due to discouragement and lack of confidence?

A—Assign subgoals which can be attained more easily than the normal learning goals.
B—Emphasize the negative aspects of poor performance by pointing out the serious consequences.
C—Raise the performance standards so the student will gain satisfaction in meeting higher standards.

A student whose slow progress is due to discouragement and a lack of confidence should be assigned "subgoals" which can be attained more easily than the normal learning goals. For this purpose, complex flight maneuvers can be separated into their elements, and each element practiced until an acceptable performance is achieved before the whole maneuver or operation is attempted. (H30) — FAA-H-8083-9, Chapter 8

LTA
5932. Should an instructor be concerned about an apt student who makes very few mistakes?

A—No. Some students have an innate, natural aptitude for flight.
B—Yes. Faulty performance may soon appear due to student overconfidence.
C—Yes. The student will lose confidence in the instructor if the instructor does not invent deficiencies in the student's performance.

Apt students can also create problems. Because they make few mistakes, they may assume that the correction of errors is unimportant. Such overconfidence soon results in faulty performance. For such students, a good

Answers

| 5927 | [A] | 5928 | [A] | 5929 | [B] | 5930 | [C] | 5931 | [A] | 5932 | [B] |

instructor will constantly raise the standard of performance for each lesson, demanding greater effort. (H30) — FAA-H-8083-9, Chapter 8

LTA
5933. When a student correctly understands the situation and knows the correct procedure for the task, but fails to act at the proper time, the student most probably

A—lacks self-confidence.
B—will be unable to cope with the demands of flying.
C—is handicapped by indifference or lack of interest.

A student may fail to act at the proper time due to lack of self-confidence, even though the situation is correctly understood. (H30) — FAA-H-8083-9, Chapter 8

LTA
5934. What should an instructor do if a student is suspected of not fully understanding the principles involved in a task, even though the student can correctly perform the task?

A—Require the student to apply the same elements to the performance of other tasks.
B—Require the student to repeat the task, as necessary, until the principles are understood.
C—Repeat demonstrating the task as necessary until the student understands the principles.

A student may perform a procedure or maneuver correctly and not fully understand the principles and objectives involved. When this is suspected by the instructor, the student should be required to vary the performance of the maneuver slightly, combine it with other operations, or apply the same elements to the performance of other maneuvers. (H30) — FAA-H-8083-9, Chapter 8

LTA
5935. When under stress, normal individuals usually react

A—with marked changes in mood on different lessons.
B—with extreme overcooperation, painstaking self-control, and laughing or singing.
C—by responding rapidly and exactly, often automatically, within the limits of their experience and training.

Normal individuals begin to respond rapidly and exactly, within the limits of their experience and training. Many responses are automatic. (H30) — FAA-H-8083-9, Chapter 8

LTA
5936. The instructor can counteract anxiety in a student by

A—treating student fear as a normal reaction.
B—allowing the student to select tasks to be performed.
C—continually citing the unhappy consequences of faulty performance.

An effective technique is to treat fears as a normal reaction. (H30) — FAA-H-8083-9, Chapter 8

LTA
5937. Which would most likely indicate that a student is reacting abnormally to stress?

A—Thinks and acts rapidly.
B—Extreme overcooperation.
C—Extreme sensitivity to surroundings.

Abnormal reaction to stress would be indicated by inappropriate reactions such as extreme over-cooperation, painstaking self-control, inappropriate laughter, singing, very rapid changes in emotions, or motion sickness. (H30) — FAA-H-8083-9, Chapter 8

LTA
5938-1. What is the primary consideration in determining the length and frequency of flight instruction periods?

A—Fatigue.
B—Mental acuity.
C—Physical conditioning.

Fatigue is the primary consideration in determining the length and frequency of flight instruction periods. (H31) — FAA-H-8083-9, Chapter 9

LTA
5938-2. What is the primary consideration in determining the length and frequency of flight instruction periods?

A—Fatigue.
B—Mental acuity.
C—Physical conditioning.

Fatigue is the primary consideration in determining the length and frequency of flight instruction periods. The amount of training which can be absorbed by one student without incurring fatigue does not necessarily indicate the capacity of another student. Fatigue which results from training operations may be either physical or mental, or both. (H31) — FAA-H-8083-9, Chapter 9

Answers

| 5933 [A] | 5934 [A] | 5935 [C] | 5936 [A] | 5937 [B] | 5938-1 [A] |

5938-2 [A]

LTA
5939. Students quickly become apathetic when they

A—understand the objectives toward which they are working.
B—are assigned goals that are difficult, but possible to attain.
C—recognize that their instructor is poorly prepared to conduct the lesson.

Students quickly become apathetic when they recognize that the instructor has made inadequate preparations for the instruction being given, or when the instruction appears to be deficient, contradictory, or insincere. (H31) — FAA-H-8083-9, Chapter 9

LTA
5940. In planning any instructional activity, the instructor's first consideration should be to

A—determine the overall objectives and standards.
B—identify the blocks of learning which make up the overall objective.
C—establish common ground between the instructor and students.

Before any important instruction can begin, a determination of standards and objectives is necessary. (H32) — FAA-H-8083-9, Chapter 10

Answers

5939 [C]   5940 [A]

Chapter 5
Procedures and Airport Operations

Airspace   *5–3*

Basic VFR Weather Minimums   *5–12*

VHF Direction Finder   *5–13*

Operations on Wet or Slippery Runways   *5–14*

Land and Hold Short Operations (LAHSO)   *5–14*

Airport Marking Aids and Signs   *5–15*

VFR Cruising Altitudes   *5–16*

Collision Avoidance   *5–17*

Fitness Physiology   *5–19*

Aeronautical Decision Making   *5–21*

Chapter 5 **Procedures and Airport Operations**

Airspace

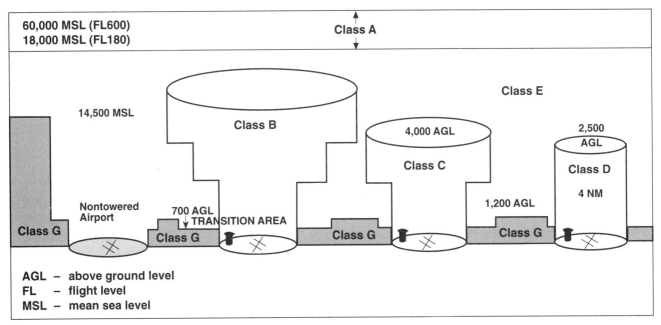

Figure 5-1. Airspace

Controlled airspace, that is, airspace within which some or all aircraft may be subject to air traffic control, consists of those areas designated as Class A, Class B, Class C, Class D, and Class E airspace.

Much of the controlled airspace begins at either 700 feet or 1,200 feet above the ground. The 700-foot lateral limits and floors of Class E airspace are defined by a magenta vignette; while the 1,200-foot lateral limits and floors are defined by a blue vignette if it abuts uncontrolled airspace. Floors other than 700 feet or 1,200 feet are shown by a number indicating the floor.

Class A—Class A airspace extends from 18,000 feet MSL up to and including FL600 and is not depicted on VFR sectional charts. No flight under visual flight rules (VFR), including VFR-On-Top, is authorized in Class A airspace.

Class B—Class B airspace consists of controlled airspace extending upward from the surface or higher to specified altitudes. Each Class B airspace sector, outlined in blue on the sectional aeronautical chart, is labeled with its delimiting altitudes. On the Terminal Area Chart, each Class B airspace sector is also outlined in blue and labeled with its delimiting arcs, radials, and altitudes. Each Class B airspace location will contain at least one primary airport. An ATC clearance is required prior to operating within Class B airspace.

A pilot landing or taking off from one of a group of 12 specific, busy airports must hold at least a Private Pilot Certificate. At other airports, a student pilot may not operate an aircraft on a solo flight within Class B airspace or to, from, or at an airport located within Class B airspace unless both ground and flight instruction has been received from an authorized instructor to operate within that Class B airspace or at that airport, and the flight and ground instruction has been received within that Class B airspace or at the specific airport for which the solo flight is authorized. The student's logbook must be endorsed within the preceding 90 days by the instructor who gave the flight training and the endorsement must specify that the student has been found competent to conduct solo flight operations in that Class B airspace or at that specific airport.

Each airplane operating within Class B airspace must be equipped with a two-way radio with appropriate ATC frequencies, and a 4096 code transponder with Mode C automatic altitude-reporting capability.

Chapter 5 **Procedures and Airport Operations**

Class C—Class C airspace is controlled airspace surrounding designated airports within which ATC provides radar vectoring and sequencing for all IFR and VFR aircraft. Each airplane operating within Class C airspace must be equipped with a two-way radio with appropriate frequencies and a 4096 code transponder with Mode C automatic altitude-reporting capability. Communications with ATC must be established prior to entering Class C airspace.

Class C airspace consists of two circles, both centered on the primary airport. The surface area has a radius of 5 NM. The airspace of the surface area normally extends from the surface of Class C airspace airport up to 4,000 feet above that airport. Some situations require different boundaries. The shelf area has a radius of 10 NM. The airspace between the 5 and 10 NM rings begins at a height of 1,200 feet and extends to the same altitude cap as the inner circle. An outer area with a normal radius of 20 NM surrounds the surface and shelf areas. Within the outer area, pilots are encouraged to participate but it is not a VFR requirement.

Class C airspace service to aircraft proceeding to a satellite airport will be terminated at a sufficient distance to allow time to change to the appropriate tower or advisory frequency. Aircraft departing satellite airports within Class C airspace shall establish two-way communication with ATC as soon as practicable after takeoff. On aeronautical charts, Class C airspace is depicted by solid magenta lines.

Class D—Class D airspace extends upward from the surface to approximately 2,500 feet AGL (the actual height is as needed). Class D airspace may include one or more airports and is normally 4 NM in radius. The actual size and shape is depicted by a blue dashed line and numbers showing the top. When the ceiling of Class D airspace is less than 1,000 feet and/or the visibility is less than 3 SM, pilots wishing to takeoff or land must hold an instrument rating, must have filed an instrument flight plan, and must have received an appropriate clearance from ATC. In addition, the aircraft must be equipped for instrument flight.

At some locations, a pilot who does not hold an instrument rating may be authorized to takeoff or land when the weather is less than that required for visual flight rules. When **special VFR** flight is prohibited, it will be depicted by "No SVFR" above the airport information on the chart. Special VFR requires the aircraft to be operated clear of clouds with flight visibility of at least 1 SM. For Special VFR operations between sunset and sunrise, the pilot must hold an instrument rating and the airplane must be equipped for instrument flight. Requests for Special VFR arrival or departure clearance should be directed to the airport traffic control tower.

Class E—Magenta shading identifies Class E airspace starting at 700 feet AGL, and an area with no shading (or blue shading if next to Class G airspace) identifies Class E airspace starting at 1,200 feet AGL. It may also start at other altitudes. All airspace from 14,500 feet to 17,999 feet is Class E airspace. It also includes the surface area of some airports with an instrument approach but no control tower. An **airway** is a corridor of controlled airspace extending from 1,200 feet above the surface (or as designated) up to and including 17,999 feet MSL, and 4 NM either side of the centerline. The airway is indicated by a centerline, shown in blue.

Class G—Class G is airspace within which Air Traffic Control has neither the authority nor responsibility to exercise any control over air traffic.

Prohibited Areas are blocks of airspace within which the flight of aircraft is prohibited.

Restricted Areas denote the presence of unusual, often invisible, hazards to aircraft such as artillery firing, aerial gunnery, or guided missiles. Penetration of Restricted Areas without authorization of the using or controlling agency may be extremely hazardous to the aircraft and its occupants.

Chapter 5 **Procedures and Airport Operations**

Warning Areas contain the same hazardous activities as those found in Restricted Areas, but are located in international airspace.

Military Operations Areas (MOAs) consist of airspace established for the purpose of separating certain military training activities from instrument flight rules (IFR) traffic. Pilots operating under VFR should exercise extreme caution while flying within an active MOA. Any Flight Service Station (FSS) within 100 miles of the area will provide information concerning MOA hours of operation. Prior to entering an active MOA, pilots should contact the controlling agency for traffic advisories.

Alert Areas may contain a high volume of pilot training activities or an unusual type of aerial activity, neither of which is hazardous to aircraft. Pilots of participating aircraft, as well as pilots transiting the area, are equally responsible for collision avoidance.

An **Airport Advisory Area** is the area within 10 statute miles of an airport where a control tower is not in operation but where a Flight Service Station (FSS) is located. The FSS provides advisory service to aircraft arriving and departing. It is not mandatory for pilots to use the advisory service, but it is strongly recommended that they do so.

Aircraft are requested to remain at least 2,000 feet above the surface of National Parks, National Monuments, Wilderness and Primitive Areas, and National Wildlife Refuges.

Military Training Routes (MTRs) have been developed for use by the military for the purpose of conducting low-altitude, high-speed training. Generally, MTRs are established below 10,000 feet MSL for operations at speeds in excess of 250 knots.

IFR Military Training Routes (IR) operations are conducted in accordance with instrument flight rules, regardless of weather conditions. VFR Military Training Routes (VR) operations are conducted in accordance with visual flight rules. IR and VR at and below 1,500 feet AGL (with no segment above 1,500) will be identified by four digit numbers, e.g., VR1351, IR1007. IR and VR above and below 1,500 feet AGL (segments of these routes may be below 1,500) will be identified by three digit numbers, e.g., IR341, VR426.

ALL, MIL
5115. After an ATC clearance has been obtained, a pilot may not deviate from that clearance, unless the pilot

A—requests an amended clearance.
B—is operating VFR on top.
C—receives an amended clearance or has an emergency.

When an ATC clearance has been obtained, no pilot-in-command may deviate from that clearance unless an amended clearance is obtained, an emergency exists, or the deviation is in response to a traffic alert and collision avoidance system resolution advisory. (B08) — 14 CFR §91.123

ALL, MIL
5043. Excluding Hawaii, the vertical limits of the Federal Low Altitude airways extend from

A—700 feet AGL up to, but not including, 14,500 feet MSL.
B—1,200 feet AGL up to, but not including, 18,000 feet MSL.
C—1,200 feet AGL up to, but not including, 14,500 feet MSL.

Each Federal airway includes that airspace extending upward from 1,200 feet above the surface up to but not including 18,000 feet MSL. Federal airways for Hawaii have no upper limits. (A66) — 14 CFR §61.87(m)

Answers

5115 [C]   5043 [B]

Chapter 5 **Procedures and Airport Operations**

ALL, MIL
5082-1. Which is true regarding flight operations in Class B airspace?

A—Flight under VFR is not authorized unless the pilot in command is instrument rated.
B—The pilot must receive an ATC clearance before operating an aircraft in that area.
C—Solo student pilot operations are not authorized.

No person may operate an aircraft within a Class B airspace area unless the operator receives an ATC clearance from the ATC facility having jurisdiction for that area before operating an aircraft in that area. (B08) — 14 CFR §91.131

Answer (A) is incorrect because a private pilot's certificate without an instrument rating is sufficient to operate in Class B airspace. Answer (C) is incorrect because with the proper training and endorsements a student pilot may operate in Class B airspace.

ALL, MIL
5082-2. Which is true regarding pilot certification requirements for operations in Class B airspace?

A—The pilot in command must hold at least a private pilot certificate with an instrument rating.
B—The pilot in command must hold at least a private pilot certificate.
C—Solo student pilot operations are not authorized.

No person may operate an aircraft within a Class B airspace area unless the pilot-in-command holds at least a private pilot certificate. (B08) — 14 CFR §91.131

Answer (A) is incorrect because a private pilot certificate without an instrument rating is sufficient to operate in Class B airspace. Answer (C) is incorrect because with the proper training and endorsements, a student pilot may operate in Class B airspace.

ALL, MIL
5082-3. Which is true regarding flight operations in Class B airspace?

A—The aircraft must be equipped with an ATC transponder and altitude reporting equipment.
B—The pilot in command must hold at least a private pilot certificate with an instrument rating.
C—The pilot in command must hold at least a student pilot certificate.

A student pilot may only operate an aircraft on a solo flight in Class B airspace if the student pilot has received both ground and flight training from an authorized instructor and has received a logbook endorsement. No person may operate an aircraft in a Class B airspace area unless the aircraft is equipped with the applicable operating transponder and automatic altitude reporting equipment. Requests for ATC authorized deviations for operation of an aircraft that is not equipped with a transponder must be made at least one hour before the proposed operation. (B08) — 14 CFR §91.131, §91.215 and §61.95

ALL, MIL
5009. What designated airspace associated with an airport becomes inactive when the control tower at that airport is not in operation?

A—Class D, which then becomes Class C.
B—Class D, which then becomes Class E.
C—Class B.

Class D airspace exists only when the control tower is operating. It reverts to Class E when the tower closes if there is an instrument approach and a weather observer. (J08) — 14 CFR §1.1

Answer (A) is incorrect because Class D airspace will revert to Class E airspace when the control tower closes. Answer (C) is incorrect because the primary airport of Class B airspace will have a control tower that operates full-time.

ALL, MIL
5564. Which is true concerning the blue and magenta colors used to depict airports on Sectional Aeronautical Charts?

A—Airports with control towers underlying Class A, B, and C airspace are shown in blue, Class D and E airspace are magenta.
B—Airports with control towers underlying Class C, D, and E airspace are shown in magenta.
C—Airports with control towers underlying Class B, C, D, and E airspace are shown in blue.

Airports having Control Towers (Class B, C, D or E airspace) are shown in blue. All others are shown in magenta. (J37) — Sectional Chart Legend

Answers

5082-1 [B]   5082-2 [B]   5082-3 [A]   5009 [B]   5564 [C]

Chapter 5 **Procedures and Airport Operations**

ALL, MIL
5117. When operating an aircraft in the vicinity of an airport with an operating control tower, in Class E airspace, a pilot must establish communications prior to

A— 8 NM, and up to and including 3,000 feet AGL.
B— 5 NM, and up to and including 3,000 feet AGL.
C— 4 NM, and up to and including 2,500 feet AGL.

Unless otherwise authorized or required by ATC, no person may operate an aircraft to, from, through, or on an airport having an operational control tower unless two-way radio communications are maintained between that aircraft and the control tower. Communications must be established prior to 4 NM from the airport, up to and including 2,500 feet AGL. (B08) — 14 CFR §91.127

ALL, MIL
5118. When approaching to land at an airport with an ATC facility, in Class D airspace, the pilot must establish communications prior to

A— 10 NM, up to and including 3,000 feet AGL.
B— 30 SM, and be transponder equipped.
C— 4 NM, up to and including 2,500 feet AGL.

Unless otherwise authorized or required by ATC, no person may operate an aircraft to, from, through, or on an airport having an operational control tower unless two-way radio communications are maintained between that aircraft and the control tower. Communications must be established prior to 4 NM from the airport, up to and including 2,500 feet AGL. (B08) — 14 CFR §91.127

ALL, MIL
5119-1. Which is true regarding flight operations to or from a satellite airport, without an operating control tower, within the Class C airspace area?

A— Prior to takeoff, a pilot must establish communication with the ATC controlling facility.
B— Aircraft must be equipped with an ATC transponder and altitude reporting equipment.
C— Prior to landing, a pilot must establish and maintain communication with an ATC facility.

No person may take off or land an aircraft at a satellite airport within a Class C airspace area except in compliance with FAA arrival and departure traffic patterns. Each person must establish two-way radio communications with the ATC facility providing air traffic services prior to entering that airspace and thereafter maintain those communications while within that airspace. (B08) — 14 CFR §91.130

Answers (A) and (B) are incorrect because from a satellite airport without an operating control tower within Class C airspace, the pilot must establish and maintain two-way radio communications with the controlling ATC facility as soon as practicable after departing; unless otherwise authorized or directed by ATC, no person may operate an aircraft in Class C airspace unless that aircraft is equipped with an operable transponder and altitude reporting equipment.

ALL, MIL
5119-2. Which is true regarding flight operations to or from a satellite airport, without an operating control tower, within the Class C airspace area?

A— Prior to entering that airspace, a pilot must establish and maintain communication with the ATC serving facility.
B— Aircraft must be equipped with an ATC transponder.
C— Prior to takeoff, a pilot must establish communication with the ATC controlling facility.

No person may take off or land an aircraft at a satellite airport within a Class C airspace area except in compliance with FAA arrival and departure traffic patterns. Each person must establish two-way radio communications with the ATC facility providing air traffic services prior to entering that airspace and thereafter maintain those communications while within that airspace. (B08) — 14 CFR §91.130

Answers (B) and (C) are incorrect because unless otherwise authorized or directed by ATC, no person may operate an aircraft in Class C airspace unless that aircraft is equipped with an operable transponder and altitude reporting equipment; from a satellite airport without an operating control tower within Class C airspace, the pilot must establish and maintain two-way radio communications with the controlling ATC facility as soon as practicable after departing.

ALL, MIL
5120-1. Which is true regarding flight operations in Class A airspace?

A— Aircraft must be equipped with approved distance measuring equipment (DME).
B— Must conduct operations under instrument flight rules.
C— Aircraft must be equipped with an approved ATC transponder.

Each person operating an aircraft in Class A airspace must conduct that operation under instrument flight rules. (B08) — 14 CFR §91.135

Answer (A) is incorrect because if VOR navigational equipment is required, no person may operate a U.S.-registered civil aircraft at or above FL240 unless that aircraft is equipped with approved distance measuring equipment (DME). Answer (C) is incorrect because all aircraft must be equipped with an approved ATC transponder and altitude reporting equipment in airspace at and above 10,000 feet MSL, excluding the airspace at and below 2,500 feet above the surface.

Answers

5117 [C]   5118 [C]   5119-1 [C]   5119-2 [A]   5120-1 [B]

Chapter 5 **Procedures and Airport Operations**

ALL, MIL
5120-2. Which is true regarding flight operations in Class A airspace?

A—Aircraft must be equipped with approved distance measuring equipment (DME).
B—Aircraft must be equipped with an ATC transponder and altitude reporting equipment.
C—May conduct operations under visual flight rules.

Unless otherwise authorized by ATC, no person may operate an aircraft within Class A airspace unless that aircraft is equipped with an approved transponder and altitude reporting equipment. (B08) — 14 CFR §91.135

AIR, GLI, LTA, MIL
5116-1. When approaching to land at an airport, without an operating control tower, in Class G airspace, the pilot should

A—make all turns to the left, unless otherwise indicated.
B—fly a left-hand traffic pattern at 800 feet AGL.
C—enter and fly a traffic pattern at 800 feet AGL.

When approaching to land at an airport without an operating control tower in a Class G airspace area each pilot of an airplane must make all turns of that airplane to the left unless the airport displays approved light signals or visual markings indicating that turns should be made to the right, in which case the pilot must make all turns to the right. (B08) — 14 CFR §91.126

RTC, MIL
5116-2. When approaching to land at an airport, without an operating control tower, in Class G airspace, a helicopter pilot should

A—avoid the flow of fixed-wing aircraft.
B—make all turns to the left, unless otherwise indicated.
C—enter and fly a traffic pattern at 800 feet AGL.

Each pilot of a helicopter must avoid the flow of fixed-wing aircraft. (B08) — 14 CFR §91.126

ALL, MIL
5572. (Refer to Figure 54, point 1.) What minimum altitude is required to avoid the Livermore Airport (LVK) Class D airspace?

A—2,503 feet MSL.
B—2,901 feet MSL.
C—3,297 feet MSL.

The Class D airspace at Livermore has a top of 2,900 feet MSL, indicated by the [29] within the blue segmented circle. Therefore, the minimum altitude to fly over and avoid the Class D airspace is 2,901 feet MSL. (J37) — AIM ¶3-2-5

Answer (A) is incorrect because 2,503 feet MSL would place you within the Class D airspace. Answer (C) is incorrect because although 3,297 feet MSL would keep you outside the Class D airspace, it is not the minimum altitude required to avoid it.

LTA
5574-1. (Refer to Figure 54, point 1.) A balloon flight over Livermore Airport (LVK) at 3,000 feet MSL

A—requires a transponder, but ATC communication is not necessary.
B—does not require a transponder or ATC communication.
C—cannot be accomplished without meeting all Class B airspace requirements.

Aircraft operating within the Mode C veil must be equipped with automatic pressure altitude reporting requirement having Mode C capability. However, aircraft that was not originally certificated with an engine-driven electrical system or which has not subsequently been certified with a system installed, may conduct operations within a Mode C veil provided the aircraft remains outside Class A, B, or C airspace, and below the altitude of the ceiling of a Class B or Class C airspace area designated for an airport or 10,000 feet MSL, whichever is lower. (J37) — AIM ¶3-2-1

RTC, MIL
5574-2. (Refer to Figure 54, point 1.) A helicopter flight over Livermore Airport (LVK) at 3,000 feet MSL

A—requires a transponder, but ATC communication is not necessary.
B—does not require a transponder or ATC communication.
C—cannot be accomplished without meeting all Class B airspace requirements.

Answers

5120-2   [B]   5116-1   [A]   5116-2   [A]   5572   [B]   5574-1   [B]   5574-2   [A]

At 3,000 feet MSL, the flight is above the Class D airspace, but within the 30-mile ring of San Francisco. Therefore, a transponder is required since the flight is in Class E airspace, but no communication is required since the flight is outside the Class B airspace (J37) — AIM ¶3-2-1

ALL, MIL
5583. (Refer to Figure 52, point 6.) Mosier Airport is

A—an airport restricted to use by private and recreational pilots.
B—a restricted military stage field within restricted airspace.
C—a nonpublic use airport.

The "Pvt" after the airport names indicates Mosier airport is a restricted or non-public use airport. (J37) — Sectional Chart Legend

Answer (A) is incorrect because the R in the circle indicates the airport is a nonpublic-use airport. Answer (B) is incorrect because the blue box near Mosier airport indicates an alert area.

ALL, MIL
5584. (Refer to Figure 54, point 2.) After departing from Byron Airport (C83) with a northeast wind, you discover you are approaching Livermore Class D airspace and flight visibility is approximately 2-1/2 miles. You must

A—stay below 700 feet to remain in Class G and land.
B—stay below 1,200 feet to remain in Class G.
C—contact Livermore ATCT on 119.65 and advise of your intentions.

The magenta shading indicates Class E airspace begins at 700 feet. The VFR minimum in controlled airspace below 10,000 feet is 3 SM. Therefore, with 2-1/2 miles visibility, you must stay below 700 feet to remain in Class G airspace. (J37) — AIM ¶3-2-1

Answer (B) is incorrect because Class E space begins at 1,200 feet when surrounded by blue shading. Answer (C) is incorrect because 119.65 is ATIS for Livermore, not ATCT.

ALL, MIL
5587. (Refer to Figure 54, point 6.) The Class C airspace at Metropolitan Oakland International (OAK) which extends from the surface upward has a ceiling of

A—both 2,100 feet and 3,000 feet MSL.
B—8,000 feet MSL.
C—2,100 feet AGL.

The letter "T" denotes the ceiling of the Class C airspace which extends up to, but does not include the floor of the overlying Class B airspace. The Class C airspace normally extends upward to 4,000 feet AGL. However, in this case the Class C airspace extends upward to the base of the Class B airspace. The overlying Class B airspace has bases of 2,100 feet MSL, and 3,000 feet MSL. (J37) — Sectional Chart Legend

Answer (B) is incorrect because 8,000 feet is the ceiling of the Class B airspace over OAK. Answer (C) is incorrect because the Class C airspace ceiling on the west side of OAK is 2,100 feet MSL, and the ceiling on the east side is 3,000 feet MSL.

ALL, MIL
5576. (Refer to Figure 54, point 4.) The thinner outer magenta circle depicted around San Francisco International Airport is

A—the outer segment of Class B airspace.
B—an area within which an appropriate transponder must be used from outside of the Class B airspace from the surface to 10,000 feet MSL.
C—a Mode C veil boundary where a balloon may penetrate without a transponder provided it remains below 8,000 feet MSL.

In all airspace within 30 NM of a Class B airspace primary airport from the surface up to 10,000 feet MSL, an appropriate transponder must be used. (J37) — Sectional Chart Legend

ALL, MIL
5569. (Refer to Figure 53, point 1.) This thin black shaded line is most likely

A—an arrival route.
B—a military training route.
C—a state boundary line.

The thin black shaded line is most likely a military training route (MTR). MTRs are normally labeled on sectional charts with either IR (IFR operations) or VR (VFR operations), followed by either three or four numbers. (J37) — Sectional Chart Legend

Answer (A) is incorrect because arrival routes are found on IFR charts. Answer (C) is incorrect because state boundaries are indicated by a thin black broken line.

Answers

| 5583 | [C] | 5584 | [A] | 5587 | [A] | 5576 | [B] | 5569 | [B] |

Chapter 5 Procedures and Airport Operations

ALL, MIL
5575. (Refer to Figure 52, point 9.) The alert area depicted within the blue lines is an area in which

A—the flight of aircraft, while not prohibited, is subject to restriction.
B—the flight of aircraft is prohibited.
C—there is a high volume of pilot training activities or an unusual type of aerial activity, neither of which is hazardous to aircraft.

Alert Areas inform pilots of airspace that may contain a high volume of pilot training or an unusual type of aerial activity. While pilots should be particularly alert in these areas, there are no restrictions on flying through them. (J37) — Pilot/Controller Glossary

Answer (A) is incorrect because this describes a Restricted Area. Answer (B) is incorrect because this describes a Prohibited Area.

ALL, MIL
5565. (Refer to Figure 52, point 1.) The floor of the Class E airspace above Georgetown Airport (Q61) is at

A—the surface.
B—700 feet AGL.
C—3,823 feet MSL.

Georgetown Airport is outside the magenta shaded area, which indicates the floor of Class E airspace is at 1,200 feet AGL. The airport elevation is given in the airport data as 2,623 feet MSL. Therefore, the Class E airspace above Georgetown Airport is 3,823 feet MSL (2,623 + 1,200). (J37) — Sectional Chart Legend

Answer (A) is incorrect because Class E airspace only begins at the surface when surrounded by a magenta segmented circle. Answer (B) is incorrect because Class E airspace begins at 700 feet AGL inside the magenta shaded areas.

ALL, MIL
5566. (Refer to Figure 52, point 7.) The floor of Class E airspace over the town of Woodland is

A—700 feet AGL over part of the town and no floor over the remainder.
B—1,200 feet AGL over part of the town and no floor over the remainder.
C—both 700 feet and 1,200 feet AGL.

Woodland has magenta shading over part of the town. Inside this magenta shading, Class E airspace begins at 700 feet AGL. Outside the magenta area, Class E airspace begins at 1,200 feet AGL. (J37) — Sectional Chart Legend

ALL, MIL
5567. (Refer to Figure 52, point 5.) The floor of the Class E airspace over University Airport (O05) is

A—the surface.
B—700 feet AGL.
C—1,200 feet AGL.

University Airport is within the magenta shading, which indicates the floor of the Class E airspace begins at 700 feet AGL. (J37) — Sectional Chart Legend

Answer (A) is incorrect because the Class E airspace would begin at the surface if the airport were surrounded by a magenta segmented circle. Answer (C) is incorrect because the Class E airspace would begin at 1,200 feet AGL if the airport were outside the magenta shaded area.

ALL, MIL
5568. (Refer to Figure 52, point 8.) The floor of the Class E airspace over the town of Auburn is

A—1,200 feet MSL.
B—700 feet AGL.
C—1,200 feet AGL.

Auburn is inside the magenta shading, which indicates the Class E airspace begins at 700 feet AGL. (J37) — Sectional Chart Legend

Answers (A) and (C) are incorrect because the Class E airspace would begin at 1,200 feet AGL (not MSL) if the airport were outside the magenta shaded area.

ALL, MIL
5570. (Refer to Figure 53, point 2.) The 1^6 indicates

A—an antenna top at 1,600 feet AGL.
B—the maximum elevation figure for that quadrangle.
C—the minimum safe sector altitude for that quadrangle.

The number 1^6 is a maximum elevation figure (MEF) which approximate and round-up from the highest known feature within each quadrangle. (J37) — Sectional Chart Legend

Answer (A) is incorrect because antennas are identified by obstruction symbols with the height above ground given in parentheses. Answer (C) is incorrect because minimum safe altitudes are not depicted on sectional charts.

ALL, MIL
5581. (Refer to Figure 52, point 4.) The highest obstruction with high intensity lighting within 10 NM of Lincoln Airport (LHM) is how high above the ground?

A—1,254 feet.
B—662 feet.
C—299 feet.

Answers

| 5575 | [C] | 5565 | [C] | 5566 | [C] | 5567 | [B] | 5568 | [B] | 5570 | [B] |
| 5581 | [C] | | | | | | | | | | |

Obstacles with high-intensity lighting are depicted by symbols with lightning bolts radiating from the top. The highest obstruction with high intensity lighting within 10 nautical miles of Lincoln Airport is directly south of the airport. The obstruction is 299 feet above the ground, which is the number in parentheses. (J37) — Sectional Chart Legend

Answer (A) is incorrect because 1,254 feet is the height above sea level of the obstruction 8.5 NM east of the airport, which does not have high-intensity lights. Answer (B) is incorrect because 662 feet is the height above ground of the obstructions 8 NM southwest of the airport, which do not have high-intensity lights.

ALL, MIL
5585. (Refer to Figure 52, point 4.) The terrain at the obstruction approximately 8 NM east southeast of the Lincoln Airport is approximately how much higher than the airport elevation?

A—376 feet.
B—835 feet.
C—1,135 feet.

```
  1,254 feet   MSL obstruction height
–   300 feet   AGL
    954 feet   MSL terrain height at obstruction
–   119 feet   MSL airport elevation
    835 feet   terrain height higher than airport elevation
```

(J37) — Sectional Chart Legend

AIR, MIL
5088. When operating an airplane for the purpose of landing or takeoff within Class D under special VFR, what minimum distance from clouds and what visibility are required?

A—Remain clear of clouds, and the ground visibility must be at least 1 SM.
B—500 feet beneath clouds, and the ground visibility must be at least 1 SM.
C—Remain clear of clouds, and the flight visibility must be at least 1 NM.

No person may operate an airplane within Class D airspace under Special VFR unless they remain clear of clouds and the ground visibility must be at least 1 SM. (B09) — 14 CFR §91.157(b)

Answer (B) is incorrect because the cloud clearance for Special VFR is clear of clouds. Answer (C) is incorrect because if ground visibility is not reported, flight visibility during landing or takeoff must be at least 1 SM.

AIR, MIL
5089. At some airports located in Class D airspace where ground visibility is not reported, takeoffs and landings under special VFR are

A—not authorized.
B—authorized by ATC if the flight visibility is at least 1 SM.
C—authorized only if the ground visibility is observed to be at least 3 SM.

No person may operate an airplane within Class D airspace under Special VFR unless they remain clear of clouds and the ground visibility must be at least 1 SM. If ground visibility is not reported at that airport, flight visibility during landing or takeoff must be at least 1 SM. (B09) — 14 CFR §91.157(b)

Answer (A) is incorrect because Special VFR is authorized if flight visibility is at least 1 SM. Answer (C) is incorrect because the visibility requirement for Special VFR is 1 SM.

AIR, MIL
5090. To operate an airplane under SPECIAL VFR (SVFR) within Class D airspace at night, which is required?

A—The pilot must hold an instrument pilot rating, but the airplane need not be equipped for instrument flight, as long as the weather will remain at or above SVFR minimums.
B—The Class D airspace must be specifically designated as a night SVFR area.
C—The pilot must hold an instrument pilot rating and the airplane must be equipped for instrument flight.

No person may operate an airplane in Class D airspace under Special VFR at night unless that person is instrument rated, and the airplane is equipped for instrument flight. (B09) — 14 CFR §91.157(b)

Answer (A) is incorrect because the airplane must be equipped for instrument flight. Answer (B) is incorrect because there is no such designation as "night SVFR area."

ALL, MIL
5577. When a dashed blue circle surrounds an airport on a sectional aeronautical chart, it will depict the boundary of

A—Special VFR airspace.
B—Class B airspace
C—Class D airspace.

Class D airspace areas are depicted on Sectional and Terminal charts with blue segmented lines. (J37) — AIM ¶3-2-5

Answer (A) is incorrect because no special VFR airspace is designated by a "NO SVFR" notation in the airport data block of the sectional. Answer (B) is incorrect because Class B airspace is depicted by a solid blue line.

Basic VFR Weather Minimums

Rules governing flight under visual flight rules (VFR) have been adopted to assist the pilot in meeting his/her responsibility to see and avoid other aircraft. Minimum weather conditions and distance from clouds required for VFR flight are listed in Figure 5-2.

When operating within a Class B, C, D, or E airspace designated for an airport, the ceiling must not be less than 1,000 feet. If the pilot intends to land, take off, or enter a traffic pattern at an airport within the lateral boundaries of Class B, C, D, or E airspace designated for an airport, the ground visibility must be at least 3 miles at that airport. If ground visibility is not reported, 3 miles flight visibility is required.

Airspace	Flight Visibility	Distance from Clouds
Class A	Not Applicable	Not Applicable
Class B	3 statute miles	Clear of clouds
Class C	3 statute miles	500 feet below 1,000 feet above 2,000 feet horizontal
Class D	3 statute miles	500 feet below 1,000 feet above 2,000 feet horizontal
Class E Less than 10,000 feet MSL	3 statute miles	500 feet below 1,000 feet above 2,000 feet horizontal
At or above 10,000 feet MSL	5 statute miles	1,000 feet below 1,000 feet above 1 statute mile horizontal
Class G 1,200 feet or less above the surface (regardless of MSL altitude) Day, except as provided in § 91.155(b)	1 statute mile	Clear of clouds
Night, except as provided in § 91.155(b)	3 statute miles	500 feet below 1,000 feet above 2,000 feet horizontal
More than 1,200 feet above the surface but less than 10,000 foot MSL Day	1 statute mile	500 feet below 1,000 feet above 2,000 feet horizontal
Night	3 statute miles	500 feet below 1,000 feet above 2,000 feet horizontal
More than 1,200 feet above the surface and at or above 10,000 feet MSL	5 statute miles	1,000 feet below 1,000 feet above 1 statute mile horizontal

Figure 5-2. Basic VFR weather minimums

ALL, MIL
5083. The minimum flight visibility for VFR flight increases to 5 statute miles beginning at an altitude of

A—14,500 feet MSL.
B—10,000 feet MSL if above 1,200 feet AGL.
C—10,000 feet MSL regardless of height above ground.

The only area requiring 5 statute miles visibility is 10,000 feet MSL and up (when above 1,200 feet AGL). (B08) — 14 CFR §91.155(a)

AIR, LTA, GLI, MIL
5085. What is the minimum flight visibility and proximity to cloud requirements for VFR flight, at 6,500 feet MSL, in Class C, D, and E airspace?

A—1 mile visibility; clear of clouds.
B—3 miles visibility; 1,000 feet above and 500 feet below.
C—5 miles visibility; 1,000 feet above and 1,000 feet below.

In Class C, D, or E airspace at 6,500 feet MSL, the VFR flight visibility requirement is 3 SM. The distance from cloud requirement is 500 feet below, 1,000 feet above, and 2,000 feet horizontal. (B08) — 14 CFR §91.155(a)

Answer (A) is incorrect because 1 SM visibility and clear of clouds is the VFR weather minimum when at or below 1,200 feet AGL in Class G airspace during the day. Answer (C) is incorrect because 5 SM visibility and cloud clearance of 1,000 feet above and below is the VFR weather minimum in Class E airspace at or above 10,000 feet MSL.

Answers

5577 [C]   5083 [B]   5085 [B]

Chapter 5 **Procedures and Airport Operations**

ALL, MIL
5121. When weather information indicates that abnormally high barometric pressure exists, or will be above _____ inches of mercury, flight operations will not be authorized contrary to the requirements published in NOTAMs.

A—31.00
B—32.00
C—30.50

Special flight restrictions exist when any information indicates that barometric pressure on the route of flight currently exceeds or will exceed 31 inches of mercury, and no person may operate an aircraft or initiate a flight contrary to the requirements established by the Administrator and published in a Notice to Airmen. (B08) — 14 CFR §91.144

AIR, LTA, GLI, MIL
5588. (Refer to Figure 53.)
GIVEN:

Location .... Madera Airport (MAE)
Altitude .... 1,000 ft AGL
Position .... 7 NM north of Madera (MAE)
Time .... 3 p.m. local
Flight visibility .... 1 SM

You are VFR approaching Madera Airport for a landing from the north. You

A—are in violation of the CFRs; you need 3 miles of visibility under VFR.
B—are required to descend to below 700 feet AGL to remain clear of Class E airspace and may continue for landing.
C—may descend to 800 feet AGL (Pattern Altitude) after entering Class E airspace and continue to the airport.

At 7 NM north of Madera, you are outside the magenta shading, which indicates the floor of the Class E airspace is 1,200 feet AGL. At 1,000 feet, you are in Class G airspace. During daylight hours, the minimum flight visibility for VFR flight is 1 SM. Inside the magenta shading, the floor of the Class E airspace drops to 700 feet. Therefore, to remain VFR, you must remain in Class G airspace, which requires you to descend below 700 feet before entering the Class E airspace to continue for landing. (J37) — 14 CFR §91.155

Answer (A) is incorrect because only 1 SM visibility is necessary to remain VFR in Class G airspace at 1,000 feet AGL during daylight hours. Answer (C) is incorrect because you must descend below 700 feet AGL before entering Class E airspace to remain VFR.

VHF Direction Finder

Many flight service stations have equipment that determines the direction to an aircraft which is transmitting a particular VHF frequency. Availability of this service at a particular airport is shown by the notation VHF/DF in the Airport/Facility Directory (A/FD). The only airborne equipment required to obtain VHF/DF service is an operating VHF transmitter and receiver.

ALL
5504. To use VHF/DF facilities for assistance in locating your position, you must have an operative VHF

A—transmitter and receiver.
B—transmitter and receiver, and an operative ADF receiver.
C—transmitter and receiver, and an operative VOR receiver.

VHF/DF is a ground-based radio receiver with an azimuth display. The pilot needs only two-way VHF radio communications capability to make use of the services available. (J22) — AC 61-23C, Chapter 8

Answers (B) and (C) are incorrect because an ADF or VOR receiver is not required to use VHF/DF facilities.

Answers

5121 [A]   5588 [B]   5504 [A]

Chapter 5 **Procedures and Airport Operations**

Operations on Wet or Slippery Runways

When taking off from a slippery runway, delay full-power checks until the aircraft is lined up on the runway and ready for takeoff.

After takeoff from a slushy runway, the landing gear should be cycled up and down to minimize the possibility of the gear being frozen in the up position.

AIR
5768. If necessary to take off from a slushy runway, the freezing of landing gear mechanisms can be minimized by

A—recycling the gear.
B—delaying gear retraction.
C—increasing the airspeed to V_{LE} before retraction.

After takeoff from a slushy runway, recycle the landing gear several times. This is a preventative measure against mud and slush freezing, causing gear operational problems. (L52) — AC 91-13C

Answer (B) is incorrect because delaying gear retraction will slow the climb out and may result in the landing gear freezing in the extended position. Answer (C) is incorrect because the landing gear should always be retracted below V_{LE} (maximum landing gear extended speed).

Land and Hold Short Operations (LAHSO)

LAHSO is an acronym for "Land And Hold Short Operations." These operations include landing and holding short of an intersecting runway, an intersecting taxiway, or some other designated point on a runway other than an intersecting runway or taxiway. LAHSO is an air traffic control procedure that requires pilot participation to balance the needs for increased airport capacity and system efficiency, consistent with safety. Student pilots or pilots not familiar with LAHSO should not participate in the program. The pilot-in-command has the final authority to accept or decline any land and hold short clearance. The safety and operation of the aircraft remain the responsibility of the pilot. Pilots are expected to decline a LAHSO clearance if they determine it will compromise safety. Available Landing Distance (ALD) data are published in the special notices section of the Airport/Facility Directory (A/FD) and in the U.S. Terminal Procedures Publications. Pilots should only receive a LAHSO clearance when there is a minimum ceiling of 1,000 feet and 3 statute miles visibility. The intent of having "basic" VFR weather conditions is to allow pilots to maintain visual contact with other aircraft and ground vehicle operations.

ALL
5138. Who has the final authority to accept or decline any "land and hold short" (LAHSO) clearance?

A—ATC tower controller.
B—Airplane owner/operator.
C—Pilot-in-Command.

The pilot-in-command has the final authority to accept or decline any land and hold short clearance. The safety and operation of the aircraft remain the responsibility of the pilot. (J13) — AIM ¶4-3-11

ALL
5139. When should pilots decline a "land and hold short" (LAHSO) clearance?

A—When it will compromise safety.
B—If runway surface is contaminated.
C—Only when the tower controller concurs.

Pilots are expected to decline a LAHSO clearance if they determine it will compromise safety. (J13) — AIM ¶4-3-11

Answers

5768 [A]   5138 [C]   5139 [A]

ALL
5140. What is the minimum visibility and ceiling required for a pilot to receive a "land and hold short" clearance?

A—3 statute miles and 1,000 feet.
B—3 nautical miles and 1,000 feet.
C—3 statute miles and 1,500 feet.

Pilots should only receive a LAHSO clearance when there is a minimum ceiling of 1,000 feet and 3 statute miles visibility. The intent of having "basic" VFR weather conditions is to allow pilots to maintain visual contact with other aircraft and ground vehicle operations. (J13) — AIM ¶4-3-11

Airport Marking Aids and Signs

Airport marking aids and signs provide information that is useful to a pilot during takeoff, taxiing, and landing. The top symbol in FAA Figure 51 prohibits entry into an area. The middle symbol is a runway boundary sign, and the bottom symbol is an ILS critical area boundary sign.

Pilots are encouraged to turn on the aircraft rotating beacon anytime the engines are in operation.

AIR, RTC, MIL
5748. Pilots are encouraged to turn on the aircraft rotating beacon

A—just prior to engine start.
B—anytime they are in the cockpit.
C—anytime an engine is in operation.

Pilots of aircraft equipped with rotating beacons are encouraged to turn them on just prior to engine start when intending to fly, as an alert to other aircraft and ground personnel. (J13) — AIM ¶4-3-23

Answer (B) is incorrect because lights are not needed if the pilot is simply sitting in the cockpit. Answer (C) is incorrect because lights are not needed in all instances an engine is in operation, such as if a mechanic is working on the aircraft and does not intend to taxi or fly.

AIR, RTC
5657. (Refer to Figure 51.) The pilot generally calls ground control after landing when the aircraft is completely clear of the runway. This is when the aircraft

A—passes the red symbol shown at the top of the figure.
B—is on the dashed-line side of the middle symbol.
C—is past the solid-line side of the middle symbol.

After landing, the pilot generally calls ground control when the aircraft is completely clear of the runway. This is when the aircraft is on the solid-line side of the middle symbol. The solid lines always indicate the side on which the aircraft is to hold. (J05) — AIM ¶2-3-8, 2-3-9

Answer (A) is incorrect because the top symbol prohibits aircraft entry into an area. Answer (B) is incorrect because you are still on the runway if you are on the dashed-line side of the middle symbol.

AIR, RTC
5658. (Refer to Figure 51.) The red symbol at the top would most likely be found

A—upon exiting all runways prior to calling ground control.
B—at an intersection where a roadway may be mistaken as a taxiway.
C—near the approach end of ILS runways.

This sign prohibits an aircraft from entering an area. Typically, this sign would be located on a taxiway intended to be used in only one direction or at the intersection of vehicle roadways with runways, taxiways, or aprons where the roadway may be mistaken as a taxiway or other aircraft movement surface. (J05) — AIM ¶2-3-8, 2-3-9

Answer (A) is incorrect because this refers to the middle symbol. Answer (C) is incorrect because this refers to the bottom symbol.

AIR, RTC
5659-1. (Refer to Figure 51.) While clearing an active runway you are most likely clear of the ILS critical area when you pass which symbol?

A—Top red.
B—Middle yellow.
C—Bottom yellow.

While clearing an active runway, you are most likely to be clear of the ILS critical area when you pass the bottom yellow sign. This is the ILS critical area boundary sign. (J05) — AIM ¶2-3-8, 2-3-9

Answer (A) is incorrect because this symbol prohibits aircraft entry into an area. Answer (B) is incorrect because the middle symbol indicates you are most likely clear of the runway.

Answers

5140 [A]   5748 [C]   5657 [C]   5658 [B]   5659-1 [C]

Chapter 5 Procedures and Airport Operations

AIR, RTC

5659-2. (Refer to Figure 51.) When taxiing up to an active runway, you are likely to be clear of the ILS critical area when short of which symbol?

A—Bottom yellow.
B—Top red.
C—Middle yellow.

The bottom yellow sign is located adjacent to the ILS holding position marking on the pavement and can be seen by pilots leaving the critical area. The sign is intended to provide pilots with another visual cue which they can use as a guide in deciding when they are clear of the ILS critical area. (J05) — AIM ¶2-3-8, 2-3-9

Answer (B) is incorrect because this is the sign prohibiting aircraft entry into an area. Answer (C) is incorrect because this is a runway boundary sign.

AIR, RTC

5660. (Refer to Figure 51.) Which symbol does not directly address runway incursion with other aircraft?

A—Top red.
B—Middle yellow.
C—Bottom yellow.

The top symbol prohibits an aircraft from entering an area. This sign would typically be located on one-way taxiways or a vehicle roadway. Thus, this sign does not directly address runway incursions with other aircraft. (J05) — AIM ¶2-3-8, 2-3-9

Answer (B) is incorrect because the middle symbol is used to indicate when you are clear of the runway. Answer (C) is incorrect because the bottom symbol is used to indicate when you are clear of the ILS critical area.

VFR Cruising Altitudes

When operating an aircraft under VFR in level cruising flight more than 3,000 feet above the surface and below 18,000 feet MSL, a pilot is required to maintain an appropriate altitude in accordance with certain rules. This requirement is sometimes called the "Hemispherical Cruising Rule," and is based on magnetic course. *See* Figure 5-3.

ALL, MIL

5091. VFR cruising altitudes are required to be maintained when flying

A—at 3,000 feet or more AGL; based on true course.
B—more than 3,000 feet AGL; based on magnetic course.
C—at 3,000 feet or more above MSL; based on magnetic heading.

In level cruise at more than 3,000 feet AGL, magnetic course determines proper altitude. (B09) — 14 CFR §91.159

Answers (A) and (C) are incorrect because VFR cruising altitudes are based on magnetic course and apply for flights above 3,000 feet AGL.

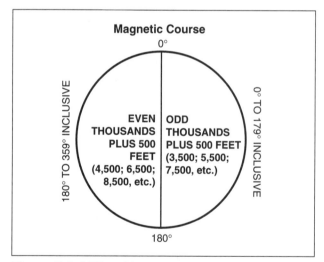

Figure 5-3. VFR cruising altitudes

Answers

5659-2 [A]   5660 [A]   5091 [B]

Collision Avoidance

Vision is the most important physical sense for safe flight. Two major factors that determine how effectively vision can be used are the level of illumination, and the technique of scanning the sky for other aircraft.

Scanning the sky for other aircraft is a key factor in collision avoidance. Pilots must develop an effective scanning technique, one that maximizes visual capabilities. Because the eyes focus on only a narrow viewing area, effective scanning is accomplished by systematically focusing with a series of short, regularly-spaced eye movements. Each movement should not exceed 10°, and each area should be observed for at least one second. At night, scan slowly to permit off-center viewing (peripheral vision). Prior to starting any maneuver, a pilot should visually scan the entire area for other aircraft. Any aircraft that appears to have no relative motion and stays in one scan quadrant is likely to be on a collision course. If a target shows neither lateral or vertical motion, but increases in size, take evasive action.

When climbing or descending VFR on an airway, execute gentle banks, right and left, to provide for visual scanning of the airspace. Particular vigilance should be exercised when operating in areas where aircraft tend to converge, such as near airports and over VOR stations.

Atmospheric haze reduces the ability to see traffic or terrain during flight, making all features appear to be farther away than their actual distance.

In preparation for a **night flight**, the pilot should avoid bright white lights for at least 30 minutes before the flight.

ALL
5272. How can you determine if another aircraft is on a collision course with your aircraft?

A—The nose of each aircraft is pointed at the same point in space.
B—The other aircraft will always appear to get larger and closer at a rapid rate.
C—There will be no apparent relative motion between your aircraft and the other aircraft.

It is essential to remember that if another aircraft appears to have no relative motion, it is likely to be on a collision course with you. If the other aircraft shows no lateral or vertical motion, but increases in size, take evasive action. (L34) — AC 90-48

ALL, MIL
5749. When in the vicinity of a VOR which is being used for navigation on VFR flights, it is important to

A—make 90° left and right turns to scan for other traffic.
B—exercise sustained vigilance to avoid aircraft that may be converging on the VOR from other directions.
C—pass the VOR on the right side of the radial to allow room for aircraft flying in the opposite direction on the same radial.

Pilots should exercise increased caution when entering high use airspace; this includes the airspace around VORs. (J14) — AIM ¶8-1-8

Answer (A) is incorrect because 90° turns are not appropriate while en route. Answer (C) is incorrect because you should try to maintain the center of the radial.

ALL, MIL
5758. To scan properly for traffic, a pilot should

A—slowly sweep the field of vision from one side to the other at intervals.
B—concentrate on any peripheral movement detected.
C—use a series of short, regularly spaced eye movements that bring successive areas of the sky into the central visual field.

Because the eyes can focus on only a narrow viewing area, effective scanning is accomplished with a series of short, regularly-spaced eye movements that bring successive areas of the sky into the central visual field. (J31) — AIM ¶8-1-6

Answer (A) is incorrect because a pilot should systematically concentrate on different segments. Answer (B) is incorrect because peripheral movement is not easily detected, especially under adverse conditions; therefore this would not be an effective scanning technique.

Answers

5272 [C]   5749 [B]   5758 [C]

Chapter 5 Procedures and Airport Operations

ALL
5130. For night flying operations, the best night vision is achieved when the

A—pupils of the eyes have become dilated in approximately 10 minutes.
B—rods in the eyes have become adjusted to the darkness in approximately 30 minutes.
C—cones in the eyes have become adjusted to the darkness in approximately 5 minutes.

The rods in the eyes make night vision possible. About 30 minutes is needed to adjust the eyes to maximum efficiency after exposure to a bright light. (H564) — FAA-H-8083-3, Chapter 10

ALL
5133. When planning a night cross-country flight, a pilot should check for

A—availability and status of en route and destination airport lighting systems.
B—red en route course lights.
C—location of rotating light beacons.

Prior to a night flight, and particularly a cross-country night flight, pilots should check the availability and status of lighting systems at the destination airport. (H568) — FAA-H-8083-3, Chapter 10

ALL
5134. Light beacons producing red flashes indicate

A—end of runway warning areas.
B—instructions for the pilot to remain clear of an airport traffic pattern and continue circling.
C—obstructions or areas considered hazardous to aerial navigation.

Beacons producing red flashes indicate obstructions or areas considered hazardous to aerial navigation. (H568) — FAA-H-8083-3, Chapter 10

ALL
5135. What is the first indication of flying into restricted visibility conditions when operating VFR at night?

A—Ground lights begin to take on an appearance of being surrounded by a halo or glow.
B—A gradual disappearance of lights on the ground.
C—Cockpit lights begin to take on an appearance of a halo or glow around them.

Generally, at night it is difficult to see clouds and restrictions to visibility, particularly on dark nights or under overcast. Usually, the first indication of flying into restricted visibility conditions is the gradual disappearance of lights on the ground. (H572) — FAA-H-8083-3, Chapter 10

Answers (A) and (C) are incorrect because ground (not cockpit) lights taking on the appearance of a halo or glow indicate ground fog.

ALL
5136. After experiencing a powerplant failure at night, one of the primary considerations should include

A—turning off all electrical switches to save battery power for the landing.
B—planning the emergency approach and landing to an unlighted portion of an area.
C—maneuvering to, and landing on a lighted highway or road.

If the condition of the nearby terrain is known, turn towards an unlighted portion of the area. Plan an emergency approach to an unlighted portion. (H574) — FAA-H-8083-3, Chapter 10

Answer (A) is incorrect because you should keep the battery on so you have lights available. Answer (C) is incorrect because you need to land away from congested areas.

ALL
5137. When planning for an emergency landing at night, one of the primary considerations should include

A—selecting a landing area close to public access, if possible.
B—landing without flaps to ensure a nose-high landing attitude at touchdown.
C—turning off all electrical switches to save battery power for the landing.

Consider an emergency landing area close to public access if possible. This may facilitate rescue or help, if needed. (H574) — FAA-H-8083-3, Chapter 10

Answer (B) is incorrect because the landing should be completed in the normal landing attitude at the slowest possible speed. Answer (C) is incorrect because you should keep the battery on so you have lights available.

Answers

| 5130 | [B] | 5133 | [A] | 5134 | [C] | 5135 | [B] | 5136 | [B] | 5137 | [A] |

Chapter 5 **Procedures and Airport Operations**

Fitness Physiology

Pilot performance can be seriously degraded by a number of physiological factors. While some of the factors may be beyond the control of the pilot, awareness of cause and effect will minimize any adverse effects. The body has no built-in alarm system to alert the pilot of many of these factors.

Hypoxia, a state of oxygen deficiency (insufficient supply), impairs functions of the brain and other organs. Headache, sleepiness, dizziness, and euphoria are all symptoms of hypoxia. For optimum protection, pilots should avoid flying above 10,000 feet MSL for prolonged periods without breathing supplemental oxygen. Federal Aviation Regulations, Part 91 require that when operating an aircraft at cabin pressure altitudes above 12,500 feet MSL up to and including 14,000 feet MSL, supplemental oxygen shall be used by the minimum flight crew during that time in excess of 30 minutes at those altitudes. Every occupant of the aircraft must be provided with supplemental oxygen above 15,000 feet. If under the effects of hypoxia, time of useful consciousness decreases with altitude.

If rapid decompression occurs in a pressurized aircraft above 30,000 feet, a pilot's time of useful consciousness is about 30 seconds. During a rapid decompression at high altitudes, the pilot should don the oxygen mask and begin a rapid descent to an appropriate lower altitude.

Aviation breathing oxygen should be used to replenish an aircraft oxygen system for high altitude flight. Oxygen used for medical purposes or welding should not be used because it may contain too much water. The excess water could condense and freeze in oxygen lines when flying at high altitudes, and this could block oxygen flow. Also, constant use of oxygen containing too much water may cause corrosion in the system. Specifications for "aviator's breathing oxygen" are 99.5% pure oxygen and not more than .005 mg. of water per liter of oxygen. Never use grease- or oil-covered hands, rags or tools while working with oxygen systems.

Hyperventilation, a deficiency (insufficient supply) of carbon dioxide within the body, can be the result of rapid or extra deep breathing due to emotional tension, anxiety, or fear. The common symptoms of hyperventilation include drowsiness, and tingling of the hands, legs and feet. A pilot should be able to overcome the symptoms or avoid future occurrences of hyperventilation by talking aloud, breathing into a bag, or slowing the breathing rate.

Carbon monoxide is a colorless, odorless, and tasteless gas contained in exhaust fumes. Symptoms of **carbon monoxide poisoning** include headache, drowsiness, or dizziness. Large accumulations of carbon monoxide in the human body result in a loss of muscular power. Susceptibility to hypoxia due to inhalation of carbon monoxide increases as altitude increases. A pilot who detects symptoms of carbon monoxide poisoning should immediately shut off the heater and open the air vents.

Various complex motions, forces, and visual scenes encountered in flight may result in various sensory organs sending misleading information to the brain. **Spatial disorientation** may result if these body signals are used to interpret flight attitude. The best way to overcome spatial disorientation is by relying on aircraft instrument indications rather than taking a chance on the sensory organs.

Extensive research has provided a number of facts about the hazards of alcohol consumption and flying. Even a small amount of **alcohol** present in the human body can impair flying skills, judgment and decision-making abilities. Alcohol also renders a pilot much more susceptible to disorientation and hypoxia. The regulations prohibit pilots from performing crew member duties within 8 hours after drinking any alcoholic beverage (bottle to throttle) or while under the influence of alcohol. However, due to the slow destruction of alcohol, a pilot may still be under influence more than 8 hours after drinking a moderate amount of alcohol.

Chapter 5 **Procedures and Airport Operations**

ALL, MIL
5757. As hyperventilation progresses, a pilot can experience

A—decreased breathing rate and depth.
B—heightened awareness and feeling of well being.
C—symptoms of suffocation and drowsiness.

The common symptoms of hyperventilation are dizziness, nausea, hot and cold sensations, tingling of the hands, legs and feet, sleepiness, and finally, unconsciousness. (J31) — AIM ¶8-1-3

Answer (A) is incorrect because hyperventilation is an increase of the breathing rate and depth. Answer (B) is incorrect because heightened awareness and feeling of well-being are symptoms of hypoxia.

ALL, MIL
5759. Which is a common symptom of hyperventilation?

A—Drowsiness.
B—Decreased breathing rate.
C—Euphoria – sense of well-being.

The common symptoms of hyperventilation are dizziness, nausea, hot and cold sensations, tingling of the hands, legs and feet, sleepiness and finally unconsciousness. (J31) — AIM ¶8-1-3

Answer (B) is incorrect because hyperventilation is an increase of the breathing rate. Answer (C) is incorrect because a feeling of well-being, or euphoria, is a symptom of hypoxia.

ALL, MIL
5760. Which would most likely result in hyperventilation?

A—Insufficient oxygen.
B—Excessive carbon monoxide.
C—Insufficient carbon dioxide.

As hyperventilation "blows off" excessive carbon dioxide from the body, a pilot can experience symptoms of lightheadedness, suffocation, drowsiness, tingling of the extremities, and coolness and react to them with even greater hyperventilation. (J31) — AIM ¶8-1-3

Answer (A) is incorrect because insufficient oxygen is a symptom of hypoxia. Answer (B) is incorrect because excessive carbon monoxide will lead to carbon monoxide poisoning.

ALL, MIL
5761. Hypoxia is the result of which of these conditions?

A—Excessive oxygen in the bloodstream.
B—Insufficient oxygen reaching the brain.
C—Excessive carbon dioxide in the bloodstream.

Hypoxia is the result of insufficient oxygen in the bloodstream going to the brain. (J31) — AIM ¶8-1-2

Answer (A) is incorrect because hypoxia is a lack of oxygen in the bloodstream. Answer (C) is incorrect because excessive carbon dioxide in the bloodstream is not a symptom of hypoxia.

ALL, MIL
5762. To overcome the symptoms of hyperventilation, a pilot should

A—swallow or yawn.
B—slow the breathing rate.
C—increase the breathing rate.

Hyperventilation can be relieved by consciously slowing the breathing rate. Talking loudly or breathing into a bag to restore carbon dioxide will effectively slow the breathing rate. (J31) — AIM ¶8-1-3

Answer (A) is incorrect because swallowing or yawning is used to relieve ear block. Answer (C) is incorrect because the breathing rate should be slowed to increase the amount of carbon dioxide in the blood.

ALL, MIL
5763. Which is true regarding the presence of alcohol within the human body?

A—A small amount of alcohol increases vision acuity.
B—An increase in altitude decreases the adverse effect of alcohol.
C—Judgment and decision-making abilities can be adversely affected by even small amounts of alcohol.

As little as one ounce of liquor, one bottle of beer, or four ounces of wine can impair flying skills. (J31) — AIM ¶8-1-1

Answer (A) is incorrect because all mental and physical activities will be decreased with even small amounts of alcohol in the bloodstream. Answer (B) is incorrect because the adverse effects of alcohol are increased as altitude is increased.

Answers

| 5757 | [C] | 5759 | [A] | 5760 | [C] | 5761 | [B] | 5762 | [B] | 5763 | [C] |

ALL, MIL
5764. Hypoxia susceptibility due to inhalation of carbon monoxide increases as

A—humidity decreases.
B—altitude increases.
C—oxygen demand increases.

Carbon monoxide inhaled in smoking or from exhaust fumes, lowered hemoglobin (anemia), and certain medications can reduce the oxygen-carrying capacity of the blood to the degree that the amount of oxygen provided to body tissues will already be equivalent to the oxygen provided to the tissues when exposed to a cabin pressure altitude of several thousand feet. (J31) — AIM ¶8-1-4

Answer (A) is incorrect because the humidity level does not have a bearing on carbon monoxide or oxygen levels. Answer (C) is incorrect because oxygen demand does not change.

AIR, RTC, LTA, MIL
5765. To best overcome the effects of spatial disorientation, a pilot should

A—rely on body sensations.
B—increase the breathing rate.
C—rely on aircraft instrument indications.

Spatial disorientation can be prevented only by visual reference to reliable fixed points on the ground or to flight instruments. (H351) — AC 61-23C, Chapter 9

Answer (A) is incorrect because body sensations must be ignored. Answer (B) is incorrect because an increase in breathing rate could cause hyperventilation.

Aeronautical Decision Making

Aeronautical decision making (ADM) is a systematic approach to the mental process used by aircraft pilots to consistently determine the best course of action in response to a given set of circumstances.

Risk Management is the part of the decision making process which relies on situational awareness, problem recognition, and good judgment to reduce risks associated with each flight.

The ADM process addresses all aspects of decision making in the cockpit and identifies the steps involved in good decision making. Steps for good decision making are:

1. Identifying personal attitudes hazardous to safe flight.
2. Learning behavior modification techniques.
3. Learning how to recognize and cope with stress.
4. Developing risk assessment skills.
5. Using all resources in a multicrew situation.
6. Evaluating the effectiveness of one's ADM skills.

There are a number of classic behavioral traps into which pilots have been known to fall. Pilots, particularly those with considerable experience, as a rule always try to complete a flight as planned, please passengers, meet schedules, and generally demonstrate that they have the "right stuff." These tendencies ultimately may lead to practices that are dangerous and often illegal, and may lead to a mishap. All experienced pilots have fallen prey to, or have been tempted by, one or more of these tendencies in their flying careers. These dangerous tendencies or behavior patterns, which must be identified and eliminated, include:

Peer Pressure. Poor decision making based upon emotional response to peers rather than evaluating a situation objectively.

Mind Set. The inability to recognize and cope with changes in the situation different from those anticipated or planned.

Continued

Answers

5764 [B]   5765 [C]

Chapter 5 **Procedures and Airport Operations**

Get-There-Itis. This tendency, common among pilots, clouds the vision and impairs judgment by causing a fixation on the original goal or destination combined with a total disregard for any alternative course of action.

Duck-Under Syndrome. The tendency to sneak a peek by descending below minimums during an approach. Based on a belief that there is always a built-in "fudge" factor that can be used or on an unwillingness to admit defeat and shoot a missed approach.

Scud Running. Pushing the capabilities of the pilot and the aircraft to the limits by trying to maintain visual contact with the terrain while trying to avoid physical contact with it. This attitude is characterized by the old pilot's joke: "If it's too bad to go IFR, we'll go VFR."

Continuing Visual Flight Rules (VFR) into instrument conditions often leads to spatial disorientation or collision with ground/obstacles. It is even more dangerous if the pilot is not instrument qualified or current.

Getting Behind the Aircraft. Allowing events or the situation to control your actions rather than the other way around. Characterized by a constant state of surprise at what happens next.

Loss of Positional or Situation Awareness. Another case of getting behind the aircraft which results in not knowing where you are, an inability to recognize deteriorating circumstances, and/or the misjudgment of the rate of deterioration.

Operating Without Adequate Fuel Reserves. Ignoring minimum fuel reserve requirements, either VFR or Instrument Flight Rules (IFR), is generally the result of overconfidence, lack of flight planning, or ignoring the regulations.

Descent Below the Minimum Enroute Altitude. The duck-under syndrome (mentioned above) manifesting itself during the enroute portion of an IFR flight.

Flying Outside the Envelope. Unjustified reliance on the (usually mistaken) belief that the aircraft's high performance capability meets the demands imposed by the pilot's (usually overestimated) flying skills.

Neglect of Flight Planning, Preflight Inspections, Checklists, Etc. Unjustified reliance on the pilot's short and long term memory, regular flying skills, repetitive and familiar routes, etc.

Each ADM student should take the Self-Assessment Hazardous Attitude Inventory Test in order to gain a realistic perspective on his/her attitudes toward flying. The inventory test requires the pilot to provide a response which most accurately reflects the reasoning behind his/her decision. The pilot must choose one of the five given reasons for making that decision, even though the pilot may not consider any of the five choices acceptable. The inventory test presents extreme cases of incorrect pilot decision making in an effort to introduce the five types of hazardous attitudes.

ADM addresses the following five hazardous attitudes:

1. **Antiauthority (don't tell me!).** This attitude is found in people who do not like anyone telling them what to do. In a sense they are saying "no one can tell me what to do." They may be resentful of having someone tell them what to do or may regard rules, regulations, and procedures as silly or unnecessary. However, it is always your prerogative to question authority if you feel it is in error. The antidote for this attitude is: Follow the rules. They are usually right.

2. **Impulsivity (do something quickly!)** is the attitude of people who frequently feel the need to do something — *anything* — immediately. They do not stop to think about what they are about to do, they do not select the best alternative, and they do the first thing that comes to mind. The antidote for this attitude is: Not so fast. Think first.

3. **Invulnerability (it won't happen to me).** Many people feel that accidents happen to others, but never to them. They know accidents can happen, and they know that anyone can be affected. They never really feel or believe that they will be personally involved. Pilots who think this way are more likely to take chances and increase risk. The antidote for this attitude is: It could happen to me.

4. **Macho (I can do it).** Pilots who are always trying to prove that they are better than anyone else are thinking "I can do it—I'll show them." Pilots with this type of attitude will try to prove themselves by taking risks in order to impress others. While this pattern is thought to be a male characteristic, women are equally susceptible. The antidote for this attitude is: taking chances is foolish.

5. **Resignation (what's the use?).** Pilots who think "what's the use?" do not see themselves as being able to make a great deal of difference in what happens to them. When things go well, the pilot is apt to think that's good luck. When things go badly, the pilot may feel that "someone is out to get me," or attribute it to bad luck. The pilot will leave the action to others, for better or worse. Sometimes, such pilots will even go along with unreasonable requests just to be a "nice guy." The antidote for this attitude is: I'm not helpless. I can make a difference.

Hazardous attitudes which contribute to poor pilot judgment can be effectively counteracted by redirecting that hazardous attitude so that appropriate action can be taken. Recognition of hazardous thoughts is the first step in neutralizing them in the ADM process. Pilots should become familiar with a means of counteracting hazardous attitudes with an appropriate antidote thought. When a pilot recognizes a thought as hazardous, the pilot should label that thought as hazardous, then correct that thought by stating the corresponding antidote.

If you hope to succeed at reducing stress associated with crisis management in the air or with your job, it is essential to begin by making a personal assessment of stress in all areas of your life. Good **cockpit stress management** begins with good life stress management. Many of the stress coping techniques practiced for life stress management are not usually practical in flight. Rather, you must condition yourself to relax and think rationally when stress appears. The following checklist outlines some thoughts on cockpit stress management.

1. Avoid situations that distract you from flying the aircraft.
2. Reduce your workload to reduce stress levels. This will create a proper environment in which to make good decisions.
3. If an emergency does occur, be calm. Think for a moment, weigh the alternatives, then act.
4. Maintain proficiency in your aircraft; proficiency builds confidence. Familiarize yourself thoroughly with your aircraft, its systems, and emergency procedures.
5. Know and respect your own personal limits.
6. Do not let little mistakes bother you until they build into a big thing. Wait until after you land, then "debrief" and analyze past actions.
7. If flying is adding to your stress, either stop flying or seek professional help to manage your stress within acceptable limits.

The DECIDE Model, comprised of a six-step process, is intended to provide the pilot with a logical way of approaching decision making. The six elements of the DECIDE Model represent a continuous loop decision process which can be used to assist a pilot in the decision making process when he/she is faced with a change in a situation that requires a judgment. This DECIDE Model is primarily focused on the intellectual component, but can have an impact on the motivational component of judgment as well. If a pilot practices the DECIDE Model in all decision making, its use can become very natural and could result in better decisions being made under all types of situations.

1. **D**etect. The decisionmaker detects the fact that change has occurred.
2. **E**stimate. The decisionmaker estimates the need to counter or react to the change.
3. **C**hoose. The decisionmaker chooses a desirable outcome (in terms of success) for the flight.
4. **I**dentify. The decisionmaker identifies actions which could successfully control the change.
5. **D**o. The decisionmaker takes the necessary action.
6. **E**valuate. The decisionmaker evaluates the effect(s) of his/her action countering the change.

Chapter 5 **Procedures and Airport Operations**

ALL, MIL
5941. Risk management, as part of the Aeronautical Decision Making (ADM) process, relies on which features to reduce the risks associated with each flight?

A— The mental process of analyzing all information in a particular situation and making a timely decision on what action to take.
B— Application of stress management and risk element procedures.
C— Situational awareness, problem recognition, and good judgment.

Risk Management is the part of the decision making process which relies on situational awareness, problem recognition, and good judgment to reduce risks associated with each flight. (L05) — AC 60-22

ALL, MIL
5942. Aeronautical Decision Making (ADM) is a

A— systematic approach to the mental process used by pilots to consistently determine the best course of action for a given set of circumstances.
B— decision making process which relies on good judgment to reduce risks associated with each flight.
C— mental process of analyzing all information in a particular situation and making a timely decision on what action to take.

ADM is a systematic approach to the mental process used by aircraft pilots to consistently determine the best course of action in response to a given set of circumstances. (L05) — AC 60-22

ALL, MIL
5943. The Aeronautical Decision Making (ADM) process identifies the steps involved in good decision making. One of these steps includes a pilot

A— making a rational evaluation of the required actions.
B— developing the "right stuff" attitude.
C— identifying personal attitudes hazardous to safe flight.

Steps for good decision making are: identifying personal attitudes hazardous to safe flight, learning behavior modification techniques, learning how to recognize and cope with stress, developing risk assessment skills, using all resources in a multicrew situation, and evaluating the effectiveness of one's ADM skills. (L05) — AC 60-22

ALL, MIL
5944. Examples of classic behavioral traps that experienced pilots may fall into are: trying to

A— assume additional responsibilities and assert PIC authority.
B— promote situational awareness and then necessary changes in behavior.
C— complete a flight as planned, please passengers, meet schedules, and demonstrate the "right stuff."

There are a number of classic behavioral traps into which pilots have been known to fall. Pilots, particularly those with considerable experience, as a rule always try to complete a flight as planned, please passengers, meet schedules, and generally demonstrate that they have the "right stuff." (L05) — AC 60-22

Answers (A) and (B) are incorrect because promoting situation awareness and then necessary changes in behavior and asserting PIC authority are positive pilot behaviors.

ALL, MIL
5945. The basic drive for a pilot to demonstrate the "right stuff" can have an adverse effect on safety, by

A— a total disregard for any alternative course of action.
B— generating tendencies that lead to practices that are dangerous, often illegal, and may lead to a mishap.
C— allowing events, or the situation, to control his or her actions.

Pilots, particularly those with considerable experience, as a rule always try to complete a flight as planned, please passengers, meet schedules, and generally demonstrate that they have the "right stuff." These tendencies ultimately may lead to practices that are dangerous and often illegal, and may lead to a mishap. (L05) — AC 60-22

ALL, MIL
5946. Most pilots have fallen prey to dangerous tendencies or behavior problems at some time. Some of these dangerous tendencies or behavior patterns which must be identified and eliminated include:

A— Deficiencies in instrument skills and knowledge of aircraft systems or limitations.
B— Performance deficiencies from human factors such as, fatigue, illness or emotional problems.
C— Peer pressure, get-there-itis, loss of positional or situation awareness, and operating without adequate fuel reserves.

Answers
5941 [C]   5942 [A]   5943 [C]   5944 [C]   5945 [B]   5946 [C]

There are a number of classic behavioral traps into which pilots have been known to fall. These dangerous tendencies or behavior patterns, which must be identified and eliminated, include: peer pressure, mind set, get-there-itis, duck-under syndrome, scud running, continuing visual flight rules into instrument conditions, getting behind the aircraft, loss of positional or situation awareness, operating without adequate fuel reserves, descent below the minimum enroute altitude, flying outside the envelope, neglect of flight planning, preflight inspections, checklists, etc. (L05) — AC 60-22

ALL, MIL
5947. An early part of the Aeronautical Decision Making (ADM) process involves

A— taking a self-assessment hazardous attitude inventory test.
B— understanding the drive to have the "right stuff."
C— obtaining proper flight instruction and experience during training.

Each ADM student should take the Self-Assessment Hazardous Attitude Inventory Test in order to gain a realistic perspective on his/her attitudes toward flying. (L05) — AC 60-22

ALL, MIL
5948. Hazardous attitudes which contribute to poor pilot judgment can be effectively counteracted by

A— early recognition of hazardous thoughts.
B— taking meaningful steps to be more assertive with attitudes.
C— redirecting that hazardous attitude so that appropriate action can be taken.

Pilots should become familiar with a means of counteracting hazardous attitudes with an appropriate antidote thought. (L05) — AC 60-22

ALL, MIL
5949. What are some of the hazardous attitudes dealt with in Aeronautical Decision Making (ADM)?

A— Antiauthority (don't tell me), impulsivity (do something quickly without thinking), macho (I can do it).
B— Risk management, stress management, and risk elements.
C— Poor decision making, situational awareness, and judgment.

ADM addresses the following five hazardous attitudes: Antiauthority (don't tell me!), Impulsivity (do something quickly!), Invulnerability (it won't happen to me), Macho (I can do it), Resignation (what's the use?). (L05) — AC 60-22

ALL, MIL
5950. When a pilot recognizes a hazardous thought, he or she then should correct it by stating the corresponding antidote. Which of the following is the antidote for MACHO?

A— Follow the rules. They are usually right.
B— Not so fast. Think first.
C— Taking chances is foolish.

Macho (I can do it). Pilots who are always trying to prove that they are better than anyone else are thinking "I can do it — I'll show them." Pilots with this type of attitude will try to prove themselves by taking risks in order to impress others. While this pattern is thought to be a male characteristic, women are equally susceptible. The antidote for this attitude is: taking chances is foolish. (L05) — AC 60-22

Answer (A) is incorrect because this is the antidote for an anti-authority attitude. Answer (B) is incorrect because this is the antidote for an impulsivity attitude.

ALL, MIL
5951. What is the first step in neutralizing a hazardous attitude in the ADM process?

A— Recognition of invulnerability in the situation.
B— Dealing with improper judgment.
C— Recognition of hazardous thoughts.

Hazardous attitudes which contribute to poor pilot judgment can be effectively counteracted by redirecting that hazardous attitude so that appropriate action can be taken. Recognition of hazardous thoughts is the first step in neutralizing them in the ADM process. (L05) — AC 60-22

Answers
5947 [A]   5948 [C]   5949 [A]   5950 [C]   5951 [C]

Chapter 5 Procedures and Airport Operations

ALL, MIL
5952. What should a pilot do when recognizing a thought as hazardous?

A—Avoid developing this hazardous thought.
B—Develop this hazardous thought and follow through with modified action.
C—Label that thought as hazardous, then correct that thought by stating the corresponding learned antidote.

When a pilot recognizes a thought as hazardous, the pilot should label that thought as hazardous, then correct that thought by stating the corresponding antidote. (L05) — AC 60-22

ALL, MIL
5953. To help manage cockpit stress, pilots must

A—be aware of life stress situations that are similar to those in flying.
B—condition themselves to relax and think rationally when stress appears.
C—avoid situations that will degrade their abilities to handle cockpit responsibilities.

Good cockpit stress management begins with good life stress management. Many of the stress coping techniques practiced for life stress management are not usually practical in flight. Rather, you must condition yourself to relax and think rationally when stress appears. (L05) — AC 60-22

ALL, MIL
5954. What does good cockpit stress management begin with?

A—Knowing what causes stress.
B—Eliminating life and cockpit stress issues.
C—Good life stress management.

If you hope to succeed at reducing stress associated with crisis management in the air or with your job, it is essential to begin by making a personal assessment of stress in all areas of your life. (L05) — AC 60-22

ALL, MIL
5955. The passengers for a charter flight have arrived almost an hour late for a flight that requires a reservation. Which of the following alternatives best illustrates the ANTIAUTHORITY reaction?

A—Those reservation rules do not apply to this flight.
B—If the pilot hurries, he or she may still make it on time.
C—The pilot can't help it that the passengers are late.

The antiauthority attitude is found in people who do not like anyone telling them what to do. In a sense, they are saying "no one can tell me what to do." They may be resentful of having someone tell them what to do or may regard rules, regulations, and procedures as silly or unnecessary. (L05) — AC 60-22

ALL, MIL
5956. While conducting an operational check of the cabin pressurization system, the pilot discovers that the rate control feature is inoperative. He knows that he can manually control the cabin pressure, so he elects to disregard the discrepancy and departs on his trip. He will just handle the system himself. Which of the following alternatives best illustrates the INVULNERABILITY reaction?

A—What is the worst that could happen.
B—He can handle a little problem like this.
C—It's too late to fix it now.

The invulnerability attitude is found in people who feel accidents happen to others, but never to them. They know accidents can happen, and they know that anyone can be affected, but they never really feel or believe that they will be personally involved. Pilots who think this way are more likely to take chances and increase risk. (L05) — AC 60-22

ALL, MIL
5957. The pilot and passengers are anxious to get to their destination for a business presentation. Level IV thunderstorms are reported to be in a line across their intended route of flight. Which of the following alternatives best illustrates the IMPULSIVITY reaction?

A—They want to hurry and get going, before things get worse.
B—A thunderstorm won't stop them.
C—They can't change the weather, so they might as well go.

Impulsivity is the attitude of people who frequently feel the need to do something, anything, immediately. They do not stop to think about what they are about to do, they do not select the best alternative, and they do the first thing that comes to mind. (L05) — AC 60-22

Answers

5952 [C]   5953 [B]   5954 [C]   5955 [A]   5956 [A]   5957 [A]

ALL, MIL
5958. While on an IFR flight, a pilot emerges from a cloud to find himself within 300 feet of a helicopter. Which of the following alternatives best illustrates the "MACHO" reaction?

A—He is not too concerned; everything will be alright.
B—He flies a little closer, just to show him.
C—He quickly turns away and dives, to avoid collision.

The Macho attitude is found in people who are always trying to prove they are better than anyone else. They are always thinking "I can do it, I'll show them." Pilots with this type of attitude will try to prove themselves by taking risks in order to impress others. (L05) — AC 60-22

ALL, MIL
5959. When a pilot recognizes a hazardous thought, he or she then should correct it by stating the corresponding antidote. Which of the following is the antidote for ANTIAUTHORITY?

A—Not so fast. Think first.
B—It won't happen to me. It could happen to me.
C—Don't tell me. Follow the rules. They are usually right.

The antiauthority (don't tell me!) attitude is found in people who do not like anyone telling them what to do. The antidote for this attitude is: follow the rules, they are usually right. (L05) — AC 60-22

Answer (A) is incorrect because this is the antidote for the impulsivity attitude. Answer (B) is incorrect because this is the antidote for the invulnerability attitude.

ALL, MIL
5960. A pilot and friends are going to fly to an out-of-town football game. When the passengers arrive, the pilot determines that they will be over the maximum gross weight for takeoff with the existing fuel load. Which of the following alternatives best illustrates the RESIGNATION reaction?

A—Well, nobody told him about the extra weight.
B—Weight and balance is a formality forced on pilots by the FAA.
C—He can't wait around to de-fuel, they have to get there on time.

The resignation attitude is found in pilots who think, "what's the use?" They do not see themselves as being able to make a great deal of difference in what happens to them. When things go well, the pilot is apt to think that's good luck. When things go badly, the pilot may feel someone is out to get them, or attribute it to bad luck. The pilot will leave the action to others, for better or worse. Sometimes, such pilots will even go along with unreasonable requests just to be a "nice guy." (L05) — AC 60-22

ALL, MIL
5961. Which of the following is the final step of the Decide Model for effective risk management and Aeronautical Decision Making?

A—Estimate.
B—Evaluate.
C—Eliminate.

The DECIDE Model, comprised of a six-step process, is intended to provide the pilot with a logical way of approaching decision making: Detect, Estimate, Choose, Identify, Do, and Evaluate. (L05) — AC 60-22

ALL, MIL
5962. Which of the following is the first step of the Decide Model for effective risk management and Aeronautical Decision Making?

A—Detect.
B—Identify.
C—Evaluate.

The DECIDE Model, comprised of a six-step process, is intended to provide the pilot with a logical way of approaching decision making: Detect, Estimate, Choose, Identify, Do, and Evaluate. (L05) — AC 60-22

ALL, MIL
5963. The Decide Model is comprised of a 6-step process to provide a pilot a logical way of approaching Aeronautical Decision Making. These steps are:

A—Detect, estimate, choose, identify, do, and evaluate.
B—Determine, evaluate, choose, identify, do, and eliminate.
C—Determine, eliminate, choose, identify, detect, and evaluate.

The DECIDE Model, comprised of a six step process, is intended to provide the pilot with a logical way of approaching decision making: Detect, Estimate, Choose, Identify, Do, and Evaluate. (L05) — AC 60-22

Answers

| 5958 | [B] | 5959 | [C] | 5960 | [A] | 5961 | [B] | 5962 | [A] | 5963 | [A] |

Chapter 6
Weather

The Earth's Atmosphere   *6-3*

Temperature   *6-6*

Wind   *6-7*

Moisture   *6-9*

Stable and Unstable Air   *6-10*

Clouds   *6-11*

Air Masses and Fronts   *6-16*

Turbulence   *6-17*

Icing   *6-18*

Thunderstorms   *6-20*

Common IFR Producers   *6-23*

Wind Shear   *6-25*

Soaring Weather   *6-28*

The Earth's Atmosphere

The major source of all weather is the sun. Every physical process of weather is accompanied by, or is a result of, unequal heating (heat exchange) of the Earth's surface. The heating of the Earth (and therefore the heating of the air in contact with the Earth) is unequal around the entire planet. Either north, south, east or west of a point directly under the sun, one square foot of sunrays is not concentrated over one square foot of the surface, but over a larger area. This lower concentration of sunrays produces less radiation (absorption) of heat over a given surface area and therefore, less atmospheric heating takes place in that area.

The unequal heating of the Earth's atmosphere creates a large air-cell circulation pattern (wind) because the warmer air has a tendency to rise (low pressure) and the colder air has a tendency to settle or descend (high pressure) and replace the rising warmer air. This unequal heating, which causes pressure variations, will also cause variations in altimeter settings between weather reporting points.

Because the Earth rotates, this large, simple air-cell circulation pattern is greatly distorted by a phenomenon known as the **Coriolis force**. When the wind, which is created by pressure differences, horizontal pressure gradient, and high pressure trying to flow into low pressure, first begins to move at higher altitudes, the Coriolis force deflects it to the right (in the Northern Hemisphere). This causes it to flow parallel to the isobars (lines of equal pressure). The Coriolis force prevents air from flowing directly from high-pressure areas to low-pressure areas because it tends to counterbalance the horizontal pressure gradient. These deflections of the large-cell circulation pattern create general wind patterns as depicted in Figure 6-1.

The **jet stream** is a river of high-speed winds (by definition, 50 knots or more) associated with a layer of atmosphere called the tropopause. The tropopause is actually the boundary layer between the troposphere and the stratosphere. Within the troposphere, temperature decreases with altitude, while the stratosphere is characterized by relatively small temperature changes. The tropopause itself is found between the two layers and is marked by an abrupt change in the temperature lapse rate.

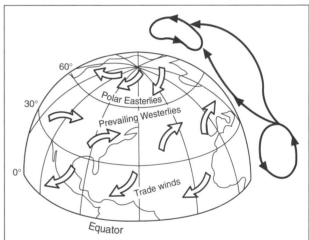

Figure 6-1. Prevailing wind systems

The troposphere varies in height from around 65,000 feet at the equator to about 20,000 feet over the poles, averaging about 37,000 feet in the mid-latitudes. It also is higher in the summer than in the winter. The height of the tropopause does not change uniformly, but rather tends to change in "steps." The jet stream is often found at or near these steps. Since the tropopause height also changes with the seasons, the location of the jet stream changes seasonally. In the winter, the jet stream moves south and increases in speed, and during the summer, the jet stream moves north and decreases in speed.

The strong winds of the jet stream create narrow zones of wind shear which often generate hazardous turbulence. The jet stream maximum is not constant; rather, it is broken into segments, shaped something like a boomerang. Jet stream segments move with pressure ridges and troughs in the upper atmosphere.

A common location of **clear air turbulence (CAT)** and strong wind shear exists with a curving jet stream. This curve is created by an upper or lower low-pressure trough. The wind speed, shown by isotachs (lines of constant wind speed), decreases outward from the jet core. The greatest rate of decrease

Continued

of wind speed is on the polar side as compared to the equatorial side. Strong wind shear and CAT can be expected on the low-pressure side or polar side of a jet stream where the speed at the core is greater than 110 knots. Air travels in a "corkscrew" path around the jet core with upward motion on the equatorial side. When high-level moisture is present, cirriform (cirrus) clouds may be visible, identifying the jet stream along with its associated turbulence.

ALL
5301. Every physical process of weather is accompanied by or is the result of

A—a heat exchange.
B—the movement of air.
C—a pressure differential.

The amount of solar energy received by any region varies with time of day, with seasons and with latitude. These differences in solar energy create temperature variation. Temperatures also vary with differences in topographical surface and with altitude. This temperature variation, or heat exchange, creates forces that drive the atmosphere in its endless motion. (I21) — AC 00-6A, Chapter 2

Answers (B) and (C) are incorrect because the movement of air and pressure differentials are caused by heat exchanges.

ALL
5310. What causes wind?

A—The Earth's rotation.
B—Air mass modification.
C—Pressure differences.

Differences in temperature create differences in pressure. These pressure differences drive a complex system of winds in a never-ending attempt to reach equilibrium. (I23) — AC 00-6A, Chapter 4

ALL
5311. In the Northern Hemisphere, the wind is deflected to the

A—right by Coriolis force.
B—right by surface friction.
C—left by Coriolis force.

The Coriolis force deflects air to the right in the Northern Hemisphere. (I23) — AC 00-6A, Chapter 4

ALL
5312. Why does the wind have a tendency to flow parallel to the isobars above the friction level?

A—Coriolis force tends to counterbalance the horizontal pressure gradient.
B—Coriolis force acts perpendicular to a line connecting the highs and lows.
C—Friction of the air with the Earth deflects the air perpendicular to the pressure gradient.

The pressure gradient force drives the wind and is perpendicular to isobars. When a pressure gradient force is first established, wind begins to blow from higher to lower pressure directly across the isobars. However, the instant air begins moving, Coriolis force deflects it to the right. Soon the wind is deflected a full 90° and is parallel to the isobars or contours. At this time, Coriolis force exactly balances pressure gradient force. With the forces in balance, wind will remain parallel to isobars or contours. (I23) — AC 00-6A, Chapter 4

ALL
5315. What prevents air from flowing directly from high-pressure areas to low-pressure areas?

A—Coriolis force.
B—Surface friction.
C—Pressure gradient force.

The pressure gradient force drives the wind and is perpendicular to isobars. When a pressure gradient force is first established, wind begins to blow from higher to lower pressure directly across the isobars. However, the instant air begins moving, Coriolis force deflects it to the right. Soon the wind is deflected a full 90° and is parallel to the isobars or contours. At this time, Coriolis force exactly balances pressure gradient force. With the forces in balance, wind will remain parallel to isobars or contours. (I23) — AC 00-6A, Chapter 4

Answer (B) is incorrect because surface friction moves air from highs to lows by decreasing wind speed, which decreases the effect of the Coriolis force. Answer (C) is incorrect because the pressure gradient force causes the initial movement from high-pressure areas to low-pressure areas.

Answers

| 5301 | [A] | 5310 | [C] | 5311 | [A] | 5312 | [A] | 5315 | [A] |

ALL
5356. Convective currents are most active on warm summer afternoons when winds are

A—light.
B—moderate.
C—strong.

Convective currents are most active on warm summer afternoons when winds are light. Heated air at the surface creates a shallow, unstable layer and the warm air is forced upward. Convection increases in strength and to greater heights as surface heating increases. (I28) — AC 00-6A, Chapter 9

Answers (B) and (C) are incorrect because moderate and strong winds disrupt the vertical movement of convective currents.

ALL
5381. Which feature is associated with the tropopause?

A—Constant height above the Earth.
B—Abrupt change in temperature lapse rate.
C—Absolute upper limit of cloud formation.

As the thin boundary layer between the troposphere and the stratosphere, the tropopause signals an abrupt change in temperature lapse rate. (I32) — AC 00-6A, Chapter 13

Answer (A) is incorrect because the tropopause is farther away from the Earth's surface at the equator than the poles. Answer (C) is incorrect because clouds may form above the tropopause.

ALL
5382. A common location of clear air turbulence is

A—in an upper trough on the polar side of a jet stream.
B—near a ridge aloft on the equatorial side of a high-pressure flow.
C—south of an east/west oriented high-pressure ridge in its dissipating stage.

Clear air turbulence (CAT) is greatest near the wind speed maxima, usually on the polar sides where there is a combination of strong wind shear, curvature in the flow, and cold air advection associated with sharply curved contours of strong lows, troughs and ridges aloft. A frequent location of CAT is in an upper trough on the cold, or polar side of the jet stream. (I32) — AC 00-6A, Chapter 13

ALL
5383. The jet stream and associated clear air turbulence can sometimes be visually identified in flight by

A—dust or haze at flight level.
B—long streaks of cirrus clouds.
C—a constant outside air temperature.

Long streaks of cirrus clouds can sometimes help the pilot to visually identify the jet stream and associated clear air turbulence (CAT). (I32) — AC 00-6A, Chapter 13

Answer (A) is incorrect because dust or haze indicates there is not enough wind or air movement to dissipate the particles. Answer (C) is incorrect because CAT is caused by mixing different air temperatures at different pressure levels.

ALL
5384. During the winter months in the middle latitudes, the jet stream shifts toward the

A—north and speed decreases.
B—south and speed increases.
C—north and speed increases.

In middle latitudes, the wind speed of the jet stream averages considerably higher in the winter months as it shifts farther south. (I32) — AC 00-6A, Chapter 13

ALL
5385. The strength and location of the jet stream is normally

A—weaker and farther north in the summer.
B—stronger and farther north in the winter.
C—stronger and farther north in the summer.

The jet stream is considerably weaker in the middle latitudes during the summer months, and is further north than in the winter. (I32) — AC 00-6A, Chapter 13

Answers

5356 [A]   5381 [B]   5382 [A]   5383 [B]   5384 [B]   5385 [A]

Chapter 6 Weather

ALL
5447. Which type of jetstream can be expected to cause the greater turbulence?

A—A straight jetstream associated with a low-pressure trough.
B—A curving jetstream associated with a deep low-pressure trough.
C—A jetstream occurring during the summer at the lower latitudes.

Curving jet streams, especially those which curve around a deep pressure trough, are more apt to have turbulent edges than straight jet streams. (K02) — AC 00-30

Answer (A) is incorrect because a curving jet stream is stronger than a straight jet stream. Answer (C) is incorrect because the jet stream is weaker in the summer.

Temperature

In aviation, temperature is measured in degrees Celsius (°C). The standard temperature at sea level is 15°C (59°F). The average decrease in temperature with altitude (standard lapse rate) is 2°C (3.5°F) per 1,000 feet. Since this is an average, the exact value seldom exists; in fact, temperature sometimes increases with altitude—this is known as an inversion. The most frequent type of ground- or surface-based temperature inversion is one that is produced on clear, cool nights, with calm or light wind. *See* Figure 6-2.

ALL
5304. Which conditions are favorable for the formation of a surface based temperature inversion?

A—Clear, cool nights with calm or light wind.
B—Area of unstable air rapidly transferring heat from the surface.
C—Broad areas of cumulus clouds with smooth, level bases at the same altitude.

An inversion often develops near the ground on clear, cool nights when the wind is light. (I21) — AC 00-6A, Chapter 2

Answer (B) is incorrect because the air near the surface must be stable to permit the cool ground to lower the temperature of the surrounding air. Answer (C) is incorrect because cumulus clouds are well above the surface.

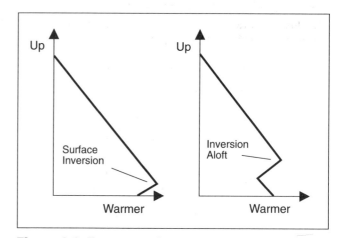

Figure 6-2. Temperature inversions

Answers
5447 [B]   5304 [A]

Wind

The circulation patterns for high- and low-pressure areas are caused by the Coriolis force.

The general circulation and wind rules in the Northern Hemisphere are:

- Air circulates in a clockwise direction around a high.
- Air circulates in a counterclockwise direction around a low.
- The closer the isobars are together, the greater the pressure gradient force, and the stronger the wind speed.
- Due to surface friction (up to about 2,000 feet AGL), surface winds do not exactly parallel the isobars, but move outward from the center of the high toward lower pressure. *See* Figure 6-3.

For preflight planning, it is useful to know that air flows out and downward (or descends) from a high-pressure area in a clockwise direction and flows upward (rises) and into a low-pressure area in a counterclockwise direction. Assume a flight from point A to point B as shown in Figure 6-4. Going direct would involve fighting the wind flowing around the low. However, by traveling south of the low-pressure area, the circulation pattern could help instead of hinder. Generally speaking, in the Northern Hemisphere, when traveling west to east, the most favorable winds can be found by flying north of high-pressure areas and south of low-pressure areas. Conversely, when flying east to west, the most favorable winds can be found south of high-pressure areas and north of low-pressure areas. If flying directly into a low-pressure area in the Northern Hemisphere, the wind direction and speed will be from the left and increasing.

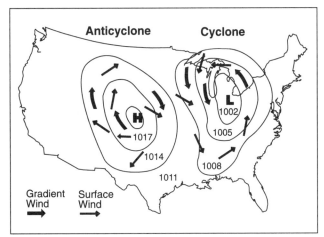

Figure 6-3. Gradient and surface wind

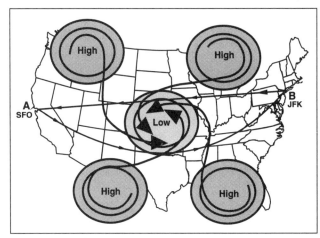

Figure 6-4. Circulation and wind

5313. The wind system associated with a low-pressure area in the Northern Hemisphere is

A—an anticyclone and is caused by descending cold air.
B—a cyclone and is caused by Coriolis force.
C—an anticyclone and is caused by Coriolis force.

The storms that develop between high-pressure systems are characterized by low pressure. As winds try to blow inward toward the center of low pressure, they are also deflected to the right. Thus, the wind around a low moves in a counterclockwise direction. The low pressure and its wind system is a cyclone. (I23) — AC 00-6A, Chapter 4

Answers (A) and (C) are incorrect because they describe a high-pressure system.

Answers

5313 [B]

Chapter 6 Weather

ALL
5314. With regard to windflow patterns shown on surface analysis charts; when the isobars are

A—close together, the pressure gradient force is slight and wind velocities are weaker.
B—not close together, the pressure gradient force is greater and wind velocities are stronger.
C—close together, the pressure gradient force is greater and wind velocities are stronger.

The closer the spacing of isobars, the stronger is the pressure gradient force. The stronger the pressure gradient force, the stronger is the wind. Thus, closely spaced isobars mean strong winds; widely spaced isobars mean lighter wind. (I23) — AC 00-6A, Chapter 4

ALL
5316. While flying cross-country, in the Northern Hemisphere, you experience a continuous left crosswind which is associated with a major wind system. This indicates that you

A—are flying toward an area of generally unfavorable weather conditions.
B—have flown from an area of unfavorable weather conditions.
C—cannot determine weather conditions without knowing pressure changes.

When flying in the Northern Hemisphere experiencing a continuous left crosswind indicates that you are entering a low-pressure system. Wind blows counterclockwise around a low which accounts for the left crosswind. In general, a low-pressure system is associated with bad weather. (I23) — AC 00-6A, Chapter 4

Answer (B) is incorrect because if you have flown from an area of unfavorable weather conditions, you are flying out of the low, which means you would have a right crosswind. Answer (C) is incorrect because the wind can provide an indication of pressure changes and weather.

ALL
5317. Which is true with respect to a high- or low-pressure system?

A—A high-pressure area or ridge is an area of rising air.
B—A low-pressure area or trough is an area of descending air.
C—A high-pressure area or ridge is an area of descending air.

Air moving out of a high or ridge depletes the quantity of air. Highs and ridges, therefore, are areas of descending air. (I23) — AC 00-6A, Chapter 4

Answer (A) is incorrect because high-pressure air descends. Answer (B) is incorrect because low-pressure air rises.

ALL
5318. Which is true regarding high- or low-pressure systems?

A—A high-pressure area or ridge is an area of rising air.
B—A low-pressure area or trough is an area of rising air.
C—Both high- and low-pressure areas are characterized by descending air.

At the surface when air converges into a low, it cannot go outward against the pressure gradient, nor can it go downward into the ground. It must go upward. Therefore, a low or trough is an area of rising air. (I23) — AC 00-6A, Chapter 4

Answers (A) and (C) are incorrect because high-pressure air descends and low-pressure air rises.

ALL
5319. When flying into a low-pressure area in the Northern Hemisphere, the wind direction and velocity will be from the

A—left and decreasing.
B—left and increasing.
C—right and decreasing.

In the Northern Hemisphere the wind around a low is counterclockwise. Thus, when flying to the center of a low, the wind will be from the left. When flying into a pressure system, spacing between isobars will decrease with increasing wind velocity. (I23) — AC 00-6A, Chapter 4

ALL
5321. The general circulation of air associated with a high-pressure area in the Northern Hemisphere is

A—outward, downward, and clockwise.
B—outward, upward, and clockwise.
C—inward, downward, and clockwise.

As the air tries to blow outward from the high pressure, it is deflected to the right by the Coriolis force. Thus, the wind around a high blows clockwise. Air moving out of a high depletes the quantity of air. Highs and ridges are areas of descending air. (I24) — AC 00-6A, Chapter 5

Answers
5314 [C] 5316 [A] 5317 [C] 5318 [B] 5319 [B] 5321 [A]

Moisture

Air has invisible water vapor in it. The water vapor content or air can be expressed in two different ways—relative humidity and dew point.

Relative humidity relates the actual water vapor present in the air to that which could be present in the air. Temperature largely determines the maximum amount of water vapor air can hold. Warm air can hold more water vapor than cold air. *See* Figure 6-5. Air with 100% relative humidity is said to be saturated, and air with less than 100% is unsaturated.

Dew point is the temperature to which air must be cooled to become saturated by the water already present in the air. *See* Figure 6-6.

Water vapor can be added to the air by either evaporation or sublimation. Water vapor is removed from the air by either condensation or sublimation.

When water vapor condenses on large objects, such as leaves, windshields, or airplanes, it will form dew, and when it condenses on microscopic particles (condensation nuclei), such as salt, dust, or combustion by-products, it will form clouds or fog.

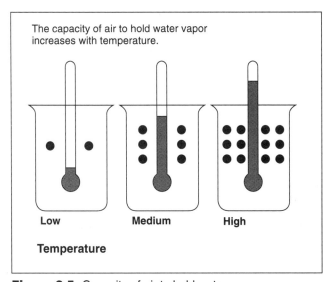

Figure 6-5. Capacity of air to hold water

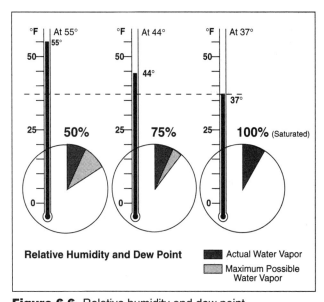

Figure 6-6. Relative humidity and dew point

If the temperature and dewpoint spread is small and decreasing, condensation is about to occur. If the temperature is above freezing, the weather most likely to develop will be fog or low clouds.

To summarize, relative humidity can be increased either by lowering the air temperature or by increasing the amount of water vapor in the air. This causes a decreased air temperature and temperature/dewpoint spread as the relative humidity increases.

Chapter 6 Weather

ALL
5320. Which is true regarding actual air temperature and dewpoint temperature spread? The temperature spread

A—decreases as the relative humidity decreases.
B—decreases as the relative humidity increases.
C—increases as the relative humidity increases.

The difference between air temperature and dewpoint temperature is called the "spread." As the spread becomes less, relative humidity increases. (I24) — AC 00-6A, Chapter 5

ALL
5323. Moisture is added to a parcel of air by

A—sublimation and condensation.
B—evaporation and condensation.
C—evaporation and sublimation.

Evaporation is the changing of liquid water to invisible water vapor. Sublimation is the changing of ice directly to water vapor. (I24) — AC 00-6A, Chapter 5

Answers (A) and (B) are incorrect because condensation removes moisture from the air.

Stable and Unstable Air

Atmospheric stability is defined as the resistance of the atmosphere to vertical motion. A stable atmosphere resists an upward or downward movement. An unstable atmosphere allows an upward or downward disturbance to grow into a vertical (convective) current.

Determining the stability of the atmosphere requires measuring the difference between the actual existing (ambient) temperature lapse rate of a given parcel of air and the dry adiabatic rate (a constant 3°C per 1,000 feet lapse rate).

A stable layer of air would be associated with a temperature inversion. Warming from below, on the other hand, would decrease the stability of an air mass.

The conditions shown in Figure 6-7 are characteristic of stable or unstable air masses.

Unstable Air	Stable Air
Cumuliform clouds	Stratiform clouds and fog
Showery precipitation	Continuous precipitation
Rough air (turbulence)	Smooth air
Good visibility except in blowing obstructions	Fair to poor visibility in haze and smoke

Figure 6-7. Characteristics of air masses

ALL
5333. Which would decrease the stability of an air mass?

A—Warming from below.
B—Cooling from below.
C—Decrease in water vapor.

A change in ambient temperature lapse rate of an air mass will determine its stability. Surface heating or cooling aloft can make the air more unstable. (I25) — AC 00-6A, Chapter 6

Answer (B) is incorrect because cooling from below increases stability. Answer (C) is incorrect because a decrease in water vapor lowers the dew point of the air, but does not affect stability.

Answers
5320 [B]   5323 [C]   5333 [A]

ALL
5336. Which would increase the stability of an air mass?

A—Warming from below.
B—Cooling from below.
C—Decrease in water vapor.

A change in ambient temperature lapse rate of an air mass will determine its stability. Surface cooling or warming aloft often tips the balance toward greater stability. (I25) — AC 00-6A, Chapter 6

Answer (A) is incorrect because warming from below decreases stability. Answer (C) is incorrect because a decrease in water vapor lowers the dew point of the air, but does not affect stability.

ALL
5334. From which measurement of the atmosphere can stability be determined?

A—Atmospheric pressure.
B—The ambient lapse rate.
C—The dry adiabatic lapse rate.

A change in ambient temperature lapse rate of an air mass will determine its stability. Surface heating or cooling aloft can make the air more unstable. On the other hand, surface cooling or warming aloft often tips the balance toward greater stability. (I25) — AC 00-6A, Chapter 6

Answer (A) is incorrect because atmospheric pressure affects temperature and air movements, but does not determine the stability of the atmosphere. Answer (C) is incorrect because the dry adiabatic lapse rate is a constant rate.

Clouds

Stability determines which of two types of clouds will be formed: cumuliform or stratiform.

Cumuliform clouds are the billowy-type clouds having considerable vertical development, which enhances the growth rate of precipitation. They are formed in unstable conditions, and they produce showery precipitation made up of large water droplets. *See* Figure 6-8.

Stratiform clouds are the flat, more evenly based clouds formed in stable conditions. They produce steady, continuous light rain and drizzle made up of much smaller raindrops. *See* Figure 6-9.

Steady precipitation (in contrast to showery) preceding a front is an indication of stratiform clouds with little or no turbulence.

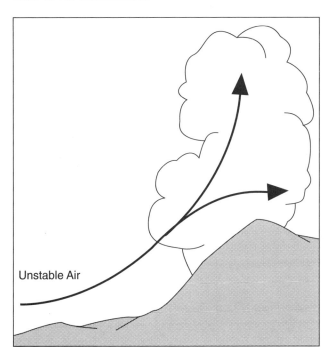

Figure 6-8. Cumulus clouds

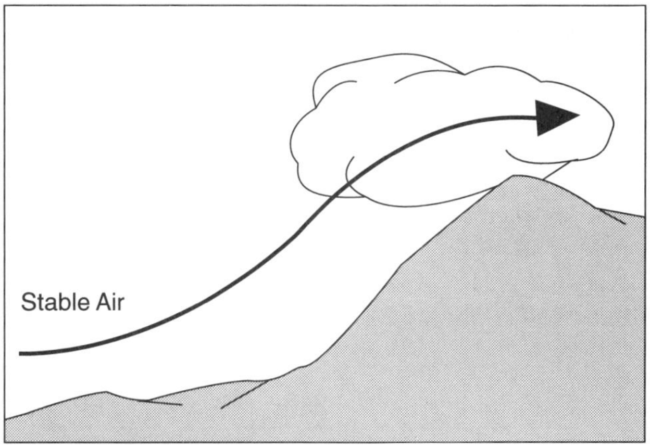

Figure 6-9. Stratiform clouds

Answers
5336 [B]   5334 [B]

Chapter 6 **Weather**

Clouds are divided into four families according to their height range: low, middle, high, and clouds with extensive vertical development. *See* Figure 6-10.

The first three families—low, middle, and high—are further classified according to the way they are formed. Clouds formed by vertical currents (unstable) are cumulus (heap) and are billowy in appearance. Clouds formed by the cooling of a stable layer are stratus (layered) and are flat and sheet-like in appearance. A further classification is the prefix "nimbo-" or suffix "-nimbus," which means raincloud. High clouds, called cirrus, are composed mainly of ice crystals; therefore, they are least likely to contribute to structural icing (since it requires water droplets).

The base of a cloud (AGL) that is formed by vertical currents (cumuliform clouds) can be roughly calculated by dividing the difference between the surface temperature and dew point by 4.4 and multiplying the remainder by 1,000. The convergence of the temperature and the dewpoint lapse rate is 4.4°F per 1,000 feet.

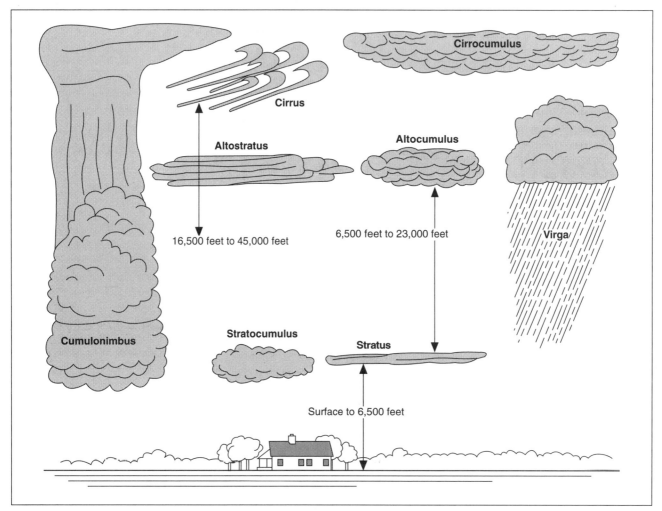

Figure 6-10. Cloud families

Chapter 6 Weather

Problem:

What is the approximate base of the cumulus clouds if the surface temperature at 3,000 feet MSL is 75°F and the dew point is 66°F?

Solution:

1.
     75°F   Temperature
   − 66°F   Dew point
      9°F   Difference

2. 9°F ÷ 4.4 = 2.05 or 2

3. 2 × 1,000 = 2,000 feet AGL (base of cloud)

4.
     2,000   feet AGL (base of cloud)
   + 3,000   feet MSL (altitude of temperature and dew point)
     5,000   feet MSL (base of cloud)

ALL
5330. What determines the structure or type of clouds which will form as a result of air being forced to ascend?

A—The method by which the air is lifted.
B—The stability of the air before lifting occurs.
C—The relative humidity of the air after lifting occurs.

Whether the air is stable or unstable within a layer largely determines cloud structure. When stable air is forced upward the air tends to retain horizontal flow and any cloudiness is flat and stratified. When unstable air is forced upward, the disturbance grows and any resulting cloudiness shows extensive vertical development. (I25) — AC 00-6A, Chapter 6

Answer (A) is incorrect because the stability determines the type of clouds that form. Answer (C) is incorrect because the relative humidity determines the amount of clouds that form.

ALL
5340. The formation of either predominantly stratiform or predominantly cumuliform clouds is dependent upon the

A—source of lift.
B—stability of the air being lifted.
C—temperature of the air being lifted.

When stable air is forced upward, the air tends to retain horizontal flow. Any cloudiness is flat and stratified. When unstable air is forced upward, the disturbance grows, and any resulting cloudiness shows extensive vertical development. (I27) — AC 00-6A, Chapter 8

Answer (A) is incorrect because the stability of the air determines the type of clouds that form. Answer (C) is incorrect because the temperature of the air determines the altitude of clouds that form.

ALL
5327. When conditionally unstable air with high-moisture content and very warm surface temperature is forecast, one can expect what type of weather?

A—Strong updrafts and stratonimbus clouds.
B—Restricted visibility near the surface over a large area.
C—Strong updrafts and cumulonimbus clouds.

Characteristics of unstable air include cumuliform clouds, showery precipitation, turbulence, and good visibility, except in blowing obstructions. (I25) — AC 00-6A, Chapter 6

Answer (A) is incorrect because stratonimbus clouds are characteristic of stable air. Answer (B) is incorrect because restricted visibility is characteristic of stable air.

ALL
5329. If clouds form as a result of very stable, moist air being forced to ascend a mountain slope, the clouds will be

A—cirrus type with no vertical development or turbulence.
B—cumulus type with considerable vertical development and turbulence.
C—stratus type with little vertical development and little or no turbulence.

Stable air resists upward movement; therefore, stratified clouds are produced. (I25) — AC 00-6A, Chapter 6

Answer (A) is incorrect because cirrus clouds are high and composed of ice crystals. Answer (B) is incorrect because unstable air causes vertical development.

Answers

5330 [B]   5340 [B]   5327 [C]   5329 [C]

ALL
5335. What type weather can one expect from moist, unstable air, and very warm surface temperatures?

A—Fog and low stratus clouds.
B—Continuous heavy precipitation.
C—Strong updrafts and cumulonimbus clouds.

Characteristics of unstable air include cumuliform clouds, showery precipitation, turbulence, and good visibility, except in blowing obstructions. (I25) — AC 00-6A, Chapter 6

Answer (A) is incorrect because fog and stratus clouds are characteristics of stable air. Answer (B) is incorrect because continuous precipitation is characteristic of stable air.

ALL
5337. The conditions necessary for the formation of stratiform clouds are a lifting action and

A—unstable, dry air.
B—stable, moist air.
C—unstable, moist air.

Stable, moist air and adiabatic cooling is necessary to form stratiform clouds. (I26) — AC 00-6A, Chapter 7

ALL
5338. Which cloud types would indicate convective turbulence?

A—Cirrus clouds.
B—Nimbostratus clouds.
C—Towering cumulus clouds.

Billowy fair weather cumulus clouds, usually seen on sunny afternoons, are signposts in the sky indicating convective turbulence. Vertical heights range from the shallow fair weather cumulus to the giant thunderstorm cumulonimbus. (I26) — AC 00-6A, Chapter 7

Answer (A) is incorrect because cirrus clouds are high clouds made of ice crystals, and are not generated by any convective activity. Answer (B) is incorrect because nimbostratus clouds are flat rain clouds, formed in stable air and do not produce convective activity or turbulence.

ALL
5341. Which combination of weather-producing variables would likely result in cumuliform-type clouds, good visibility, and showery rain?

A—Stable, moist air and orographic lifting.
B—Unstable, moist air and orographic lifting.
C—Unstable, moist air and no lifting mechanism.

Characteristics of unstable, moist air include cumuliform clouds, showery precipitation, turbulence, and good visibility, except in blowing obstructions. "Orographic lifting" is the lifting action produced by a physical object, such as a mountain slope, forcing air upward. (I27) — AC 00-6A, Chapter 8

Answer (A) is incorrect because if the air is stable, steady precipitation and stratiform clouds will form. Answer (C) is incorrect because a lifting mechanism must exist to form cumuliform clouds and showery rain.

ALL
5332. What are the characteristics of stable air?

A—Good visibility; steady precipitation; stratus clouds.
B—Poor visibility; steady precipitation; stratus clouds.
C—Poor visibility; intermittent precipitation; cumulus clouds.

Characteristics of stable air include stratiform clouds and fog, continuous precipitation, smooth air, and fair to poor visibility in haze and smoke. (I25) — AC 00-6A, Chapter 6

Answer (A) is incorrect because good visibility is characteristic of unstable air. Answer (C) is incorrect because intermittent precipitation and cumulus clouds are characteristic of unstable air.

ALL
5342. What is a characteristic of stable air?

A—Stratiform clouds.
B—Fair weather cumulus clouds.
C—Temperature decreases rapidly with altitude.

Characteristics of stable air include stratiform clouds and fog, continuous precipitation, smooth air, and fair to poor visibility in haze and smoke. (I27) — AC 00-6A, Chapter 8

Answer (B) is incorrect because cumulus clouds are characteristic of unstable air. Answer (C) is incorrect because a rapid temperature decrease with altitude indicates a high lapse rate and is characteristic of unstable air.

Answers

5335 [C]   5337 [B]   5338 [C]   5341 [B]   5332 [B]   5342 [A]

ALL
5343. A moist, unstable air mass is characterized by

A—poor visibility and smooth air.
B—cumuliform clouds and showery precipitation.
C—stratiform clouds and continuous precipitation.

Characteristics of unstable air include cumuliform clouds, showery precipitation, turbulence, and good visibility, except in blowing obstructions. (I27) — AC 00-6A, Chapter 8

Answer (A) is incorrect because poor visibility and smooth air are characteristic of stable air. Answer (C) is incorrect because stratiform clouds and continuous precipitation are characteristic of stable air.

ALL
5344. When an air mass is stable, which of these conditions are most likely to exist?

A—Numerous towering cumulus and cumulonimbus clouds.
B—Moderate to severe turbulence at the lower levels.
C—Smoke, dust, haze, etc., concentrated at the lower levels with resulting poor visibility.

Characteristics typical of a stable air mass are:

- *Stratiform clouds and fog*
- *Continuous precipitation*
- *Smooth air*
- *Fair to poor visibility in haze and smoke*

(I27) — AC 00-6A, Chapter 8

Answers (A) and (B) are incorrect because towering cumulus, cumulonimbus clouds, and turbulence are characteristic of an unstable air mass.

ALL
5345. Which is a characteristic of stable air?

A—Cumuliform clouds.
B—Excellent visibility.
C—Restricted visibility.

Characteristics typical of a stable air mass are:

- *Stratiform clouds and fog*
- *Continuous precipitation*
- *Smooth air*
- *Fair to poor visibility in haze and smoke*

(I27) — AC 00-6A, Chapter 8

Answers (A) and (B) are incorrect because cumuliform clouds and excellent visibility are characteristic of an unstable air mass.

ALL
5346. Which is a characteristic typical of a stable air mass?

A—Cumuliform clouds.
B—Showery precipitation.
C—Continuous precipitation.

Characteristics typical of a stable air mass are:

- *Stratiform clouds and fog*
- *Continuous precipitation*
- *Smooth air*
- *Fair to poor visibility in haze and smoke*

(I27) — AC 00-6A, Chapter 8

Answers (A) and (B) are incorrect because cumuliform clouds and showery precipitation are characteristic of an unstable air mass.

ALL
5348. Which are characteristics of a cold air mass moving over a warm surface?

A—Cumuliform clouds, turbulence, and poor visibility.
B—Cumuliform clouds, turbulence, and good visibility.
C—Stratiform clouds, smooth air, and poor visibility.

Cool air moving over a warm surface is heated from below, generating instability and increasing the possibility of showers. Unstable air is characterized by cumuliform clouds, turbulence and good visibility. (I27) — AC 00-6A, Chapter 8

ALL
5349. The conditions necessary for the formation of cumulonimbus clouds are a lifting action and

A—unstable, dry air.
B—stable, moist air.
C—unstable, moist air.

For cumulonimbus clouds to form, the air must have sufficient water vapor, an unstable lapse rate, and an initial upward boost (lifting) to start the storm process in motion. (I27) — AC 00-6A, Chapter 8

Answers

5343 [B]   5344 [C]   5345 [C]   5346 [C]   5348 [B]   5349 [C]

ALL
5328. What is the approximate base of the cumulus clouds if the temperature at 2,000 feet MSL is 10°C and the dewpoint is 1°C?

A—3,000 feet MSL.
B—4,000 feet MSL.
C—6,000 feet MSL.

In a convection current, the temperature and dew point converge at about 2.5°C per 1,000 feet. An estimate of convective cloud bases can be found by dividing the convergence into the temperature spread.

1. $(10 - 1) \div 2.5 = 3.6 \times 1,000 = 3,600$ feet base
2.
   ```
      2,000   feet MSL
   + 3,600   feet AGL
     5,600   feet MSL
   ```

(I25) — AC 00-6A, Chapter 6

ALL
5331. Refer to the excerpt from the following METAR report:

KTUS.....08004KT 4SM HZ .....26/04 A2995 RMK RAE36

At approximately what altitude AGL should bases of convective-type cumuliform clouds be expected?

A—4,400 feet.
B—8,800 feet.
C—17,600 feet.

The reported temperature is 26°C, and the dew point is 4°C. In a convection current, the temperature and dew point converge at about 4.4°F (2.5°C) per 1,000 feet. An estimate of convective cloud bases can be found by dividing the convergence into the temperature spread.

$(26 - 4) \div 2.5 = 8.8 \times 1,000 = 8,800$ feet base

(I25) — AC 00-6A, Chapter 6

Air Masses and Fronts

When a body of air comes to rest on, or moves slowly over, an extensive area having fairly uniform properties of temperature and moisture, the air takes on these properties. The area from which the air mass acquires its identifying distribution of temperature and moisture is its "source region." As this air mass moves from its source region, it tends to take on the properties of the new underlying surface. The trend toward change is called air mass modification.

A **ridge** is an elongated area of high pressure. A **trough** is an elongated area of low pressure. All fronts lie in troughs. A **cold front** is the leading edge of an advancing cold air mass. A **warm front** is the leading edge of an advancing warm air mass. Warm fronts move about half as fast as cold fronts. Frontal waves and cyclones (areas of low pressure) usually form on slow-moving cold fronts or stationary fronts. Figure 6-11 shows the symbols that would appear on a weather map.

The physical manifestations of a warm or cold front can be different with each front. They vary with the speed of the air mass on the move and the degree of stability of the air mass being overtaken. A stable air mass forced aloft will continue to exhibit stable characteristics, while an unstable air mass forced to ascend will continue to be characterized by cumulus clouds, turbulence, showery precipitation, and good visibility.

Occlusions form because cold fronts move faster than warm fronts. In a cold front occlusion, the coldest air is under the cold front. When it overtakes the warm front, it lifts the warm front aloft and the cold air replaces cool air at the surface.

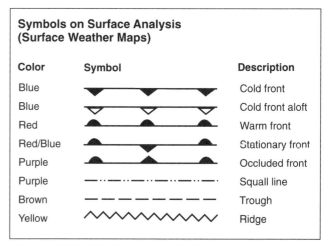

Figure 6-11. Weather map symbols

Answers
5328   [C]   5331   [B]

Frontal passage will be indicated by the following discontinuities:

1. A temperature change (the most easily recognizable discontinuity);
2. A continuous decrease in pressure followed by an increase as the front passes; and
3. A shift in the wind direction, speed, or both.

ALL
5347. Which is true regarding a cold front occlusion? The air ahead of the warm front

A—is colder than the air behind the overtaking cold front.
B—is warmer than the air behind the overtaking cold front.
C—has the same temperature as the air behind the overtaking cold front.

In the cold front occlusion, the coldest air is under the cold front. When it overtakes the warm front, it lifts the warm front aloft and cold air replaces cool air at the surface. (I27) — AC 00-6A, Chapter 8

Turbulence

Cumulus clouds are formed by convective currents (heating from below). Therefore, a pilot can expect turbulence below or inside cumulus clouds, especially towering cumulus clouds. The greatest turbulence could be expected inside cumulonimbus clouds. Strong winds (35+ knots) across ridges and mountain ranges can also cause severe turbulence and severe downdrafts on the lee side. The greatest potential danger from turbulent air currents exists when flying into the wind while on the leeward side of ridges and mountain ranges. *See* Figure 6-12.

Winds blowing across a mountain may produce an almond- or lens-shaped cloud (lenticular cloud), which appears stationary, but which may contain winds of 50 knots or more. The presence of these clouds is an indication of very strong turbulence. The stationary crests of standing mountain waves downwind of a mountain also resemble the almond or lens shape and are referred to as standing lenticular clouds. Favorable conditions for a strong mountain wave consist of a stable layer of air being disturbed by the mountains with winds of at least 20 knots across the ridge. One of the most dangerous features of mountain waves is the turbulent areas in and below rotor clouds that form under lenticular clouds.

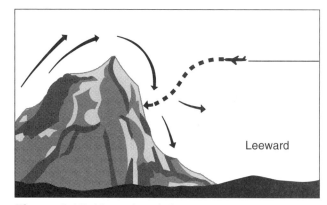

Figure 6-12. Mountain turbulence

ALL
5339. The presence of standing lenticular altocumulus clouds is a good indication of

A—lenticular ice formation in calm air.
B—very strong turbulence.
C—heavy icing conditions.

Standing lenticular and/or rotor clouds suggest a mountain wave; expect turbulence many miles to the lee of mountains. (I26) — AC 00-6A, Chapter 7

Answers
5347 [B]   5339 [B]

Chapter 6 **Weather**

ALL
5357. When flying low over hilly terrain, ridges, or mountain ranges, the greatest potential danger from turbulent air currents will usually be encountered on the

A—leeward side when flying with a tailwind.
B—leeward side when flying into the wind.
C—windward side when flying into the wind.

Dangerous downdrafts may be encountered on the lee side. (I28) — AC 00-6A, Chapter 9

Answer (A) is incorrect because with a tailwind you would be flying away from the mountain with the wind. Answer (C) is incorrect because you would be flying in air that is rising up on the windward side.

ALL
5393. The conditions most favorable to wave formation over mountainous areas are a layer of

A—stable air at mountaintop altitude and a wind of at least 20 knots blowing across the ridge.
B—unstable air at mountaintop altitude and a wind of at least 20 knots blowing across the ridge.
C—moist, unstable air at mountaintop altitude and a wind of less than 5 knots blowing across the ridge.

A strong mountain wave requires:

1. Marked stability in the airstream disturbed by the mountains;

2. Wind speed at the level of the summit should exceed a minimum which varies from 15 to 25 knots, depending on the height of the range; and

3. Wind direction within 30° to the range. Lift diminishes as winds more closely parallel the range.

(I35) — AC 00-6A, Chapter 16

ALL
5450. One of the most dangerous features of mountain waves is the turbulent areas in and

A—below rotor clouds.
B—above rotor clouds.
C—below lenticular clouds.

"Rotor clouds" appear to remain stationary, parallel the range, and stand a few miles leeward of the mountains. Turbulence is most frequent and most severe in and below the standing rotors just beneath the wave crests at or below mountain-top levels. (N33) — Soaring Flight Manual

Icing

Structural icing occurs on an aircraft whenever supercooled droplets of water make contact with any part of the aircraft that is also at a temperature below freezing. An in-flight condition necessary for structural icing to form is visible moisture (clouds or raindrops).

Icing in precipitation (rain) is of concern to the VFR pilot because it can occur outside of clouds. Aircraft structural ice will most likely have the highest accumulation in freezing rain, which indicates warmer temperature (more than 32°F) at a higher altitude. *See* Figures 6-13 and 6-14. But the air temperature at the point where freezing precipitation is encountered is 32°F or less, causing the supercooled droplet to freeze on impact with the aircraft's surface.

If rain falling through colder air freezes during descent, **ice pellets** form. The presence of ice pellets at the surface is evidence that there is freezing rain at a higher altitude, while wet snow indicates that the temperature at your altitude is above freezing.

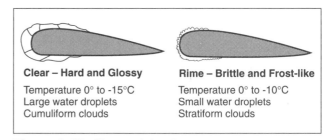

Figure 6-13. Clear and rime ice

Clear – Hard and Glossy
Temperature 0° to -15°C
Large water droplets
Cumuliform clouds

Rime – Brittle and Frost-like
Temperature 0° to -10°C
Small water droplets
Stratiform clouds

Chances for structural icing increase in the vicinity of fronts.

Frost is described as ice deposits formed by sublimation on a surface when the temperature of the collecting surface is at or below the dew point of the adjacent air, and the dew point is below freezing. Frost causes early airflow separation on an airfoil that results in a loss of lift, causing the airplane to stall at an angle of attack lower than normal. Therefore, all frost should be removed from the lifting surfaces of an airplane before flight, or it may prevent the airplane from becoming airborne.

Answers

5357 [B]   5393 [A]   5450 [A]

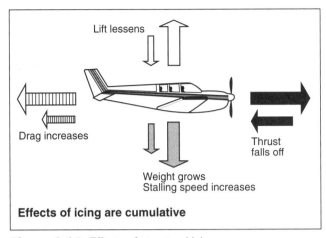

Figure 6-14. Effects of structural icing

ALL
5324. Ice pellets encountered during flight normally are evidence that

A—a warm front has passed.
B—a warm front is about to pass.
C—there are thunderstorms in the area.

Rain falling from warm air above through colder air below may freeze during its descent, falling as ice pellets. This can happen any time a warmer layer of air exists above a colder layer (i.e., a warm front or a cold front). (I24) — AC 00-6A, Chapter 5

Answer (A) is incorrect because after the warm front has passed there will no longer be a layer of warm air above a layer of cold air, which is required for the formation of ice pellets. Answer (C) is incorrect because ice pellets do not necessarily come from thunderstorms, but from rain freezing at a higher altitude.

ALL
5325. What is indicated if ice pellets are encountered at 8,000 feet?

A—Freezing rain at higher altitude.
B—You are approaching an area of thunderstorms.
C—You will encounter hail if you continue your flight.

Rain falling from warm air above through colder air below may freeze during its decent, falling as ice pellets. This can happen any time a warmer layer of air exists above a colder layer (i.e., a warm front or a cold front). (I24) — AC 00-6A, Chapter 5

Answer (B) is incorrect because freezing rain can be encountered without thunderstorms. Answer (C) is incorrect because ice pellets are a form of hail.

ALL
5326. Ice pellets encountered during flight are normally evidence that

A—a cold front has passed.
B—there are thunderstorms in the area.
C—freezing rain exists at higher altitudes.

Rain falling from warm air above through colder air below may freeze during its decent, falling as ice pellets. This can happen any time a warmer layer of air exists above a colder layer (i.e., a warm front or a cold front). (I24) — AC 00-6A, Chapter 5

ALL
5360. Which situation would most likely result in freezing precipitation? Rain falling from air which has a temperature of

A—32°F or less into air having a temperature of more than 32°F.
B—0°C or less into air having a temperature of 0°C or more.
C—more than 32°F into air having a temperature of 32°F or less.

Rain falling through colder air may become supercooled, freezing on impact as freezing rain, or it may freeze during its descent, falling as ice pellets. Water can freeze at 0°C or 32°F. (I29) — AC 00-6A, Chapter 10

AIR
5739. Frost covering the upper surface of an airplane wing usually will cause

A—the airplane to stall at an angle of attack that is higher than normal.
B—the airplane to stall at an angle of attack that is lower than normal.
C—drag factors so large that sufficient speed cannot be obtained for takeoff.

The frost on the wing causes airflow disturbances. This will cause airflow separation (stall) at a lower angle of attack, resulting in a tendency to stall during takeoff. (I29) — AC 00-6A, Chapter 10

Answer (A) is incorrect because frost on the wing surface will usually cause the airplane to stall at a lower angle of attack. Answer (C) is incorrect because the drag will usually not be enough to prevent the aircraft from obtaining takeoff speed.

Answers

5324 [B]   5325 [A]   5326 [C]   5360 [C]   5739 [B]

Thunderstorms

Thunderstorms present many hazards to flying. Three conditions necessary to the formation of a thunderstorm are:

- Sufficient water vapor
- An unstable lapse rate
- An initial upward boost (lifting)

The initial upward boost can be caused by heating from below, frontal lifting, or by mechanical lifting (wind blowing air upslope on a mountain). There are three stages of a thunderstorm: the cumulus, mature, and dissipating stages. *See* Figure 6-15.

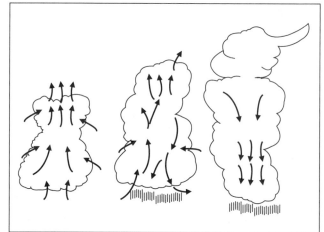

Figure 6-15. Stages of a thunderstorm

The cumulus stage consists of continuous updrafts, and these updrafts create low-pressure areas. Thunderstorms reach their greatest intensity during the mature stage, which is characterized by updrafts and downdrafts inside the cloud. Precipitation inside the cloud assists the development of these downdrafts, and the start of rain at the Earth's surface signals the beginning of the mature stage. Precipitation that evaporates before it reaches the ground is called virga.

When **lightning** occurs, the cloud is classified as a thunderstorm. Very frequent lightning, cumulonimbus clouds, and roll clouds indicate extreme turbulence in a thunderstorm. The dissipating stage of a thunderstorm features mainly downdrafts. Lightning is always associated with a thunderstorm.

Hail is formed inside thunderstorms (or cumulonimbus clouds) by the constant freezing, melting, and refreezing of water as it is carried about by the up- and downdrafts. Hailstones may be thrown outward from a storm cloud for several miles.

A pilot should always expect the hazardous and invisible atmospheric phenomena called **wind shear turbulence** when operating anywhere near a thunderstorm (within 20 NM). Wind shear is thought to be the most hazardous condition associated with a thunderstorm.

Thunderstorms that generally produce the most intense hazard to aircraft are called **squall-line thunderstorms**. These non-frontal, narrow bands of thunderstorms often contain severe steady-state thunderstorms that develop ahead of a cold front. The intense hazards found in these storms include destructive winds, heavy hail, and tornadoes. Embedded thunderstorms are those that are obscured by massive cloud layers and cannot be seen visually.

Airborne weather avoidance radar detects only precipitation drops; it does not detect minute cloud droplets. Therefore, the radar scope provides no assurance of avoiding instrument weather in clouds and fog. Weather radar precisely measures rainfall density which can be related to turbulence associated with the radar echoes. The most intense echoes are severe thunderstorms, and should be avoided by at least 20 miles. You should avoid flying between these intense echoes unless they are separated by at least 40 miles.

ALL
5322. Virga is best described as

A—streamers of precipitation trailing beneath clouds which evaporates before reaching the ground.
B—wall cloud torrents trailing beneath cumulonimbus clouds which dissipate before reaching the ground.
C—turbulent areas beneath cumulonimbus clouds.

"Virga" refers to the streamers of precipitation trailing beneath clouds that evaporate before reaching the ground. (I24) — AC 00-6A, Chapter 5

Answer (B) is incorrect because virga is usually thin and wispy. Answer (C) is incorrect because virga is a form of precipitation.

ALL
5361. Which statement is true concerning the hazards of hail?

A—Hail damage in horizontal flight is minimal due to the vertical movement of hail in the clouds.
B—Rain at the surface is a reliable indication of no hail aloft.
C—Hailstones may be encountered in clear air several miles from a thunderstorm.

Hailstones can fall some distance from the storm core. Hail has been observed in clear air several miles from the parent thunderstorm. (I30) — AC 00-6A, Chapter 11

Answer (A) is incorrect because hail is one of the greatest hazards to aircraft. Answer (B) is incorrect because rain at the surface does not mean the absence of hail aloft.

ALL
5362. Hail is most likely to be associated with

A—cumulus clouds.
B—cumulonimbus clouds.
C—stratocumulus clouds.

You should anticipate possible hail with any thunderstorm, especially beneath the anvil of a large cumulonimbus. (I30) — AC 00-6A, Chapter 11

ALL
5363. The most severe weather conditions, such as destructive winds, heavy hail, and tornadoes, are generally associated with

A—slow-moving warm fronts which slope above the tropopause.
B—squall lines.
C—fast-moving occluded fronts.

A squall line is a non-frontal, narrow band of active thunderstorms. It often contains severe steady-state thunderstorms and presents the single most intense weather hazard to aircraft. (I30) — AC 00-6A, Chapter 11

Answer (A) is incorrect because warm fronts do not usually produce severe weather. Answer (C) is incorrect because the weather produced by occluded fronts is not as severe as a squall line.

ALL
5364. Of the following, which is accurate regarding turbulence associated with thunderstorms?

A—Outside the cloud, shear turbulence can be encountered 50 miles laterally from a severe storm.
B—Shear turbulence is encountered only inside cumulonimbus clouds or within a 5-mile radius of them.
C—Outside the cloud, shear turbulence can be encountered 20 miles laterally from a severe storm.

Hazardous turbulence is present in all thunderstorms, and a severe thunderstorm can damage an airframe. Strongest turbulence within the clouds occurs with shear between updrafts and downdrafts. Outside the cloud, shear turbulence has been encountered several thousand feet above and 20 miles laterally from a severe storm. (I30) — AC 00-6A, Chapter 11

ALL
5365. If airborne radar is indicating an extremely intense thunderstorm echo, this thunderstorm should be avoided by a distance of at least

A—20 miles.
B—10 miles.
C—5 miles.

If the use of airborne radar indicates extremely intense echoes, they should be avoided by at least 20 miles. (I30) — AC 00-6A, Chapter 11

ALL
5366. Which statement is true regarding squall lines?

A—They are always associated with cold fronts.
B—They are slow in forming, but rapid in movement.
C—They are nonfrontal and often contain severe, steady-state thunderstorms.

A squall line is a non-frontal, narrow band of active thunderstorms. It often contains severe steady-state thunderstorms and presents the single most intense weather hazard to aircraft. (I30) — AC 00-6A, Chapter 11

Answer (A) is incorrect because squall lines can form in any area of unstable air, but usually are found ahead of cold fronts. Answer (B) is incorrect because squall lines usually form quickly.

Answers

5322 [A]	5361 [C]	5362 [B]	5363 [B]	5364 [C]	5365 [A]
5366 [C]					

Chapter 6 Weather

ALL
5367. Which statement is true concerning squall lines?

A—They form slowly, but move rapidly.
B—They are associated with frontal systems only.
C—They offer the most intense weather hazards to aircraft.

A squall line is a non-frontal, narrow band of active thunderstorms. It often contains severe steady-state thunderstorms and presents the single most intense weather hazard to aircraft. (I30) — AC 00-6A, Chapter 11

Answer (A) is incorrect because squall lines usually form rapidly. Answer (B) is incorrect because squall lines can form in any area of unstable air, but usually are found ahead of cold fronts.

ALL
5368. Select the true statement pertaining to the life cycle of a thunderstorm.

A—Updrafts continue to develop throughout the dissipating stage of a thunderstorm.
B—The beginning of rain at the Earth's surface indicates the mature stage of the thunderstorm.
C—The beginning of rain at the Earth's surface indicates the dissipating stage of the thunderstorm.

The mature stage of a thunderstorm starts when precipitation begins to fall from the cloud base. The downdrafts reach speeds that may exceed 2,500 feet per minute. Meanwhile, updrafts reach a maximum with speeds possibly exceeding 6,000 feet per minute. Updrafts and downdrafts in close proximity create strong vertical shear and a very turbulent environment. (I30) — AC 00-6A, Chapter 11

Answer (A) is incorrect because updrafts do not continue through the dissipating stage of a thunderstorm. Answer (C) is incorrect because this indicates the beginning of the mature stage.

ALL
5369. What visible signs indicate extreme turbulence in thunderstorms?

A—Base of the clouds near the surface, heavy rain, and hail.
B—Low ceiling and visibility, hail, and precipitation static.
C—Cumulonimbus clouds, very frequent lightning, and roll clouds.

Cumulonimbus clouds represent an unstable air mass which indicates turbulent conditions. The more frequent the lightning, the more severe the thunderstorm. The roll cloud is most prevalent with cold frontal or squall line thunderstorms and signifies an extremely turbulent zone. (I30) — AC 00-6A, Chapter 11

ALL
5370. Which weather phenomenon signals the beginning of the mature stage of a thunderstorm?

A—The start of rain.
B—The appearance of an anvil top.
C—Growth rate of cloud is maximum.

The mature stage of a thunderstorm starts when precipitation begins to fall from the cloud base. The downdrafts reach speeds that may exceed 2,500 feet per minute. Meanwhile, updrafts reach a maximum with speeds possibly exceeding 6,000 feet per minute. (I30) — AC 00-6A, Chapter 11

Answer (B) is incorrect because the anvil top appears during the mature stage, but not necessarily at the beginning. Answer (C) is incorrect because maximum cloud growth rate occurs in the middle to the end of the mature stage.

ALL
5371. What feature is normally associated with the cumulus stage of a thunderstorm?

A—Roll cloud.
B—Continuous updraft.
C—Beginning of rain at the surface.

The key feature of the cumulus stage of a thunderstorm is a continuous updraft. (I30) — AC 00-6A, Chapter 11

Answers (A) and (C) are incorrect because the roll cloud and the beginning of rain at the surface are features of the mature stage.

ALL
5372. During the life cycle of a thunderstorm, which stage is characterized predominately by downdrafts?

A—Mature.
B—Developing.
C—Dissipating.

Downdrafts characterize the dissipating stage of the thunderstorm cell. (I30) — AC 00-6A, Chapter 11

Answer (A) is incorrect because the mature stage has both updrafts and downdrafts. Answer (B) is incorrect because the developing stage primarily has updrafts.

Answers

| 5367 | [C] | 5368 | [B] | 5369 | [C] | 5370 | [A] | 5371 | [B] | 5372 | [C] |

ALL
5373. What minimum distance should exist between intense radar echoes before any attempt is made to fly between these thunderstorms?

A—20 miles.
B—30 miles.
C—40 miles.

A pilot should avoid flying between very intense echoes unless they are separated by at least 40 miles. (I30) — AC 00-6A, Chapter 11

ALL
5375. Which is true regarding the use of airborne weather-avoidance radar for the recognition of certain weather conditions?

A—The radarscope provides no assurance of avoiding instrument weather conditions.
B—The avoidance of hail is assured when flying between and just clear of the most intense echoes.
C—The clear area between intense echoes indicates that visual sighting of storms can be maintained when flying between the echoes.

Weather radar detects only precipitation drops. It does not detect minute cloud droplets. Therefore, the radar scope provides no assurance of avoiding instrument weather in clouds and fog. (I31) — AC 00-6A, Chapter 12

Answer (B) is incorrect because hail can be thrown several miles from the intense echoes. Answer (C) is incorrect because clouds without precipitation may exist between the echoes.

Common IFR Producers

Fog is a surface-based cloud (restricting visibility) composed of either water droplets or ice crystals. Fog may form by cooling the air to its dew point or by adding moisture to the air near the ground. A small temperature/dewpoint spread is essential to the formation of fog. An abundance of condensation nuclei from combustion products makes fog prevalent in industrial areas.

Fog is classified by the way it is formed:

Radiation fog (ground fog) is formed when terrestrial radiation cools the ground (land areas only), which in turn cools the air in contact with it. When the air is cooled to its dew point, or within a few degrees, fog will form. This fog will form most readily in warm, moist air over low, flatland areas on clear, calm (no wind) nights.

Advection fog (sea fog) is formed when warm, moist air moves (wind is required) over colder ground or water; for example, an air mass moving inland from the coast in winter. Advection fog is usually more extensive and much more persistent than radiation fog. It can move in rapidly regardless of the time of day or night. This fog deepens as wind speed increases up to about 15 knots. Winds much stronger than 15 knots lift the fog into a layer of low stratus clouds.

Upslope fog is formed when moist, stable air is cooled to its dew point as it moves up sloping terrain (wind is required). Cooling will be at the dry adiabatic lapse rate of approximately 3°C per 1,000 feet.

Precipitation-induced fog (frontal fog) is formed when relatively warm rain or drizzle falls through cool air; evaporation from the precipitation saturates the cool air and forms fog. It is most commonly associated with warm fronts, but can occur with slow moving cold fronts and with stationary fronts.

Steam fog forms in winter when cold, dry air passes from land areas over comparatively warm ocean waters. Condensation takes place just above the surface of the water and appears as "steam" rising from the ocean.

Answers
5373 [C]   5375 [A]

Chapter 6 Weather

ALL
5350. Fog produced by frontal activity is a result of saturation due to

A—nocturnal cooling.
B—adiabatic cooling.
C—evaporation of precipitation.

When relatively warm rain or drizzle falls through cool air, evaporation from the precipitation saturates the cool air and forms fog. (I27) — AC 00-6A, Chapter 8

Answer (A) is incorrect because nocturnal cooling produces radiation fog. Answer (B) is incorrect because adiabatic cooling produces upslope fog.

ALL
5374. Which in-flight hazard is most commonly associated with warm fronts?

A—Advection fog.
B—Radiation fog.
C—Precipitation-induced fog.

When relatively warm rain or drizzle falls through cool air, evaporation from the precipitation saturates the cool air and forms fog. Precipitation-induced fog can become quite dense and continue for an extended period of time. This fog may cover large areas, completely suspending air operations. It is most commonly associated with warm fronts, but can occur with slow moving cold fronts and with stationary fronts. (I31) — AC 00-6A, Chapter 12

Answer (A) is incorrect because advection fog forms from the movement of warm, humid air over a cold water surface. Answer (B) is incorrect because radiation fog forms from terrestrial cooling of the Earth's surface on clear, calm nights.

ALL
5376. A situation most conducive to the formation of advection fog is

A—a light breeze moving colder air over a water surface.
B—an air mass moving inland from the coastline during the winter.
C—a warm, moist air mass settling over a cool surface under no-wind conditions.

Advection fog forms when warm, moist air moves over colder ground or water. The fog forms offshore and is then carried inland by the wind. It is most common along coastal areas but often develops deep in continental areas. (I31) — AC 00-6A, Chapter 12

Answer (A) is incorrect because this describes steam fog. Answer (C) is incorrect because this describes radiation fog.

ALL
5377. Advection fog has drifted over a coastal airport during the day. What may tend to dissipate or lift this fog into low stratus clouds?

A—Nighttime cooling.
B—Surface radiation.
C—Wind 15 knots or stronger.

Advection fog deepens as wind speed increases up to about 15 knots. Winds much stronger than 15 knots lift the fog into a layer of low stratus clouds or stratocumulus. (I31) — AC 00-6A, Chapter 12

Answers (A) and (B) are incorrect because nighttime cooling and surface radiation form radiation fog.

ALL
5378. What lifts advection fog into low stratus clouds?

A—Nighttime cooling.
B—Dryness of the underlying land mass.
C—Surface winds of approximately 15 knots or stronger.

Advection fog deepens as wind speed increases up to about 15 knots. Winds much stronger than 15 knots lift the fog into a layer of low stratus clouds or stratocumulus. (I31) — AC 00-6A, Chapter 12

Answers (A) and (B) are incorrect because nighttime cooling and dryness of the underlying land mass lead to radiation fog.

ALL
5379. In what ways do advection fog, radiation fog, and steam fog differ in their formation or location?

A—Radiation fog is restricted to land areas; advection fog is most common along coastal areas; steam fog forms over a water surface.
B—Advection fog deepens as windspeed increases up to 20 knots; steam fog requires calm or very light wind; radiation fog forms when the ground or water cools the air by radiation.
C—Steam fog forms from moist air moving over a colder surface; advection fog requires cold air over a warmer surface; radiation fog is produced by radiational cooling of the ground.

Radiation fog is restricted to land areas because water surfaces cool little from nighttime radiation. Advection fog is most common along coastal areas but often develops deep in continental areas. Steam fog, also known as "sea smoke," forms in the winter when cold, dry air passes from land areas over comparatively warm ocean waters. (I31) — AC 00-6A, Chapter 12

Answers

| 5350 | [C] | 5374 | [C] | 5376 | [B] | 5377 | [C] | 5378 | [C] | 5379 | [A] |

ALL
5380. With respect to advection fog, which statement is true?

A—It is slow to develop, and dissipates quite rapidly.
B—It forms almost exclusively at night or near daybreak.
C—It can appear suddenly during day or night, and it is more persistent than radiation fog.

Advection fog is more persistent than radiation fog and can move in rapidly regardless of the time of day or night. (I31) — AC 00-6A, Chapter 12

Answer (A) is incorrect because advection fog can move in rapidly regardless of the time of day or night and is persistent. Answer (B) is incorrect because this describes radiation fog.

Wind Shear

Wind shear is defined as a change in wind direction and/or speed in a very short distance in the atmosphere. This can occur at any level of the atmosphere and can exist in both horizontal and vertical direction. The amount of wind shear can be detected by the pilot as a sudden change in airspeed.

Low-level (low-altitude) wind shear can be expected during strong temperature inversions, on all sides of a thunderstorm and directly below the cell. Low-level wind shear can also be found near frontal activity because winds can be significantly different in the two air masses which meet to form the front.

In warm front conditions, the most critical period is before the front passes. Warm front shear may exist below 5,000 feet for about 6 hours before surface passage of the front. The wind shear associated with a warm front is usually more extreme than that found in cold fronts.

The shear associated with cold fronts is usually found behind the front. If the front is moving at 30 knots or more, the shear zone will be 5,000 feet above the surface 3 hours after frontal passage.

Potentially hazardous wind shear may be encountered during periods of a strong temperature inversion with calm or light surface winds, and strong winds above the inversion. Eddies (turbulence) in the shear zone cause airspeed fluctuation as an aircraft climbs or descends through the inversions. During an approach, the most easily recognized means of detecting possible windshear conditions includes monitoring the rate of descent (vertical velocity) and power required. The power needed to hold the glide slope will be different from a no-shear situation.

There are two potentially hazardous shear situations:

1. **Loss of Tailwind**—A tailwind may shear to either a calm or headwind component. In this instance, initially the airspeed will increase by an amount equal to the change of wind velocity, the aircraft pitches up, and altitude increases. Lower than normal power would be required initially, followed by a further decrease as the shear is encountered, and then an increase as glide slope is regained. *See* Figure 6-16.

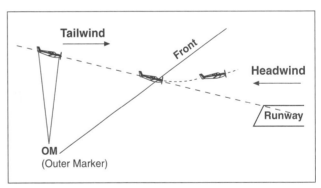

Figure 6-16. Tailwind shearing to headwind or calm

Answers
5380 [C]

2. **Loss of Headwind**—A headwind may shear to a calm or tailwind component. The decrease in headwind will cause a loss in airspeed equal to the decrease in wind velocity. Initially, the airspeed decreases, the aircraft pitches down, and altitude decreases. *See* Figure 6-17.

Some airports can report **boundary winds** as well as the wind at the tower. When a tower reports a boundary wind which is significantly different from the airport wind, there is a possibility of hazardous wind shear.

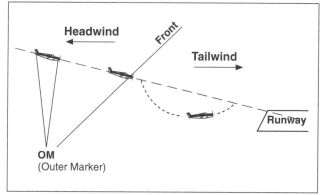

Figure 6-17. Headwind shearing to tailwind or calm

ALL
5351. What is an important characteristic of wind shear?

A—It is present at only lower levels and exists in a horizontal direction.
B—It is present at any level and exists in only a vertical direction.
C—It can be present at any level and can exist in both a horizontal and vertical direction.

Wind shear may be associated with either a wind shift or a wind speed gradient at any level in the atmosphere. It may be associated with a low-level temperature inversion, in a frontal zone, or clear air turbulence (CAT) at high levels associated with a jet stream or strong circulation. (I28) — AC 00-6A, Chapter 9

Answers (A) and (B) are incorrect because wind shear occurs both vertically and horizontally, and at all altitudes.

ALL
5352. Hazardous wind shear is commonly encountered

A—near warm or stationary frontal activity.
B—when the wind velocity is stronger than 35 knots.
C—in areas of temperature inversion and near thunderstorms.

Often there is a strong wind just above the top of an inversion layer. Flying into or out of this wind induces a shear situation. The most prominent meteorological phenomena that cause significant low-level wind shear problems are thunderstorms and certain frontal systems at or near the airport. (I28) — AC 00-6A, Chapter 9

Answer (A) is incorrect because hazardous wind shear is more commonly found near inversions and thunderstorms. Answer (B) is incorrect because strong wind does not mean that there will always be wind shear; the wind must be in different directions.

ALL
5449. The low-level wind shear Alert System (LLWAS) provides wind data and software process to detect the presence of a

A—rotating column of air extending from a cumulonimbus cloud.
B—change in wind direction and/or speed within a very short distance in the atmosphere.
C—downward motion of the air associated with continuous winds blowing with an easterly component due to the rotation of the Earth.

Wind shear may be associated with either a wind shift or a wind speed gradient at any level in the atmosphere. It may be associated with a low-level temperature inversion, in a frontal zone, or clear air turbulence (CAT) at high levels associated with a jet stream or strong circulation. (J25) — AC 00-6A, Chapter 9

Answer (A) is incorrect because this describes a tornado. Answer (C) is incorrect because this wind shear is a change in wind direction and/or speed within a very short distance in the atmosphere.

ALL
5353. Low-level wind shear may occur when

A—surface winds are light and variable.
B—there is a low-level temperature inversion with strong winds above the inversion.
C—surface winds are above 15 knots and there is no change in wind direction and windspeed with height.

When taking off or landing in calm wind under clear skies within a few hours before or after sunrise, be prepared for a temperature inversion near the ground. You can be relatively certain of a shear zone in the inversion if you know the wind at 2,000 to 4,000 feet is 25 knots or more. Allow a margin of airspeed above normal climb or approach speed to alleviate the danger of a stall in event of turbulence or sudden change in wind velocity. (I28) — AC 00-6A, Chapter 9

Answer (A) is incorrect because light surface winds alone would not cause wind shear. Answer (C) is incorrect because wind shear refers to an abrupt change in wind speed and/or direction.

Answers

5351 [C]   5449 [B]   5352 [C]   5353 [B]

ALL
5354. If a temperature inversion is encountered immediately after takeoff or during an approach to a landing, a potential hazard exists due to

A—wind shear.
B—strong surface winds.
C—strong convective currents.

You can be relatively certain of a shear zone in the inversion if you know the wind at 2,000 to 4,000 feet is 25 knots or more. Allow a margin of airspeed above normal climb or approach speed to alleviate the danger of a stall in event of turbulence or sudden change in wind velocity. (I28) — AC 00-6A, Chapter 9

Answer (B) is incorrect because strong surface winds do not present as great a danger as wind shear. Answer (C) is incorrect because a temperature inversion does not generate strong convective currents.

ALL
5355. GIVEN:

Winds at 3,000 feet AGL .... 30 kts
Surface winds .... Calm

While approaching for landing under clear skies a few hours after sunrise, one should

A—increase approach airspeed slightly above normal to avoid stalling.
B—keep the approach airspeed at or slightly below normal to compensate for floating.
C—not alter the approach airspeed, these conditions are nearly ideal.

When taking off or landing in calm wind under clear skies within a few hours before or after sunrise, be prepared for a temperature inversion near the ground. You can be relatively certain of a shear zone in the inversion if you know the wind at 2,000 to 4,000 feet is 25 knots or more. Increase airspeed slightly above normal climb or approach speed to alleviate the danger of a stall in event of turbulence or sudden change in wind velocity. (I28) — AC 00-6A, Chapter 9

Answer (B) is incorrect because the hazard is wind shear. Answer (C) is incorrect because these conditions are not ideal—wind shear may be present.

AIR, RTC
5358. During an approach, the most important and most easily recognized means of being alerted to possible wind shear is monitoring the

A—amount of trim required to relieve control pressures.
B—heading changes necessary to remain on the runway centerline.
C—power and vertical velocity required to remain on the proper glidepath.

Since rate of descent on the glide slope is directly related to ground speed, a high descent rate would indicate a strong tailwind. Conversely, a low descent rate indicates a strong headwind. The power needed to hold the glide slope also will be different from typical, no-shear conditions. Less power than normal will be needed to maintain the glide slope when a tailwind is present and more power is needed for strong headwind. (I28) — AC 00-6A, Chapter 9

Answer (A) is incorrect because trim adjustments are a function of power settings, airspeeds, and aircraft configuration. Answer (B) is incorrect because heading changes are due to the crosswind component.

ALL
5359. During departure, under conditions of suspected low-level wind shear, a sudden decrease in headwind will cause

A—a loss in airspeed equal to the decrease in wind velocity.
B—a gain in airspeed equal to the decrease in wind velocity.
C—no change in airspeed, but groundspeed will decrease.

The worst situation on departure occurs when the aircraft encounters a rapidly increasing tailwind, decreasing headwind, and/or downdraft. Taking off under these circumstances would lead to a decreased performance condition. An increasing tailwind or decreasing headwind, when encountered, will cause a decrease in indicated airspeed. The aircraft will initially pitch down due to the decreased lift in proportion to the airspeed loss. After encountering the shear, if the wind remains constant, aircraft ground speed will gradually increase and indicated airspeed will return to its original value. (I28) — AC 00-6A, Chapter 9

Answer (B) is incorrect because a sudden decrease in headwind will cause a loss in airspeed. Answer (C) is incorrect because there is an initial loss of airspeed, followed by an increase in ground speed.

Answers

5354 [A]   5355 [A]   5358 [C]   5359 [A]

Chapter 6 **Weather**

ALL
5448. A strong wind shear can be expected

A—in the jetstream front above a core having a speed of 60 to 90 knots.
B—if the 5°C isotherms are spaced between 7° to 10° of latitude.
C—on the low-pressure side of a jetstream core where the speed at the core is stronger than 110 knots.

Jet streams stronger than 110 knots (at the core) are apt to have areas of significant turbulence near them in the sloping tropopause above the core, in the jet stream front below the core and on the low-pressure side of the core. In these areas there are frequently strong wind shears. (K02) — AC 00-30

Answer (A) is incorrect because 60 to 90 knots is common for the jet stream and if turbulence were to be found it would be to the sides and bottom of the core. Answer (B) is incorrect because these conditions do not exclusively create wind shear.

Soaring Weather

GLI
5386. Select the true statement concerning thermals.

A—Thermals are unaffected by winds aloft.
B—Strong thermals have proportionately increased sink in the air between them.
C—A thermal invariably remains directly above the surface area from which it developed.

For every rising current there is a compensating downward current. The downward currents frequently occur over broader areas than do the upward currents; therefore, they have a slower vertical speed than do rising currents. A thermal is simply the updraft in a small-scale convective current. (I35) — AC 00-6A, Chapter 16

GLI
5387. A thermal column is rising from an asphalt parking lot and the wind is from the south at 12 knots. Which statement would be true?

A—As altitude is gained, the best lift will be found directly above the parking lot.
B—As altitude is gained, the center of the thermal will be found farther north of the parking lot.
C—The slowest rate of sink would be close to the thermal and the fastest rate of sink farther from it.

Wind causes a thermal to lean with altitude. When seeking the thermal supporting soaring birds or aircraft, you must make allowance for the wind. The thermal at lower levels usually is upwind from your high-level visual cue. (I35) — AC 00-6A, Chapter 16

GLI
5388. Which is true regarding the development of convective circulation?

A—Cool air must sink to force the warm air upward.
B—Warm air is less dense and rises on its own accord.
C—Warmer air covers a larger surface area than the cool air; therefore, the warmer air is less dense and rises.

Warm air does expand when heated, but the convective circulation or lifting force comes from the dense, cool air drawn to the ground by gravity and forcing the warm air upward. (I35) — AC 00-6A, Chapter 16

GLI
5389. Which is generally true when comparing the rate of vertical motion of updrafts with that of downdrafts associated with thermals?

A—Updrafts and downdrafts move vertically at the same rate.
B—Downdrafts have a slower rate of vertical motion than do updrafts.
C—Updrafts have a slower rate of vertical motion than do downdrafts.

For every rising current there is a compensating downward current. The downward currents frequently occur over broader areas than do the upward currents; therefore, they have a slower vertical speed than do rising currents. A thermal is simply the updraft in a small-scale convective current. (I35) — AC 00-6A, Chapter 16

Answers

5448   [C]   5386   [B]   5387   [B]   5388   [A]   5389   [B]

GLI
5390. Which thermal index would predict the best probability of good soaring conditions?

A— -10.
B— -5.
C— +20.

Strength of thermals is proportional to the magnitude of the negative value of the thermal index (TI). A TI of -8 or -10 predicts very good lift and a long soaring day. (I35) — AC 00-6A, Chapter 16

GLI
5391. Which is true regarding the effect of fronts on soaring conditions?

A— A slow moving front provides the strongest lift.
B— Good soaring conditions usually exist after passage of a warm front.
C— Frequently, the air behind a cold front provides excellent soaring for several days.

In the central and eastern United States, the most favorable weather for cross-country soaring occurs behind a cold front.

1. *The cold polar air is usually dry, and thermals can build to relatively high altitudes.*
2. *The polar air is colder than the ground and thus the warm ground aids solar radiation in heating the air. Thermals begin earlier in the morning and last later in the evening. On occasions, soarable lift has been found at night.*
3. *Quite often, colder air at high altitudes moves over the cold, low-level outbreak intensifying the instability and strengthening the thermals.*
4. *The wind profile frequently favors thermal streeting— a real benefit to speed and distance.*

The same four factors may occur with cold frontal passages over mountainous regions in the western United States. (I35) — AC 00-6A, Chapter 16

GLI
5394. When soaring in the vicinity of mountain ranges, the greatest potential danger from vertical and rotor-type currents will usually be encountered on the

A— leeward side when flying with a tailwind.
B— leeward side when flying into the wind.
C— windward side when flying into the wind.

Dangerous downdrafts may be encountered on the lee side. (I35) — AC 00-6A, Chapter 16

GLI
5395. Which is true regarding ridge soaring with the wind direction perpendicular to the ridge?

A— When flying between peaks along a ridge, the pilot can expect a significant decrease in wind and lift.
B— When very close to the surface of the ridge, the glider's speed should be reduced to the minimum sink speed.
C— If the glider drifts downwind from the ridge and sinks slightly lower than the crest of the ridge, the glider should be turned away from the ridge and a high speed attained.

The glider pilot would want to get out of the area of strong sink as rapidly as possible. Prompt and drastic action is required. Considerable altitude may be lost in the process. Occasionally a very strong wind will sweep down the lee side of the ridge. (I35) — AC 00-6A, Chapter 16

GLI
5396. (Refer to Figure 6.) With regard to the soundings taken at 1400 hours, between what altitudes could optimum thermalling be expected at the time of the sounding?

A— From 2,500 to 6,000 feet.
B— From 6,000 to 10,000 feet.
C— From 13,000 to 15,000 feet.

The actual lapse rate must exceed the dry adiabatic rate of cooling for air to be unstable. That is, the line representing the lapse rate must slope parallel to, or slope more than, the dry adiabats. The 1400 GMT sounding slopes more than the adiabats from 2,500 to 6,000 feet, parallels the adiabats from 6,000 to 10,000 feet and slopes less than the adiabats from 10,000 to 13,000 and from 13,000 to 15,000 feet. (I35) — AC 00-6A, Chapter 16

Answers

5390 [A]   5391 [C]   5394 [B]   5395 [C]   5396 [A]

Chapter 6 **Weather**

GLI
5397. (Refer to Figure 6.) With regard to the soundings taken at 0900 hours, from 2,500 feet to 15,000 feet, as shown on the Adiabatic Chart, what minimum surface temperature is required for instability to occur and for good thermals to develop from the surface to 15,000 feet MSL?

A—58°F.
B—68°F.
C—80°F.

The actual lapse rate must exceed the adiabatic rate for good thermals to develop. Find the intersection of 0900 GMT (Greenwich Mean Time) sounding and 15,000 feet. Draw a line parallel to the diagonals (dry adiabatic lapse rate) back to the surface at 2,500 feet MSL. The surface temperature must exceed about 80°F. (I35) — AC 00-6A, Chapter 16

GLI
5437. (Refer to Figure 7.) According to the lifted index and K-index shown on the Stability Chart, which area of the U.S. would have the least satisfactory conditions for thermal soaring on the day of the soundings?

A—Southeastern.
B—North central.
C—Western seaboard.

Lifted index is computed as if a parcel of air near the surface were lifted to 500 millibars. A positive index means very stable air. A negative index means the air is unstable and suggests the possibility of convection. The lifted index is plotted above the line and K index below the line. Largest positive, e.g., highly stable, values of K are seen in the North central area. (I49) — AC 00-45E, Chapter 12

GLI
5742. (Refer to Figure 6.) At the 0900 hours sounding and the line plotted from the surface to 10,000 feet, what temperature must exist at the surface for instability to take place between these altitudes? Any temperature

A—less than 68°F.
B—more than 68°F.
C—less than 43°F.

Note that the actual lapse rate must exceed the adiabatic rate for good thermals to develop.

1. *Locate the intersection of the 0900 GMT sounding and the 10,000-foot altitude.*
2. *Draw a line parallel to the diagonals and downward to the right to intercept the line representing the surface, 2,500 feet MSL.*
3. *From that intercept draw a line downward and read the temperature, 20°C or 68°F. Temperatures greater than this will result in instability.*

(N23) — Soaring Flight Manual

GLI
5743. (Refer to Figure 6.) At the sounding taken at 0900 hours from 2,500 feet to 15,000 feet, what minimum surface temperature is required for instability to occur and for good thermals to develop from the surface to 15,000 feet MSL?

A—58°F.
B—68°F.
C—80°F.

Note that the actual lapse rate must exceed the adiabatic rate for good thermals to develop.

1. *Locate the intersection of the 0900 GMT sounding and the 15,000-foot altitude.*
2. *Draw a line parallel to the diagonals and downward to the right to intercept the line representing the surface, 2,500 feet MSL.*
3. *From that intercept draw a line downward and read the temperature, 27°C or 80°F. Temperatures greater than this will result in instability.*

(N23) — Soaring Flight Manual

GLI
5744. (Refer to Figure 6.) At the soundings taken at 1400 hours, is the atmosphere stable or unstable and at what altitudes?

A—Stable from 6,000 to 10,000 feet.
B—Stable from 10,000 to 13,000 feet.
C—Unstable from 10,000 to 13,000 feet.

If the sounding line is parallel to, or has less slope than the diagonals, the air is stable. (N23) — Soaring Flight Manual

Answers

5397 [C]	5437 [B]	5742 [B]	5743 [C]	5744 [B]

GLI
5745. Which thermal index would predict the best probability of good soaring conditions?

A— +5.
B— -5.
C— -10.

A negative thermal index indicates unstable air. (N23) — Soaring Flight Manual

GLI
5746. Which is true regarding the effect of fronts on soaring conditions?

A—Good soaring conditions usually exist after passage of a warm front.
B—Excellent soaring conditions usually exist in the cold air ahead of a warm front.
C—Frequently the air behind a cold front provides excellent soaring for several days.

Frequently, the unstable air behind a cold front provides good soaring conditions. (N23) — AC 00-6A, Chapter 16

GLI
5747. Which is true regarding ridge soaring with the wind direction perpendicular to the ridge?

A—When very close to the surface of the ridge, the glider's speed should be reduced to the minimum sink speed.
B—When the wind and lift are very strong on the windward side of the ridge, a weak sink condition will exist on the leeward side.
C—If the glider drifts downwind from the ridge and sinks slightly lower than the crest of the ridge, the glider should be turned away from the ridge and a high speed attained.

The glider pilot would want to get out of the area of strong sink as rapidly as possible. (I35) — AC 00-6A, Chapter 16

GLI
5392. Convective circulation patterns associated with sea breezes are caused by

A—water absorbing and radiating heat faster than the land.
B—land absorbing and radiating heat faster than the water.
C—cool and less dense air moving inland from over the water, causing it to rise.

Land is warmer than the sea during the day; wind blows from the cool water to warm land. (I35) — AC 00-6A, Chapter 16

Answer (A) is incorrect because water absorbs and radiates heat slower than land. Answer (C) is incorrect because cool air is more dense, therefore it will sink.

Answers

5745 [C]   5746 [C]   5747 [C]   5392 [B]

Chapter 7
Weather Services

Aviation Routine Weather Report (METAR)   *7–3*

Pilot Report (UA)   *7–5*

Terminal Aerodrome Forecast (TAF)   *7–8*

Aviation Area Forecast (FA)   *7–9*

Winds and Temperatures Aloft Forecast (FD)   *7–10*

Inflight Weather Advisories (WA, WS, WST)   *7–10*

Transcribed Weather Broadcast (TWEB)
and En Route Flight Advisory Service (EFAS)   *7–12*

Radar Weather Report   *7–14*

Surface Analysis Chart   *7–15*

Weather Depiction Chart   *7–16*

Constant Pressure Chart   *7–18*

Tropopause Height/Vertical Wind Shear Prognostic Chart   *7–18*

Significant Weather Prognostics   *7–19*

Composite Moisture Stability Chart   *7–21*

Chapter 7 **Weather Services**

Aviation Routine Weather Report (METAR)

An international weather reporting code is used for weather reports (METAR) and forecasts (TAFs) worldwide. The reports follow the format shown in Figure 7-1.

For aviation purposes, the ceiling is the lowest broken or overcast layer, or vertical visibility into an obscuration.

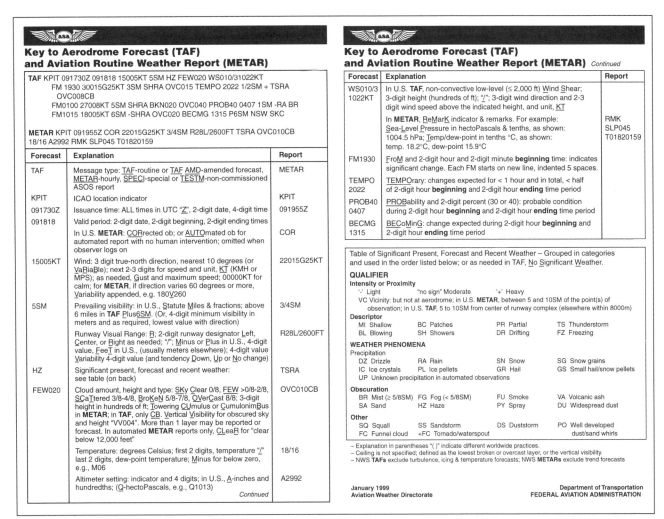

Figure 7-1. TAF/METAR weather card

ALL
5398. During preflight preparation, weather report forecasts which are not routinely available at the local service outlet (AFSS) can best be obtained by means of contacting the

A—weather forecast office (WFD).
B—air route traffic control center.
C—pilot's automatic telephone answering service.

Any reports or forecasts not routinely available at the local service outlet are available through the request/reply service at all FSS's, WSOs and WSFOs. (I40) — AC 00-45E, Chapter 1

Answer (B) is incorrect because ARTCC deals with air traffic control. Answer (C) is incorrect because PATWAS is limited in the number of route forecasts and synopses it can provide.

Answers

5398   [A]

Chapter 7 Weather Services

ALL
5399. The most current en route and destination weather information for an instrument flight should be obtained from the

A—AFSS.
B—ATIS broadcast.
C—Notices to Airman publications.

The FAA Flight Service Station (AFSS) provides more aviation weather briefing service than any other government service outlet. The AFSS provides weather briefings, scheduled and unscheduled weather broadcasts, and furnishes weather support to flight in its area. (I40) — AC 00-45E, Chapter 1

Answer (B) is incorrect because ATIS provides information for operations at a specific airport and does not provide information for enroute operations. Answer (C) is incorrect because NOTAMs is a publication containing information on airport operations, and would not have information for en route and destination weather.

ALL, MIL
5402. The remarks section of the Aviation Routine Weather Report (METAR) contains the following coded information. What does it mean?

RMK FZDZB45 WSHFT 30 FROPA

A—Freezing drizzle with cloud bases below 4,500 feet.
B—Freezing drizzle below 4,500 feet and wind shear
C—Wind shift at three zero due to frontal passage.

RMK—remarks follow

FZDZB45—freezing drizzle began 45 minutes after the hour

WSHFT 30 FROPA—wind shift at three zero due to frontal passage

(J25) — AIM ¶7-1-28

ALL, MIL
5403. What is meant by the Special METAR weather observation for KBOI?

SPECI KBOI 091854Z 32005KT 1 1/2SM RA BR OVC007 17/16 A2990 RMK RAB12

A—Rain and fog obscuring two-tenths of the sky; rain began at 1912Z.
B—Rain and mist obstructing visibility; rain began at 1812Z.
C—Rain and overcast at 1200 feet AGL.

SPECI KBOI—special report for KBOI

091854Z—the date and time the observation is taken is the 9th of the month, 1854 Zulu time

32005KT—winds are 320° true at 5 knots

1-1/2 SM RA BR—visibilities are 1-1/2 statute miles in rain and mist

OVC 007—ceiling is 700 feet overcast

17/16—temperature is 17°C and the dew point is 16°C

A2990—altimeter is 29.90

RMK RAB12—remarks, rain began 12 minutes past the hour

(J25) — AIM ¶7-1-28

ALL, MIL
5404. The station originating the following METAR observation has a field elevation of 3,500 feet MSL. If the sky cover is one continuous layer, what is the thickness of the cloud layer? (Top of overcast reported at 7,500 feet MSL).

METAR KHOB 151250Z 17006KT 4SM OVC005 13/11 A2998

A—2,500 feet.
B—3,500 feet.
C—4,000 feet.

KHOB reports a ceiling of 500 feet (OVC005). This means the bottom of the overcast layer is 4,000 feet (3,500 MSL + 500 feet AGL). The top of the overcast is reported at 7,500 feet MSL. Therefore, the overcast layer is 3,500 feet thick (7,500 – 4,000). (J25) — AIM ¶7-1-28

ALL
5405. What wind conditions would you anticipate when squalls are reported at your destination?

A—Rapid variations in windspeed of 15 knots or more between peaks and lulls.
B—Peak gusts of at least 35 knots combined with a change in wind direction of 30° or more.
C—Sudden increases in windspeed of at least 15 knots to a sustained speed of 20 knots or more for at least 1 minute.

A squall is a sudden increase in wind speed of 15 knots to a sustained speed of 20 knots or more lasting for at least 1 minute. (I41) — AC 00-45E, Chapter 2

Answer (A) is incorrect because this describes gusts. Answer (B) is incorrect because this describes wind shear.

Answers

5399 [A]   5402 [C]   5403 [B]   5404 [B]   5405 [C]

Pilot Report (UA)

Aircraft in flight are the only means of directly observing cloud tops, icing, and turbulence; therefore, no observation is more timely than one made from the cockpit. While the FAA encourages pilots to report inflight weather, a report of any unforecast weather is required by regulation. A Pilot Report, or PIREP (identified by the letters "UA") is usually transmitted in a prescribed format. *See* Figure 7-2.

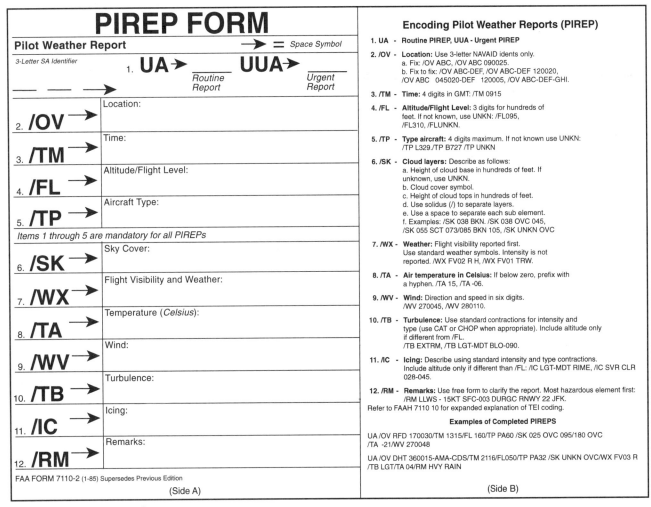

Figure 7-2. Pilot report form

Turbulence and icing should be reported by using the intensity tables shown in Figure 7-3 and Figure 7-4.

ALL, MIL
5406. What significant cloud coverage is reported by this pilot report?

KMOB
UA/OV 15NW MOB 1340Z/SK OVC 025/045 OVC 090

A—Three (3) separate overcast layers exist with bases at 250, 7,500 and 9,000 feet.
B—The top of the lower overcast is 2,500 feet; base and top of second overcast layer is 4,500 and 9,000 feet, respectively.
C—The base of the second overcast layer is 2,500 feet; top of second overcast layer is 7,500 feet; base of third layer is 9,000 feet.

UA—pilot report

/OV 15NW MOB—pilot was 15 miles northwest of Minot, at 1340 UTC time

/SK OVC 025—sky coverage, overcast layer top at 2,500 feet

/045 OVC 090—a second layer is at 4,500 feet with the top at 9,000 feet MSL

(J25) — AIM ¶7-1-19

ALL
5407. To best determine observed weather conditions between weather reporting stations, the pilot should refer to

A—pilot reports.
B—Area Forecasts.
C—prognostic charts.

Pilot weather reports are the best means to determine observed weather conditions between weather-reporting stations. It is also the only method of directly observing cloud tops, icing and turbulence. (I42) — AC 00-45E, Chapter 3

Answers (B) and (C) are incorrect because Area Forecasts and prognostic charts are forecasts and not observed weather.

ALL
5444. A pilot reporting turbulence that momentarily causes slight, erratic changes in altitude and/or attitude should report it as

A—light chop.
B—light turbulence.
C—moderate turbulence.

Icing Intensities		
Intensity	Airframe ice accumulation	Pilot report
Trace	Ice becomes perceptible. Rate of accumulation slightly greater than rate of sublimation. It is not hazardous even though deicing/anti-icing equipment is not used unless encountered for an extended period of time (over one hour).	Aircraft indentification, location, time UTC, intensity and type of icing*, altitude/FL, aircraft type, IAS
Light	The rate of accumulation may create a problem if flight is prolonged in this environment (over one hour). Occasional use of deicing/anti-icing equipment removes/prevents accumulation. It does not present a problem if the deicing/anti-icing equipment is used.	
Moderate	The rate of accumulation is such that even short encounters become potentially hazardous and use of deicing/anti-icing equipment or diversion is necessary.	
Severe	The rate of accumulation is such that deicing/anti-icing equipment fails to reduce or control the hazard. Immediate diversion is necessary.	

*Icing may be rime, clear and mixed.
Rime ice: Rough milky opaque ice formed by the instantaneous freezing of small supercooled water droplets.
Clear ice: A glossy, clear or translucent ice formed by the relatively slow freezing of large supercooled water droplets.
Mixed ice: A combination of rime and clear ice.

Figure 7-3. Icing intensities

Turbulence that momentarily causes slight, erratic changes in altitude and/or attitude (pitch, roll, yaw) is reported as light turbulence. (I53) — AC 00-45E, Chapter 14

Answer (A) is incorrect because light chop is slight, rapid, somewhat rhythmic bumpiness. Answer (C) is incorrect because moderate turbulence causes changes in altitude and/or attitude, and variations in indicated airspeed.

Answers
5406 [B]   5407 [A]   5444 [B]

Chapter 7 Weather Services

Turbulence Intensities			
Intensity	Aircraft reaction	Reaction inside aircraft	Reporting term definition
Light	Turbulence that momentarily causes slight, erratic changes in altitude and/or attitude (pitch, roll, yaw). Report as *Light Turbulence;** or Turbulence that causes slight, rapid and somewhat rhythmic bumpiness without appreciable changes in altitude or attitude. Report as *Light Chop.*	Occupants may feel a slight strain against belts or shoulder straps. Unsecured objects may be displaced slightly. Food service may be conducted and little or no difficulty is encountered in walking.	Occasional – less than 1/3 of the time. Intermittent – 1/3 to 2/3 of the time. Continuous – More than 2/3 of the time.
Moderate	Turbulence that is similar to Light Turbulence but of greater intensity. Changes in altitude and/or attitude occur but the aircraft remains in positive control at all times. It usually causes variations in indicated airspeed. Report as *Moderate Turbulence;** or Turbulence that is similar to Light Chop but of greater intensity. It causes rapid bumps or jolts without appreciable changes in aircraft altitude or attitude. Report as *Moderate Chop.*	Occupants feel definite strains against seat belts or shoulder straps. Unsecured objects are dislodged. Food service and walking are difficult.	Note: 1. Pilots should report location(s), time (UTC), intensity, whether in or near clouds, altitude, type of aircraft and, when applicable, duration of turbulence. 2. Duration may be based on time between two locations or over a single location. All locations should be readily identifiable.
Severe	Turbulence that causes large, abrupt changes in altitude and/or attitude. It usually causes large variations in indicated airspeed. Aircraft may be momentarily out of control. Report as *Severe Turbulence.**	Occupants are forced violently against seat belts or shoulder straps. Unsecured objects are tossed about. Food service and walking are impossible.	
Extreme	Turbulence in which the aircraft is violently tossed about and is practically impossible to control. It may cause structural damage. Report as *Extreme Turbulence.**		

* High level turbulence (normally above 15,000 feet AGL) that is not associated with cumuliform cloudiness, including thunderstorms, should be reported as CAT (clear air turbulence) preceded by the appropriate intensity, or light or moderate chop.

Figure 7-4. Turbulence reporting criteria

ALL
5445. When turbulence causes changes in altitude and/or attitude, but aircraft control remains positive, that should be reported as

A—light.
B—severe.
C—moderate.

Moderate turbulence is similar to light turbulence but of greater intensity. Changes in altitude and/or attitude occur but the aircraft remains in positive control at all times. It usually causes variations in indicated airspeed. Moderate chop is similar to light chop but of greater intensity. It causes rapid bumps or jolts without appreciable changes in aircraft altitude or attitude. (I53) — AC 00-45E, Chapter 14

Answer (A) is incorrect because light turbulence momentarily causes slight, erratic changes in altitude and/or attitude. Answer (B) is incorrect because severe turbulence causes large, abrupt changes in altitude and/or attitude and the aircraft may be momentarily out of control.

ALL
5446. Turbulence that is encountered above 15,000 feet AGL not associated with cumuliform cloudiness, including thunderstorms, should be reported as

A—severe turbulence.
B—clear air turbulence.
C—convective turbulence.

High-level turbulence (normally above 15,000 feet AGL) not associated with cumuliform cloudiness, including thunderstorms, should be reported as CAT (clear air turbulence) preceded by the appropriate intensity, or light or moderate chop. (I53) — AC 00-45E, Chapter 14

Answer (A) is incorrect because severe turbulence is not dependent on altitude. Answer (C) is incorrect because convective turbulence refers to the lifting action related to turbulence with cumulus clouds.

Answers

5445 [C]   5446 [B]

Terminal Aerodrome Forecast (TAF)

A Terminal Aerodrome Forecast (TAF) is a concise statement of the expected meteorological conditions at an airport during a specified period (usually 24 hours). TAFs use the same code used in the METAR weather reports (*See* Figure 7-1 on Page 7-3).

TAFs are issued in the following format:

TYPE / LOCATION / ISSUANCE TIME / VALID TIME / FORECAST

Note: The "/" above are for separation purposes and do not appear in the actual TAFs.

ALL, MIL
5409. What is the meaning of the terms PROB40 2102 +TSRA as used in a Terminal Aerodrome Forecasts (TAF)?

A—Probability of heavy thunderstorms with rain showers below 4000 feet at time 2102.
B—Between 2100Z and 0200Z there is a forty percent (40%) probability of thunderstorms with heavy rain.
C—Beginning at 2102Z forty percent (40%) probability of heavy thunderstorms and rain showers.

The TAF reports there is a 40% probability (PROB40) between 2100Z and 0200Z (2102) of thunderstorms and heavy rain (+TSRA). (J25) — AIM ¶7-1-28

ALL, MIL
5410. What does the contraction VRB in the Terminal Aerodrome Forecast (TAF) mean?

A—Wind speed is variable throughout the period.
B—Cloud base is variable.
C—Wind direction is variable.

A variable wind direction is noted by "VRB" where the three-digit direction usually appears. (J25) — AIM ¶7-1-28

ALL, MIL
5411. Which statement pertaining to the following Terminal Aerodrome Forecast (TAF) is true?

TAF
KMEM 091135Z 0915 15005KT 5SM HZ BKN060
FM1600 VRB04KT P6SM SKC

A—Wind in the valid period implies surface winds are forecast to be greater than 5 KTS.
B—Wind direction is from 160° at 4 KTS and reported visibility is 6 statute miles.
C—SKC in the valid period indicates no significant weather and sky clear.

TAF KMEM — terminal aerodrome forecast for Memphis

091135Z — observation is taken on the 9th of the month at 1135 Zulu time

15005KT — winds are from 150° at 5 knots

5SM HZ — visibility is 5 statute miles in haze

BKN060 — ceiling is 6,000 feet broken

FM1600 — from 1600Z

VRB04KT — wind is variable at 4 knots

P6SM — visibility is greater than 6 statute miles

SKC — sky is clear

(J25) — AIM ¶7-1-28

ALL, MIL
5412. The visibility entry in a Terminal Aerodrome Forecast (TAF) of P6SM implies that the prevailing visibility is expected to be greater than

A—6 nautical miles.
B—6 statute miles.
C—6 kilometers.

"P6SM" indicates the visibility is greater than 6 statute miles. (J25) — AIM ¶7-1-28

Answers
5409 [B] 5410 [C] 5411 [C] 5412 [B]

Chapter 7 **Weather Services**

ALL, MIL
5413. Terminal Aerodrome Forecasts (TAF) are issued how many times a day and cover what period of time?

A— Four times daily and are usually valid for a 24 hour period.
B— Six times daily and are usually valid for a 24 hour period including a 4-hour categorical outlook.
C— Four times daily and are valid for 12 hours including a 6-hour categorical outlook.

Terminal Aerodrome Forecast (TAF) is a concise statement of the expected meteorological conditions at an airport during a specified period (usually 24 hours). TAFs use the same codes as METAR weather reports. They are scheduled four times daily for 24-hour periods beginning at 0000Z, 0600Z, 1200Z, and 1800Z. (J25) — AIM ¶7-1-28

Aviation Area Forecast (FA)

The aviation Area Forecast is a forecast of general weather conditions over an area the size of several states. It is used to determine forecast enroute weather and to interpolate conditions at airports which do not have TAFs issued.

The Area Forecast is issued 3 times a day, and is comprised of the four sections: a communications and product header section, a precautionary statement section, and two weather sections; a SYNOPSIS section and a VFR CLOUDS/WX section.

ALL
5414. Which information section is contained in the Aviation Area Forecast (FA)?

A— Winds aloft, speed and direction.
B— VFR Clouds and Weather (VFR CLDS/WX).
C— In-Flight Aviation Weather Advisories.

The aviation Area Forecast (FA) is comprised of four sections: a communications and product header section, a precautionary statement section, and two weather sections; a SYNOPSIS section and a VFR CLOUDS/ WX section. (I43) — AC 00-45E, Chapter 4

ALL
5415. The section of the Aviation Area Forecast (FA) entitled VFR Clouds and Weather contains a summary of

A— forecast sky cover, cloud tops, visibility, and obstructions to vision along specific routes.
B— only those weather systems producing liquid or frozen precipitation, fog, thunderstorms, or IFR ceilings.
C— sky condition, cloud heights, visibility, obstructions to vision, precipitation, and sustained surface winds of 20 knots or greater.

The specific forecast section gives a general description of clouds and weather which cover an area greater than 3,000 square miles and is significant to VFR flight operations. Surface visibility and obstructions to vision are included when the forecast visibility is six statute miles or less. Precipitation, thunderstorms, and sustained wind of 20 knots or greater are always included when forecast. (I43) — AC 00-45E, Chapter 4

ALL
5416. In-Flight Aviation Weather Advisories include what type of information?

A— Forecasts for potentially hazardous flying conditions for enroute aircraft.
B— State and geographic areas with reported ceilings and visibilities below VFR minimums.
C— IFR conditions, turbulence, and icing within a valid period for the listed states.

Inflight Aviation Weather Advisories are forecasts to advise en route aircraft of development of potentially hazardous weather. (I43) — AIM ¶7-1-5

ALL
5419. The Aviation Weather Center (AWC) prepares FAs for the contiguous U.S.

A— twice each day.
B— three times each day.
C— every 6 hours unless significant changes in weather require it more often.

All Area Forecasts (FA) (6 areas) are issued by AWC. FAs are prepared 3 times a day in the conterminous U.S. and Alaska (4 times in Hawaii), and amended as required. (I43) — AIM ¶7-1-1

Answers

5413 [A]   5414 [B]   5415 [C]   5416 [A]   5419 [B]

Winds and Temperatures Aloft Forecast (FD)

The Winds and Temperatures Aloft Forecast (FD) shows wind direction, wind velocity, and the temperature that is forecast to exist at specified levels. No winds are forecast for a level within 1,500 feet of the surface, and no temperatures are forecast for the 3,000-foot level or for a level within 2,500 feet of the surface. The wind direction is shown in tens of degrees with reference to true north, and the velocity is shown in knots. Temperatures are in degrees Celsius, and are negative above 24,000 feet.

Each 6-digit group includes wind and temperature. For example, the entry "2045-26" indicates wind from 200° true north at 45 knots, temperature -26°C.

A forecast wind from 160° at 115 knots with a temperature at -34°C at the 30,000-foot level would be encoded "661534," with 50 added to the wind direction and 100 subtracted from the velocity. A wind coded "7699" would indicate wind from 260° at 199 knots or more. "9900" indicates wind light and variable, speed less than 5 knots.

ALL
5424. What values are used for Winds Aloft Forecasts?

A—True direction and MPH.
B—True direction and knots.
C—Magnetic direction and knots.

Wind direction is with reference to true north and wind speed is given in knots on winds aloft forecast reports. (I43) — AC 00-45E, Chapter 4

Inflight Weather Advisories (WA, WS, WST)

Inflight Weather Advisories advise pilots en route of the possibility of encountering hazardous flying conditions that may not have been forecast at the time of the preflight weather briefing.

AIRMETs (WA) contain information on weather that may be hazardous to single engine, other light aircraft, and VFR pilots. The items covered are moderate icing or turbulence, sustained winds of 30 knots or more at the surface, widespread areas of IFR conditions, and extensive mountain obscurement.

SIGMETs (WS) advise of weather potentially hazardous to all aircraft. The items covered are severe icing, severe or extreme turbulence, and widespread sandstorms, dust storms or volcanic ash lowering visibility to less than 3 miles.

SIGMETs and AIRMETs are broadcast upon receipt and at 30-minute intervals (H + 15 and H + 45) during the first hour. If the advisory is still in effect after the first hour, an alert notice will be broadcast. Pilots may contact the nearest FSS to ascertain whether the advisory is pertinent to their flights.

CONVECTIVE SIGMETs (WST) cover weather developments such as tornadoes, lines of thunderstorms, and embedded thunderstorms; they also imply severe or greater turbulence, severe icing, and low-level wind shear. When a SIGMET forecasts embedded thunderstorms, it indicates that the thunderstorms are obscured by massive cloud layers and cannot be seen. Convective SIGMET bulletins are issued hourly at H + 55. Unscheduled convective SIGMETs are broadcast upon receipt and at 15-minute intervals for the first hour (H + 15; H + 30; H + 45).

Answers

5424 [B]

ALL, MIL

5400. The Hazardous Inflight Weather Advisory Service (HIWAS) is a broadcast service over selected VORs that provides

A—SIGMETs and AIRMETs at 15 minutes and 45 minutes past the hour for the first hour after issuance.
B—continuous broadcast of inflight weather advisories.
C—SIGMETs, CONVECTIVE SIGMETs and AIRMETs at 15 minutes and 45 minutes past the hour.

The Hazardous Inflight Weather Advisory Service (HIWAS) is a continuous broadcast service over selected VORs of Inflight Weather Advisories; i.e. SIGMETs, CONVECTIVE SIGMETs, AIRMETs, Severe Weather Forecast Alerts (AWWs), and Center Weather Advisories (CWAs). (J25) — AIM ¶7-1-9

ALL

5422. SIGMETs are issued as a warning of weather conditions which are hazardous

A—to all aircraft.
B—particularly to heavy aircraft.
C—particularly to light airplanes.

A SIGMET advises of weather potentially hazardous to all aircraft, other than convective activity. In the conterminous U.S., items covered are:

1. *Severe icing;*
2. *Severe or extreme turbulence; and*
3. *Dust storms, sandstorms or volcanic ash lowering visibilities to less than 3 miles.*

(I43) — AC 00-45E, Chapter 4

ALL

5423. Which correctly describes the purpose of convective SIGMETs (WST)?

A—They consist of an hourly observation of tornadoes, significant thunderstorm activity, and large hailstone activity.
B—They contain both an observation and a forecast of all thunderstorm and hailstone activity. The forecast is valid for 1 hour only.
C—They consist of either an observation and a forecast or just a forecast for tornadoes, significant thunderstorm activity, or hail greater than or equal to 3/4 inch in diameter.

Convective SIGMETs are issued in the conterminous U.S. for any of the following:

1. *Severe thunderstorms due to surface winds greater than or equal to 50 knots or hail at the surface greater than or equal to 3/4 inches in diameter or tornadoes;*
2. *Embedded thunderstorms;*
3. *Line of thunderstorms; or*
4. *Thunderstorms greater than or equal to VIP level four affecting 40% or more of an area at least 3,000 square miles.*

(I43) — AC 00-45E, Chapter 4

Answer (A) is incorrect because SIGMETs are unscheduled forecasts. Answer (B) is incorrect because SIGMETs are issued only for severe thunderstorms and hail 3/4 inch or larger and they can be valid for up to 2 hours.

ALL

5417. What type of Inflight Weather Advisories provides an en route pilot with information regarding the possibility of moderate icing, moderate turbulence, winds of 30 knots or more at the surface and extensive mountain obscurement?

A—Convective SIGMETs and SIGMETs.
B—Severe Weather Forecast Alerts (AWW) and SIGMETs.
C—AIRMETs and Center Weather Advisories (CWA).

AIRMETs and Center Weather Advisories (CWA) may be of significance to any pilot or aircraft operator and are issued for all domestic airspace. They are of particular concern to operators and pilots of aircraft sensitive to the phenomena described and to pilots without instrument ratings. They are issued for the following weather phenomena which are potentially hazardous to aircraft: moderate icing, moderate turbulence, sustained winds of 30 knots or more at the surface, widespread area of ceilings less than 1,000 feet and/or visibility less than three miles, and extensive mountain obscurement. (J25) — AIM ¶7-1-5

Answers

5400 [B]   5422 [A]   5423 [C]   5417 [C]

Chapter 7 **Weather Services**

ALL
5418. What single reference contains information regarding volcanic eruption, that is occurring or expected to occur?

A—In-Flight Weather Advisories.
B—Terminal Area Forecasts (TAF).
C—Weather Depiction Chart.

In-Flight Weather Advisories (SIGMETs) provide information regarding volcanic eruption, that is occurring or expected to occur. (J25) — AIM ¶7-1-5

Answer (B) is incorrect because Terminal Area Forecasts (TAF) is a concise statement of the expected meteorological conditions at an airport during a specified period. Answer (C) is incorrect because a Weather Depiction Chart is a report of weather as of the time of observation.

Transcribed Weather Broadcast (TWEB) and En Route Flight Advisory Service (EFAS)

A TWEB is the continuous broadcast of weather information on low/medium frequency NAVAIDs and on selected VORs. TWEBs contain information concerning sky cover, cloud tops, visibility, weather, and obstructions to vision in a route format.

En Route Flight Advisory Service (EFAS) is a service specifically designed to provide en route aircraft with timely and meaningful weather advisories pertinent to the type of flight intended, route of flight, and altitude. It provides communications capabilities for aircraft flying at 5,000 feet above ground level to 17,500 feet MSL on a common frequency of 122.0 MHz. Contact Flight Watch by using the name of the ARTCC facility identification serving the area of your location, followed by your aircraft identification, and the name of the nearest VOR to your position.

ALL, MIL
5401. The Telephone Information Briefing Service (TIBS) provided by AFSS's includes

A—weather information service on a common frequency (122.0 mHz).
B—recorded weather briefing service for the local area, usually within 50 miles and route forecasts.
C—continuous recording of meteorological and/or aeronautical information available by telephone.

TIBS, provided by automated flight service stations (AFSS's) is a continuous recording of meteorological and aeronautical information, available by telephone. (J25) — AIM ¶7-1-8

Answer (A) is incorrect because TIBS is available via telephone. Answer (B) is incorrect because TIBS provides area and/or route briefings, airspace procedures, and special announcements (if applicable) concerning aviation interests.

ALL, MIL
5421. To obtain a continuous transcribed weather briefing including winds aloft and route forecasts for a cross-country flight, a pilot could monitor

A—a TWEB on a low-frequency and/or VOR receiver.
B—the regularly scheduled weather broadcast on a VOR frequency.
C—a high-frequency radio receiver tuned to En Route Flight Advisory Service.

TWEB broadcasts contain recorded route-oriented data broadcast continuously over selected low-frequency navigational aids and/or VORs. (J25) — AC 00-45E, Chapter 4

Answer (B) is incorrect because weather broadcasts only include observed weather at specific stations. Answer (C) is incorrect because Enroute Flight Advisory Service provides pilots the opportunity to discuss weather with Flight Watch, but it does not have continuous transcribed weather data.

Answers

5418 [A]   5401 [C]   5421 [A]

ALL, MIL

5560. Weather Advisory Broadcasts, including Severe Weather Forecast Alerts (AWW), Convective SIGMETs, and SIGMETs, are provided by

A—ARTCCs on all frequencies, except emergency, when any part of the area described is within 150 miles of the airspace under their jurisdiction.
B—AFSSs on 122.2 MHz and adjacent VORs, when any part of the area described is within 200 miles of the airspace under their jurisdiction.
C—selected low-frequency and/or VOR navigational aids.

ARTCCs broadcast a Severe Weather Forecast Alert (AWW), Convective SIGMET, SIGMET, or CWA alert once on all frequencies, except emergency, when any part of the area described is within 150 miles of the airspace under their jurisdiction. These broadcasts contain SIGMET or CWA (identification) and a brief description of the weather activity and general area affected. (J25) — AIM ¶7-1-9

ALL

5420. Which forecast provides specific information concerning expected sky cover, cloud tops, visibility, weather, and obstructions to vision in a route format?

A—Area Forecast.
B—Terminal Forecast.
C—Transcribed Weather Broadcast.

The TWEB Route Forecast is similar to the Area Forecast (FA) except information is contained in a route format. Forecast sky cover, cloud tops, visibility, weather and obstructions to vision are described for a corridor 25 miles either side of the route. (I43) — AC 00-45E, Chapter 4

Answers (A) and (B) are incorrect because Area Forecasts and Terminal Aerodrome Forecasts are not issued in a route format.

ALL, MIL

5559. En route Flight Advisory Service (EFAS) is a service that provides en route aircraft with timely and meaningful weather advisories pertinent to the type of flight intended, route, and altitude. This information is received by

A—listening to en route VORs at 15 and 45 minutes past the hour.
B—contacting flight watch, using the name of the ARTCC facility identification in your area, your aircraft identification, and name of nearest VOR, on 122.0 MHz below 17,500 feet MSL.
C—contacting the AFSS facility in your area, using your airplane identification, and the name of the nearest VOR.

EFAS provides communications capabilities for aircraft flying at 5,000 feet above ground level to 17,500 feet MSL on a common frequency of 122.0 MHz. Contact Flight Watch by using the name of the ARTCC facility identification serving the area of your location, followed by your aircraft identification, and the name of the nearest VOR to your position. (J25) — AIM ¶7-1-4

Answers

5560 [A] 5420 [C] 5559 [B]

Chapter 7 **Weather Services**

Radar Weather Report

Radar Weather Reports are of special interest to the pilot because they indicate the location of precipitation along with type, intensity, and trend.

The Radar Summary Chart will yield a three-dimensional view of clouds and precipitation when used in conjunction with other charts and reports. Since radar detects only drops or ice particles of precipitation size, it does not detect clouds or fog. It is the only chart that shows lines and cells of hazardous thunderstorms.

Areas from which precipitation echoes were received are shown on the Radar Summary Chart, with echo intensity indicated by contours as shown in Figure 7-5. Echo heights are displayed in hundreds of feet MSL. Tops are entered above a short line, while any available bases are entered below a short line.

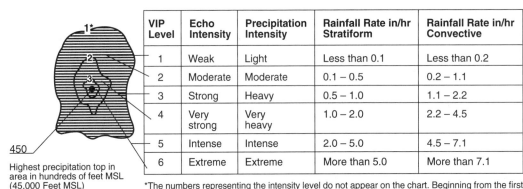

VIP Level	Echo Intensity	Precipitation Intensity	Rainfall Rate in/hr Stratiform	Rainfall Rate in/hr Convective
1	Weak	Light	Less than 0.1	Less than 0.2
2	Moderate	Moderate	0.1 – 0.5	0.2 – 1.1
3	Strong	Heavy	0.5 – 1.0	1.1 – 2.2
4	Very strong	Very heavy	1.0 – 2.0	2.2 – 4.5
5	Intense	Intense	2.0 – 5.0	4.5 – 7.1
6	Extreme	Extreme	More than 5.0	More than 7.1

450 — Highest precipitation top in area in hundreds of feet MSL (45,000 Feet MSL)

*The numbers representing the intensity level do not appear on the chart. Beginning from the first contour line, bordering the area, the intensity level is 1–2, second contour is 3–4, and the third contour is 5–6.

Symbols Used on Charts

Symbol	Meaning	Symbol	Meaning	Symbol	Meaning
R	Rain	+	Intensity increasing or new echo	)	Line of echoes
RW	Rain shower	–	Intensity decreasing	SLD	8/10 or greater coverage in a line
Hail	Hail	No Symbol	No change in intensity	WS999	Severe thunderstorm watch
S	Snow			WT999	Tornado watch
PL	Ice pellets	35	Cell movement to NE at 35 knots	LEWP	Line echo wave pattern
SW	Snow shower		Line or area movement to East at 20 knots	HOOK	Hook echo
L	Drizzle			BWER	Bounded weak echo region
T	Thunderstorm	LM	Little movement	PCLL	Persistent cell
ZR, ZL	Freezing precipitation	MA	Echoes mostly aloft	FNLN	Fine line
NE	No echoes observed	PA	Echoes partly aloft		
NA	Observations unavailbale				
OM	Out for Maintenance				
STC	STC ON – all precipitation may not be seen				
ROBEPS	Radar Operating Below Performance Standards				
RHINO	Range Height Indicator Not Operating				

Figure 7-5. Radar weather report

Chapter 7 **Weather Services**

Movement of an area of echoes is indicated by a shaft and barb combination with the shaft indicating the direction, and the barb indicating the speed. A whole barb is 10 knots, a half barb is 5 knots, and a pennant is 50 knots. Individual cell movement is indicated by an arrow that shows the direction of movement with the speed in knots entered as a number. Severe weather watch areas are outlined by heavy dashed lines, usually in the form of a large rectangular box.

ALL
5408. Which is true concerning the radar weather report (SD) for KOKC?

KOKC 1934 LN 8TRW++/+ 86/40 164/60 199/115 15W L2425 MT 570 AT 159/65 2
INCH HAIL RPRTD THIS CELL

A—There are three cells with tops at 11,500, 40,000, and 60,000 feet.
B—The line of cells is moving 060° with winds reported up to 40 knots.
C—The maximum tops of the cells is 57,000 feet located 65 NM southeast of the station.

KOKC 1934—Oklahoma City Radar Weather Report at 1934 UTC

LN—line

8—8/10 coverage

TRW++/+—thunderstorm (T), very heavy rainshowers (RW++), and increasing intensity (/+) within the line

86/40 164/60 199/115—086° at 40 NM, 164° at 60 NM, and 199° at 115 NM

15W—15 NM wide

L2425—line moving from 240° at 25 knots

MT 570 AT 159/65—maximum top 57,000 feet MSL, located 159° at 65 NM

2 INCH HAIL RPRTD THIS CELL—remarks, 2 inches of hail reported this cell

(I42) — AC 00-45E, Chapter 3

ALL
5432. What information is provided by the Radar Summary Chart that is not shown on other weather charts?

A—Lines and cells of hazardous thunderstorms.
B—Ceilings and precipitation between reporting stations.
C—Areas of cloud cover and icing levels within the clouds.

Radar summary charts show lines and cells of hazardous thunderstorms for the conterminous U.S., thus covering areas without reported observed weather. (I46) — AC 00-45E, Chapter 7

Answer (B) is incorrect because ceilings and precipitation between reporting stations comes from the Weather Depiction Chart. Answer (C) is incorrect because areas of cloud cover and icing levels are forecast on the prog charts.

Surface Analysis Chart

The Surface Analysis Chart depicts frontal positions, pressure patterns, temperature, dew point, wind, weather, and obstructions to vision as of the valid time of the chart. Pressure patterns are shown with isobar lines around the highs and lows. Isobars are solid lines depicting sea level pressure patterns. They are usually spaced at 4-millibar intervals; close spacing of the isobars indicates a strong pressure gradient. When the pressure gradient is weak, dashed isobars are sometimes inserted at 2-millibar intervals to more clearly define the pressure pattern.

ALL
5425. On a Surface Analysis Chart, the solid lines that depict sea level pressure patterns are called

A—isobars.
B—isogons.
C—millibars.

Isobars are solid lines depicting the pressure pattern. They are usually spaced at four-millibar intervals on a surface analysis chart. (I43) — AC 00-45E, Chapter 4

Answer (B) is incorrect because isogons are lines of magnetic variation, found on navigational charts. Answer (C) is incorrect because millibars are units of pressure.

Answers

5408 [C]   5432 [A]   5425 [A]

Chapter 7 **Weather Services**

ALL
5426. Dashed lines on a Surface Analysis Chart, if depicted, indicate that the pressure gradient is

A—weak.
B—strong.
C—unstable.

When a pressure gradient is weak, dashed isobars are sometimes inserted at two-millibar intervals to more clearly define the pressure pattern. (I44) — AC 00-45E, Chapter 5

Answer (B) is incorrect because strong pressure gradients are indicated by closely-spaced solid isobars at 4-mb intervals. Answer (C) is incorrect because stability is concerned with temperature lapse rates.

ALL
5427. Which chart provides a ready means of locating observed frontal positions and pressure centers?

A—Surface Analysis Chart.
B—Constant Pressure Analysis Chart.
C—Weather Depiction Chart.

Surface Analysis Charts are used to denote pressure systems and fronts. (I44) — AC 00-45E, Chapter 5

Answer (B) is incorrect because Constant Pressure Analysis Charts provide observed moisture content, temperatures, and winds aloft. Answer (C) is incorrect because the Weather Depiction Chart depicts frontal location, cloud coverage and height, and VFR/MVFR/IFR.

ALL
5428. On a Surface Analysis Chart, close spacing of the isobars indicates

A—weak pressure gradient.
B—strong pressure gradient.
C—strong temperature gradient.

Close spacing of isobars indicate that a strong pressure gradient exists. (I44) — AC 00-45E, Chapter 5

Answer (A) is incorrect because farther spaced isobars indicate a weaker pressure gradient. Answer (C) is incorrect because isotherms indicate changing temperatures.

ALL
5429. The Surface Analysis Chart depicts

A—frontal locations and expected movement, pressure centers, cloud coverage, and obstructions to vision at the time of chart transmission.
B—actual frontal positions, pressure patterns, temperature, dewpoint, wind, weather, and obstructions to vision at the valid time of the chart.
C—actual pressure distribution, frontal systems, cloud heights and coverage, temperature, dewpoint, and wind at the time shown on the chart.

The Surface Analysis Chart depicts frontal system positions, pressure patterns, and the temperature, dew point, wind, weather and obstructions to vision at individual stations on the surface chart valid for the time given. (I44) — AC 00-45E, Chapter 5

Answer (A) is incorrect because the Surface Analysis Chart reports weather as of the time of observation. Answer (C) is incorrect because Surface Analysis Charts do not indicate cloud heights.

Weather Depiction Chart

The Weather Depiction Chart, which shows conditions that existed at the valid time of the chart, allows the pilot to readily determine general weather conditions on which to base flight planning.

A legend in the lower right-hand corner of the chart describes the method of identifying areas of VFR, MVFR, and IFR conditions.

Total sky cover is depicted by shading of the station circle as shown in Figure 7-6. Cloud height is entered under the station circle in hundreds of feet above ground level. Symbols for weather and obstructions to vision are normally shown to the left of the station circle.

Visibility (in statute miles) is entered to the left of any weather symbol. If the visibility is greater than 6 miles it will be omitted. Figure 7-7 shows examples of plotted data.

Answers

5426 [A]   5427 [A]   5428 [B]   5429 [B]

Symbol	Total Sky Cover
○	Sky clear
◐ (small)	Less than 1/10 (FEW)
◐	1/10 to 5/10 Inclusive (SCATTERED)
◕	6/10 to 9/10 inclusive (BROKEN)
◎	10/10 with breaks (BINOVC)
●	10/10 (OVERCAST)
⊗	Sky obscured or partially obscured

Figure 7-6. Total sky cover symbols

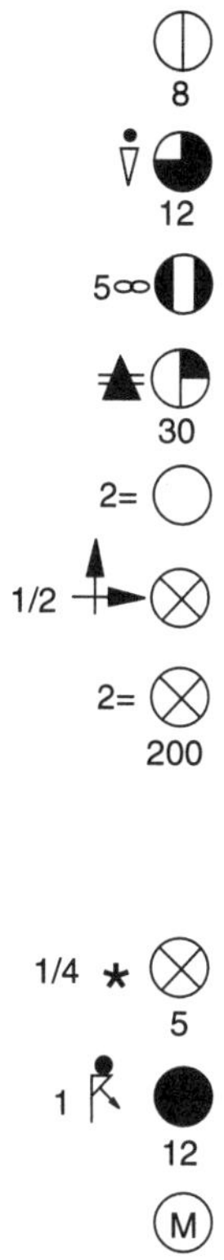

Plotted	Interpreted
◐ 8	Few clouds, base 800 feet, visibility more than 6.
↓ ◕ 12	Broken sky cover, ceiling 1,200 feet, rain shower, visibility more than 6.
5∞ ◎	Thin overcast with breaks, visibility 5 in haze.
▲ ◐ 30	Scattered at 3,000 feet, clouds topping ridges, visibility more than 6.
2= ○	Sky clear, visibility 2, ground fog or fog.
1/2 ↑← ⊗	Sky partially obscured, visibility 1/2, blowing snow, no cloud layers observed.
2= ⊗ 200	Sky partially obscured, visibility 2, fog, cloud layer at 20,000 feet. Assume sky is partially obscured since 20,000 feet cannot be vertical visibility into fog. It is questionable if 20,000 feet is lowest scattered layer or ceiling.*
1/4 * ⊗ 5	Sky obscured, ceiling 500, visibility 1/4, snow
1 ⚡ ● 12	Overcast, ceiling 1,200 feet thunderstorm, rain shower, visibility 1.
(M)	Data missing.

* Note: Since a partial and a total obscuration (x) is entered as total sky cover, it can be difficult to determine if a height entry is a cloud layer above a partial obscuration or vertical visibility into a total obscuration. Check the METAR.

Figure 7-7. Examples of plotted data on the weather depiction chart

ALL
5430. Which provides a graphic display of both VFR and IFR weather?

A—Surface Weather Map.
B—Radar Summary Chart.
C—Weather Depiction Chart.

The Weather Depiction Chart provides the pilot with a broad overview of conditions (including IFR, MVFR and VFR) valid at the time of the chart. (I45) — AC 00-45E, Chapter 6

Answer (A) is incorrect because Surface Weather Maps depict pressure systems. Answer (B) is incorrect because Weather Depiction Charts depict a collection of radar reports.

ALL
5431. When total sky cover is few or scattered, the height shown on the Weather Depiction Chart is the

A—top of the lowest layer.
B—base of the lowest layer.
C—base of the highest layer.

If the total sky cover is few or scattered, the height given is the base of the lowest layer. (I45) — AC 00-45E, Chapter 6

Answers

5430 [C]   5431 [B]

Chapter 7 **Weather Services**

Constant Pressure Chart

A Constant Pressure Analysis Chart is an upper air weather map where all the information depicted is at the specified pressure-level of the chart. Each of the Constant Pressure Analysis Charts (850 MB, 700 MB, 500 MB, 300 MB, 250 MB, and 200 MB) can provide observed temperature/dewpoint spread, wind, height of the pressure surface, and the height changes over the previous 12-hour period.

ALL
5440. Hatching on a Constant Pressure Analysis Chart indicates

A—hurricane eye.
B—windspeed 70 knots to 110 knots.
C—windspeed 110 knots to 150 knots.

To aid in identifying areas of strong winds, hatching denotes wind speeds of 70 to 110 knots, a clear area within a hatched area indicates wind of 110 to 150 knots, and an area of 150 to 190 knots of wind is hatched. (I51) — AC 00-45E, Chapter 8

Answer (A) is incorrect because the hurricane eye has very low winds. Answer (C) is incorrect because a clear area within a hatched area indicates wind of 110 to 150 knots.

ALL
5441. What flight planning information can a pilot derive from Constant Pressure Analysis Charts?

A—Winds and temperatures aloft.
B—Clear air turbulence and icing conditions.
C—Frontal systems and obstructions to vision aloft.

The Constant Pressure Analysis Chart is a source of observed temperature, temperature/dewpoint spread, moisture and wind. (I51) — AC 00-45E, Chapter 8

Answer (B) is incorrect because clear air turbulence can be found on prognostic charts and in area forecasts. Answer (C) is incorrect because frontal systems and obstructions to vision aloft can be found on Surface Analysis and Weather Depiction charts.

ALL
5442. From which of the following can the observed temperature, wind, and temperature/dewpoint spread be determined at a specified altitude?

A—Stability Charts.
B—Winds Aloft Forecasts.
C—Constant Pressure Analysis Charts.

The Constant Pressure Analysis Chart is a source of observed temperature, temperature/dewpoint spread, moisture, and wind. (I51) — AC 00-45E, Chapter 8

Answer (A) is incorrect because Stability Charts provide information on stability, freezing level, precipitation, and average relative humidity. Answer (B) is incorrect because Winds Aloft Forecasts do not provide information about temperature/dewpoint spread.

Tropopause Height/Vertical Wind Shear Prognostic Chart

The Tropopause Height/Vertical Wind Shear Prognostic Chart is a two-panel chart containing a maximum wind prog and a vertical wind shear prog. The chart is prepared for the contiguous 48 states and is available once a day with a valid time of 18Z. It shows the temperature, pressure, and wind at the tropopause.

ALL
5443. The minimum vertical wind shear value critical for probable moderate or greater turbulence is

A—4 knots per 1,000 feet.
B—6 knots per 1,000 feet.
C—8 knots per 1,000 feet.

The vertical shear critical for probable turbulence is 6 knots per 1,000 feet. (I52) — AC 00-45E, Chapter 14

Answers

5440 [B]   5441 [A]   5442 [C]   5443 [B]

Chapter 7 **Weather Services**

Significant Weather Prognostics

The Low-Level Significant Weather Prognostic Chart (surface to 24,000 feet) portrays forecast weather which may influence flight planning, including those areas or activities of most significant turbulence and icing. It is a four-panel chart; the two lower panels are 12- and 24-hour surface progs. The chart is issued four times daily. The chart uses standard weather symbols as shown in Figure 7-8.

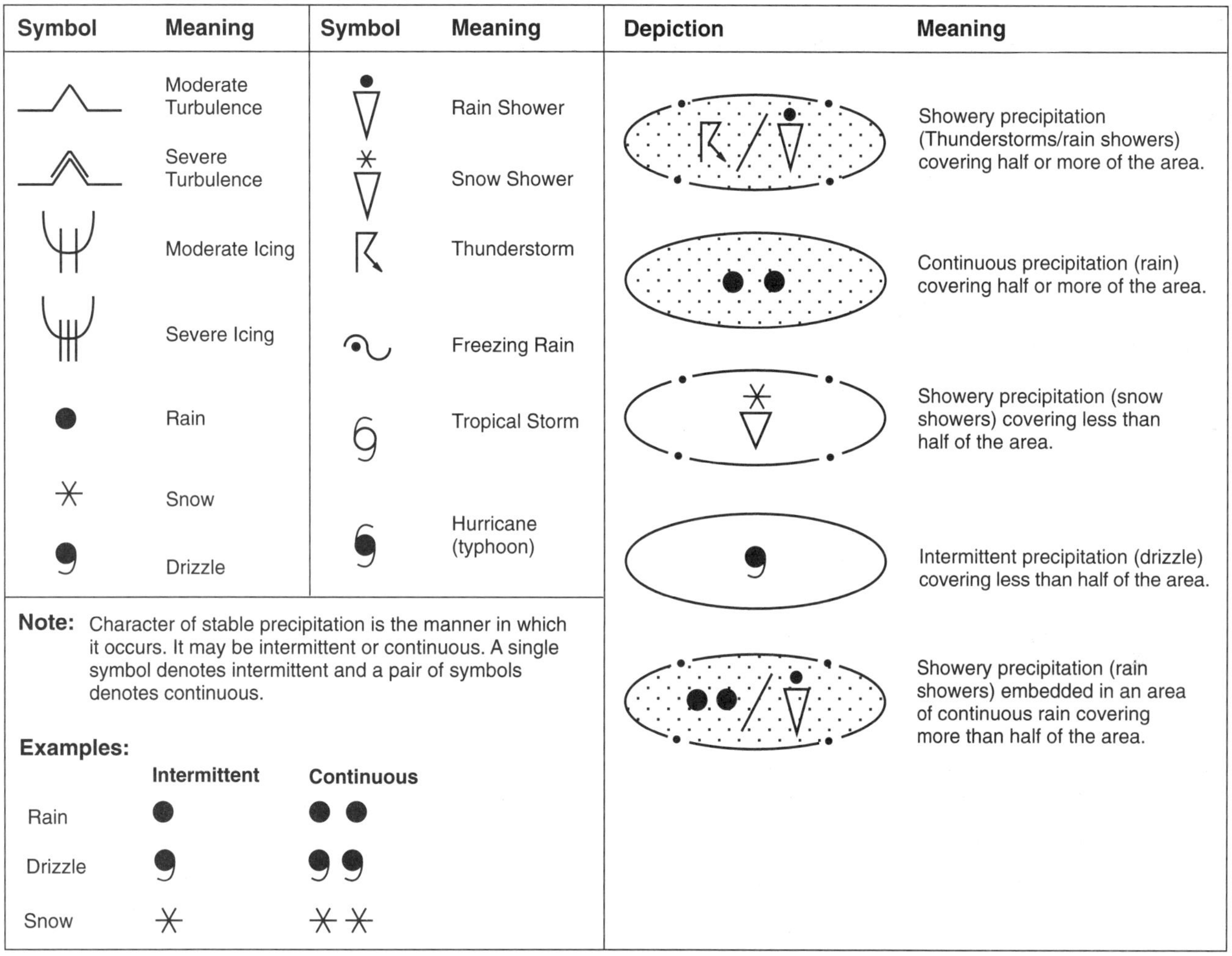

Symbol	Meaning	Symbol	Meaning	Depiction	Meaning
∧	Moderate Turbulence	▽	Rain Shower	(showers/thunderstorm symbol in dotted oval)	Showery precipitation (Thunderstorms/rain showers) covering half or more of the area.
∧∧	Severe Turbulence	⁎▽	Snow Shower		
ψ	Moderate Icing	↙	Thunderstorm	(two dots in dotted oval)	Continuous precipitation (rain) covering half or more of the area.
ψψ	Severe Icing	∼	Freezing Rain		
●	Rain	6	Tropical Storm	(snow shower symbol in oval)	Showery precipitation (snow showers) covering less than half of the area.
⁎	Snow	6 (with tail)	Hurricane (typhoon)		
9	Drizzle			(drizzle symbol in oval)	Intermittent precipitation (drizzle) covering less than half of the area.

Note: Character of stable precipitation is the manner in which it occurs. It may be intermittent or continuous. A single symbol denotes intermittent and a pair of symbols denotes continuous.

Examples:

	Intermittent	Continuous
Rain	●	● ●
Drizzle	9	9 9
Snow	⁎	⁎ ⁎

(rightmost column bottom): Showery precipitation (rain showers) embedded in an area of continuous rain covering more than half of the area.

Figure 7-8. Significant weather prognostics

The High-Level Significant Weather Prognostic Chart (24,000 feet to 63,000 feet) outlines areas of forecast turbulence and cumulonimbus clouds, shows the expected height of the tropopause, and predicts jet stream location and velocity. The chart depicts clouds and turbulence as shown in Figure 7-9 on the next page.

The height of the tropopause is depicted in hundreds of feet MSL and is enclosed in a rectangular box. Areas of forecast moderate or greater Clear Air Turbulence (CAT) are bounded by heavy dashed lines and are labeled with the appropriate symbol and the vertical extent in hundreds of feet MSL. Cumulonimbus clouds imply moderate or greater turbulence and icing.

Chapter 7 **Weather Services**

ALL
5433. Which weather chart depicts conditions forecast to exist at a specific time in the future?

A—Freezing Level Chart.
B—Weather Depiction Chart.
C—12-Hour Significant Weather Prognostication Chart.

Prog (prognostic) charts portray weather forecasts for sometime in the future for the conterminous U.S. and adjacent areas. (I47) — AC 00-45E, Chapter 11

Answer (A) is incorrect because there is no chart named the Freezing Level Chart. Answer (B) is incorrect because the Weather Depiction Charts show conditions valid only for the time given for the chart.

ALL
5436. What is the upper limit of the Low Level Significant Weather Prognostic Chart?

A—30,000 feet.
B—24,000 feet.
C—18,000 feet.

The U.S. Low-Level Significant Weather Prog includes significant weather from the surface to 24,000 feet (400 millibars). (I47) — AC 00-45E, Chapter 11

ALL
5434. What weather phenomenon is implied within an area enclosed by small scalloped lines on a U.S. High-Level Significant Weather Prognostic Chart?

A—Cirriform clouds, light to moderate turbulence, and icing.
B—Cumulonimbus clouds, icing, and moderate or greater turbulence.
C—Cumuliform or standing lenticular clouds, moderate to severe turbulence, and icing.

Small-scalloped lines enclose areas of expected cumulonimbus development. Cumulonimbus clouds imply moderate or greater turbulence and icing. (I47) — AC 00-45E, Chapter 11

ALL
5435. The U.S. High-Level Significant Weather Prognostic Chart forecasts significant weather for what airspace?

A—18,000 feet to 45,000 feet.
B—24,000 feet to 45,000 feet.
C—24,000 feet to 63,000 feet.

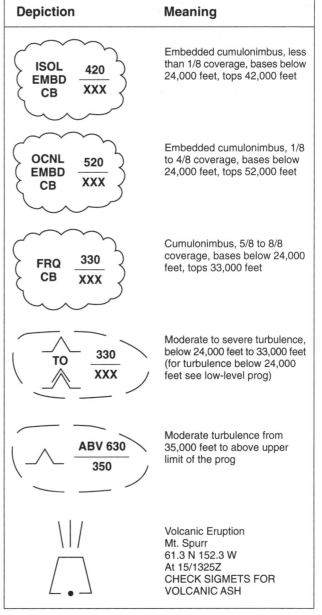

Figure 7-9. Prognostic chart depictions

The U.S. High-Level Significant Weather Prognostic Chart forecasts significant weather encompassing airspace from 24,000 feet to 63,000 feet pressure altitude (400 to 70 millibars). (I47) — AC 00-45E, Chapter 11

Answers

5433 [C]   5436 [B]   5434 [B]   5435 [C]

Chapter 7 **Weather Services**

Composite Moisture Stability Chart

Composite Moisture Stability Charts are analysis charts using observed upper air data. These charts are available twice daily with valid times of 12Z and 00Z. They contain four panels: freezing level, precipitable water, average relative humidity, stability.

The stability panel outlines areas of stable and unstable air. Two stability indices are computed for each upper air station:

Lifted index (LI)—This index indicates the stability at 500 mb (18,000 feet MSL).

K index (K)—This index examines the temperature and moisture profile of the environment. The number above is the lifted index, and the number below the line is the K index. A positive lifted index means that a parcel of air, if lifted, would be colder than existing air at 500 mb. For example, 23 over 28 indicates that the air is very stable. There will be stratified cloudiness and steady precipitation. A negative lifted index means that the low-level air, if lifted to 500 mb would be warmer than existing air at 500 mb. For example, -1 over 35 indicates that the air is unstable, and there is an air-mass thunderstorm probability of 61% to 80%.

ALL
5438. A freezing level panel of the composite moisture stability chart is an analysis of

A—forecast freezing level data from surface observations.
B—forecast freezing level data from upper air observations.
C—observed freezing level data from upper air observations.

A freezing level panel is an analysis of observed freezing level data from upper air observations, on the Composite Moisture Stability Chart. (I49) — AC 00-45E, Chapter 9

ALL
5439. The difference found by subtracting the temperature of a parcel of air theoretically lifted from the surface to 500 millibars and the existing temperature at 500 millibars is called the

A—lifted index.
B—negative index.
C—positive index.

The lifted index is computed as if a parcel of air near the surface were lifted to 500 millibars. As the air is "lifted," it cools by expansion. The temperature the parcel would have at 500 millibars is then subtracted from the existing 500 millibar (mb) temperature. The difference is the lifted index; it may be positive, zero, or negative. (I49) — AC 00-45E, Chapter 9

Answer (B) is incorrect because a positive index means the lifted parcel of air is colder than the existing air at 500 mb, and the air is stable. Answer (C) is incorrect because a negative index means the lifted parcel of air is warmer than the existing air at 500 mb, and the air is unstable.

Answers
5438 [C]   5439 [A]

Chapter 8
Aircraft Performance

Weight and Balance   *8–3*
    Computing Weight and Balance   *8–4*
    Graph Weight and Balance Problems   *8–4*
    Weight Change   *8–5*
    Weight Shift   *8–6*
Rotorcraft Weight and Balance   *8–12*
Glider Weight and Balance   *8–14*
Ground Operations During Windy Conditions   *8–16*
Headwind and Crosswind Components   *8–17*
Density Altitude   *8–19*
Takeoff and Landing Considerations   *8–20*
Takeoff and Landing Distance   *8–22*
Fuel Consumption vs. Brake Horsepower   *8–26*
Time, Fuel and Distance to Climb   *8–27*
    Table Method   *8–27*
    Graph Method   *8–28*
Cruise Performance Table   *8–32*
Cruise and Range Performance Table   *8–35*
Maximum Rate of Climb   *8–36*
Rotorcraft Performance   *8–37*
Glider Performance   *8–40*
Balloon Performance   *8–43*

Chapter 8 Aircraft Performance

Weight and Balance

Even though an aircraft has been certificated for flight at a specified maximum gross weight, it may not be safe to take off with that load under all conditions. High altitude, high temperature, and high humidity are additional factors which may require limiting the load to be carried.

In addition to considering the weight, the pilot must ensure the load is arranged to keep the aircraft in balance. The balance point, or **center of gravity (CG)**, is the point at which all of the weight of the system is considered to be concentrated. For an aircraft to be safe to fly, the center of gravity must fall between specified limits. To keep the CG within safe limits, it may be necessary to move weight toward the nose of the aircraft (forward) which moves the CG forward, or toward the tail (aft) which moves the CG aft. The aircraft and the various compartments of the aircraft are designed for specific maximum weights and critical load factors. Compartments placarded for a given weight may be loaded to the maximum allowable load only if the CG is kept within limits and the aircraft is flown safely within critical load factor limits.

The **datum** is an imaginary vertical reference line from which locations on/in an aircraft are measured. The datum is established by the manufacturer and may vary in location in different aircraft. *See* Figure 8-1.

The **arm** (or station) is the horizontal distance measured in inches from the datum line to a point on the aircraft. If measured aft toward the tail, the arm is given a positive value; if measured forward toward the nose, the arm is given a negative value. When all arms are positive, the datum is located at the nose or in front of the airplane. *See* Figure 8-2.

The **moment** is the product of the weight of an object multiplied by its arm and is expressed in pound-inches (lbs-in): Weight x Arm = Moment. The moment index is a moment divided by a constant such as 100 or 1,000. It is used to simplify computations where heavy items and long arms result in large, unmanageable numbers.

The center of gravity (CG) is the point about which an aircraft will balance, and its position is expressed in inches from datum. The center of gravity is found by dividing the total moments by the total weight. The formula is usually expressed as follows:

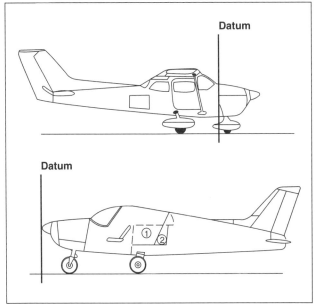

Figure 8-1. Datum lines

$$\text{CG (inches aft of datum)} = \frac{\text{Total Moment}}{\text{Total Weight}}$$

The specified forward and aft points within which the CG must be located for safe flight are called the **Center of Gravity Limits**. These limits are specified by the manufacturer. The distance between the forward and aft CG limits is called the **Center of Gravity Range**. The empty weight includes the airframe, engine(s), all items of equip-

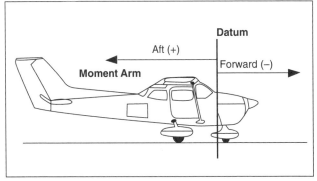

Figure 8-2. Positive and negative arms

ment, unusable fuel, hydraulic fluid, and undrainable oil or, in some aircraft, all of the oil. The useful load includes the pilot, passengers, baggage, fuel and oil. The takeoff weight is the empty weight plus the useful load. The landing weight is the takeoff weight minus any fuel used.

Computing Weight and Balance

Problem:

Calculate the weight and balance and determine where the CG would be located for three weights given the following:

Weight D—160 lbs at 45 inches aft of datum

Weight E—170 lbs at 145 inches aft of datum

Weight F—105 lbs at 185 inches aft of datum

Solution:

1. Construct a table, with column headings as follows:

Item	Weight	Arm	Moment

2. Find the moment for each item by multiplying the weight times the arm:

Item	Weight	Arm	Moment
Weight D	160	45	7,200
Weight E	170	145	24,650
Weight F	105	185	19,425

3. Total the weights and moments:

Weight	Moment
160	7,200
170	24,650
+ 105	+ 19,425
435 lbs	51,275 lbs-in

4. Determine the center of gravity by dividing the total moment by the total weight.

    $$CG = \frac{\text{Total Moment}}{\text{Total Weight}}$$

    $CG = 51,275 \div 435 = 117.8$ inches

Graph Weight and Balance Problems

Airplane manufacturers use one of several available systems to provide loading information. Weight and balance computations are greatly simplified by two graphic aids: the Loading Graph and the Center of Gravity Moment Envelope. The graphs shown in FAA Figure 38 are typical of those found in a Pilot's Operating Handbook.

Chapter 8 Aircraft Performance

Problem:

Determine if the airplane is loaded within limits, given the following:

Empty weight (oil included) .... 1,271 lbs
Empty weight moment (lbs-in/1,000) .... 102.04 lbs-in
Pilot and copilot .... 400 lbs
Rear seat passenger .... 140 lbs
Cargo .... 100 lbs
Fuel .... 37 gallons

Solution:

1. Construct a table with the Item, Weight, and Moment. The moment of each item is found by using the Loading Graph in FAA Figure 38. Locate the weight of each item on the left side of the chart, and proceed to the right to intersect the correct item line. At point of intersection, proceed downward to find the moment for that item at the specified weight.

Item	Weight	Moment/1,000
Aircraft empty weight	1,271	102.04
Pilot and copilot	400	36.00
Aft passenger	140	18.00
Cargo	100	11.50
Fuel (37.0 gal x 6 lbs/gal)	+ 222	+ 20.00
Totals	2,133	187.54

2. Compare the total weight and moment to the limits in the center of gravity envelope in FAA Figure 38. Move up the loaded aircraft weight scale to 2,133, then across to 187.54 on the moment/1,000 scale. The weight and CG are within limits.

Weight Change

There are a number of additional weight and balance formulas that may be used to determine the effect on the CG due to a weight change. Use the following formula to find the change in CG, if weight has been added or subtracted. The amount of weight changed and the new total weight must be known, in addition to the distance between the original CG and the point where the weight is being added or subtracted.

$$\frac{\text{Weight Lost or Gained}}{\text{New Total Weight}} = \frac{\text{Change in CG}}{\text{Distance Between Original CG and Point of Weight Removed or Added}}$$

Problem:

Determine the new CG location after 1 hour 45 minutes of flight time, given the following:

Total weight .... 4,037 lbs
CG location .... Station 67.8
Fuel consumption .... 14.7 GPH
Fuel CG .... Station 68.0

Solution:

1. Find the amount of weight change. The aircraft has consumed 14.7 GPH for 1 hour 45 minutes. The total fuel consumed is:

    14.7 GPH x 1.75 hr = 25.7 gal

    Which weighs:

    25.7 gal x 6 lbs/gal = 154 lbs

2. Determine the new total weight by subtracting the weight of the fuel consumed (154 lbs) from the total weight (4,037 lbs).

    4,037 − 154 = 3,883 lbs new total weight

3. Find the distance between the original CG (67.8) and the point weight removed (fuel CG = 68.0).

    68.0 − 67.8 = 0.2 inches

4. Place the three known values into the formula.

    $$\frac{154 \text{ lbs}}{3,883 \text{ lbs}} = \frac{\text{Change in CG}}{0.2 \text{ in}}$$

    $$\text{Change in CG} = \frac{154 \text{ lbs} \times 0.2 \text{ in}}{3,883 \text{ lbs}} = 0.01 \text{ in}$$

5. The CG was found to shift approximately 0.01 in. Since the weight was removed aft (68.0 in) of the CG (67.8 in), the CG shifted forward 0.01 in.

    ```
    67.80   original CG
    − 0.01  forward shift
    67.79   new CG
    ```

Weight Shift

A shift in weight may have a great affect on the CG location even though the total weight may not have changed. The change in CG caused by a weight shift, or the amount of weight to shift in order to move the CG within limits, may be found by the formula below:

$$\frac{\text{Weight Shifted}}{\text{Total Weight}} = \frac{\text{Change in CG}}{\text{Distance Shifted}}$$

Problem:

Determine the new CG after luggage has been shifted, given the following:

Gross weight .... 5,000 pounds (including three pieces of luggage)

CG .... 98 in. aft of datum (2 inches aft of limits)

2 pieces of luggage .... 100 lbs (weighed together)

The luggage is moved from the rear baggage compartment (145 inches aft of datum) to the front compartment (45 inches aft of datum).

Solution:

1. Find the distance shifted by taking the difference between the original location (baggage compartment 145 inches) and the new location (front compartment 45 inches).

    145 in. baggage compartment
    – 45 in. front compartment
    100 in. distance shifted

2. Determine the change in CG by substituting the values into the formula.

$$\frac{100 \text{ lbs}}{5,000 \text{ lbs}} = \frac{\text{Change in CG}}{100 \text{ in}}$$

$$\frac{100 \text{ lbs} \times 100 \text{ in}}{5,000 \text{ lbs}} = \text{Change in CG} = 2 \text{ inches}$$

3. The CG is moved 2 inches. Since the weight was shifted from the rear compartment (145 inches) to the front compartment (45 inches), the CG moved forward by 2 inches.

    98 in. original CG
    – 2 in. CG shift
    96 in. new CG

The new CG is located 96 inches aft of datum, which places it on the aft limit.

AIR
5164. Baggage weighing 90 pounds is placed in a normal category airplane's baggage compartment which is placarded at 100 pounds. If this airplane is subjected to a positive load factor of 3.5 Gs, the total load of the baggage would be

A—315 pounds and would be excessive.
B—315 pounds and would not be excessive.
C—350 pounds and would not be excessive.

Each cargo compartment must be designed for its placarded maximum weight of contents in relation to the critical load factors. Normal category aircraft having a gross weight of less than 4,000 pounds are designed for a maneuvering load factor of 3.8 times the aircraft weight or 3.8 Gs. Thus a load factor of 3.5 Gs would not be excessive. Load factor is the ratio of a given load to its weight. Thus:

    90.0 lbs weight
    x 3.5 Gs
    315.0 lbs loading

(H303) — AC 61-23C, Chapter 1

ALL
5632. When computing weight and balance, the basic empty weight includes the weight of the airframe, engine(s), and all installed optional equipment. Basic empty weight also includes

A—the unusable fuel, full operating fluids, and full oil.
B—all usable fuel, full oil, hydraulic fluid, but does not include the weight of pilot, passengers, or baggage.
C—all usable fuel and oil, but does not include any radio equipment or instruments that were installed by someone other than the manufacturer.

Basic empty weight includes unusable fuel, full operating fluids, and full oil. (H315) — AC 61-23C, Chapter 4

Answers
5164 [B]   5632 [A]

Chapter 8 Aircraft Performance

ALL
5633. If all index units are positive when computing weight and balance, the location of the datum would be at the

A—centerline of the main wheels.
B—nose, or out in front of the airplane.
C—centerline of the nose or tailwheel, depending on the type of airplane.

In general, the arm (index unit) of a location in the aircraft is in inches aft of the datum. A positive arm indicates the position of the object is aft of the datum. If all index units are positive, the datum must be at least at the nose of the aircraft. (H316) — AC 61-23C, Chapter 4

Answer (A) is incorrect because the engine would have a negative arm if the datum was at the centerline of the main wheels. Answer (C) is incorrect because the propeller would have a negative arm if the datum were at the centerline of the nose or tailwheel.

ALL
5634. The CG of an aircraft can be determined by which of the following methods?

A—Dividing total arms by total moments.
B—Multiplying total arms by total weight.
C—Dividing total moments by total weight.

Total moment is the weight of the airplane multiplied by the distance between the datum and the CG. Therefore:

$$CG = \frac{Total\ Moment}{Total\ Weight}$$

(H12) — FAA-H-8083-1, Chapter 3

ALL
5635. The CG of an aircraft may be determined by

A—dividing total arms by total moments.
B—dividing total moments by total weight.
C—multiplying total weight by total moments.

Total moment is the weight of the airplane multiplied by the distance between the datum and the CG. Therefore:

$$CG = \frac{Total\ Moment}{Total\ Weight}$$

(H12) — FAA-H-8083-1, Chapter 3

ALL
5636. GIVEN:

Weight A: 155 pounds at 45 inches aft of datum
Weight B: 165 pounds at 145 inches aft of datum
Weight C: 95 pounds at 185 inches aft of datum

Based on this information, where would the CG be located aft of datum?

A—86.0 inches.
B—116.8 inches.
C—125.0 inches.

1. Arm is the aft of datum value for each item:

Weight	x	Arm	=	Moment	
155	x	45	=	6,975	for A
165	x	145	=	23,925	for B
+ 95	x	185	=	+ 17,575	for C
415 lbs				48,475 lbs-in	

2. Moments ÷ Total Weight = CG Location

    48,475 ÷ 415 = 116.8 inches

(H12) — FAA-H-8083-1, Chapter 3

ALL
5637. GIVEN:

Weight A: 140 pounds at 17 inches aft of datum
Weight B: 120 pounds at 110 inches aft of datum
Weight C: 85 pounds at 210 inches aft of datum

Based on this information, the CG would be located how far aft of datum?

A—89.11 inches.
B—96.89 inches.
C—106.92 inches.

1. Arm is the aft of datum value for each item:

Weight	x	Arm	=	Moment	
140	x	17	=	2,380	for A
120	x	110	=	13,200	for B
+ 85	x	210	=	+ 17,850	for C
345 lbs				33,430 lbs-in	

2. Moments ÷ Total Weight = CG Location

    33,430 ÷ 345 = 96.89 inches

(H12) — FAA-H-8083-1, Chapter 3

Answers

5633 [B]   5634 [C]   5635 [B]   5636 [B]   5637 [B]

ALL
5638. GIVEN:

Weight A: 135 pounds at 15 inches aft of datum
Weight B: 205 pounds at 117 inches aft of datum
Weight C: 85 pounds at 195 inches aft of datum

Based on this information, the CG would be located how far aft of datum?

A—100.2 inches.
B—109.0 inches.
C—121.7 inches.

1. Arm is the aft of datum value for each item:

Weight	x	Arm	=	Moment	
135	x	15	=	2,025	for A
205	x	117	=	23,985	for B
+ 85	x	195	=	+ 16,575	for C
425 lbs				42,585 lbs-in	

2. Moments ÷ Total Weight = CG Location

   42,585 ÷ 425 = 100.2 inches

(H12) — FAA-H-8083-1, Chapter 3

ALL
5639. GIVEN:

Weight A: 175 pounds at 135 inches aft of datum
Weight B: 135 pounds at 115 inches aft of datum
Weight C: 75 pounds at 85 inches aft of datum

The CG for the combined weights would be located how far aft of datum?

A—91.76 inches.
B—111.67 inches.
C—118.24 inches.

1. Arm is the aft of datum value for each item:

Weight	x	Arm	=	Moment	
175	x	135	=	23,625	for A
135	x	115	=	15,525	for B
+ 75	x	85	=	+ 6,375	for C
385 lbs				45,525 lbs-in	

2. Moments ÷ Total Weight = CG Location

   45,525 ÷ 385 = 118.25 inches

(H12) — FAA-H-8083-1, Chapter 3

AIR
5650. (Refer to Figure 38.)
GIVEN:

Empty weight (oil is included) .... 1,271 lb
Empty weight moment (in-lb/1,000) 102.04
Pilot and copilot .... 400 lb
Rear seat passenger .... 140 lb
Cargo .... 100 lb
Fuel .... 37 gal

Is the airplane loaded within limits?

A—Yes, the weight and CG is within limits.
B—No, the weight exceeds the maximum allowable.
C—No, the weight is acceptable, but the CG is aft of the aft limit.

1. Construct a table with the Item, Weight, and Moment. The moment of each item is found by using the Loading Graph in FAA Figure 38. Locate the weight of each item on the left side of the chart, and proceed to the right to intersect the correct item line. At point of intersection, proceed downward to find the moment for that item at the specified weight.

Item	Weight	Moment
Aircraft empty weight	1,271	102.04
Pilot and copilot	400	36.00
Aft passenger	140	18.00
Cargo	100	11.50
Fuel (37.0 gal x 6 lbs/gal)	+ 222	+ 20.00
Totals	2,133	187.54

2. Compare the total weight and moment to the limits in the center of gravity envelope in FAA Figure 38. Move up the loaded aircraft weight scale to 2,133, then across to 187.54 on the moment/1,000 scale. The weight and CG are within limits.

(H15) — FAA-H-8083-1, Chapter 3

Answers

5638 [A]   5639 [C]   5650 [A]

Chapter 8 Aircraft Performance

AIR
5651. (Refer to Figure 38.)
GIVEN:

Empty weight (oil is included) .... 1,271 lb
Empty weight moment (in-lb/1,000) 102.04
Pilot and copilot .... 260 lb
Rear seat passenger .... 120 lb
Cargo .... 60 lb
Fuel .... 37 gal

Under these conditions, the CG is determined to be located

A—within the CG envelope.
B—on the forward limit of the CG envelope.
C—within the shaded area of the CG envelope.

1. Construct a table with the Item, Weight, and Moment. The moment of each item is found by using the Loading Graph in FAA Figure 38. Locate the weight of each item on the left side of the chart, and proceed to the right to intersect the correct item line. At point of intersection, proceed downward to find the moment for that item at the specified weight.

Item	Weight	Moment
Aircraft empty weight	1,271	102.04
Pilot and copilot	260	23.30
Aft passenger	120	15.00
Cargo	60	6.80
Fuel (37.0 gal x 6 lbs/gal)	+ 222	+ 20.00
Totals	1,933	167.14

2. Compare to the limits in FAA Figure 38. Using the CG Envelope Graph, move up the loaded aircraft weight scale to 1,933, then across to 167.14 on the moment/1,000 scale. The weight and CG are within limits.

(H15) — FAA-H-8083-1, Chapter 3

AIR
5652. (Refer to Figure 38.)
GIVEN:

Empty weight (oil is included) .... 1,271 lb
Empty weight moment (in-lb/1,000) 102.04
Pilot and copilot .... 360 lb
Cargo .... 340 lb
Fuel .... 37 gal

Will the CG remain within limits after 30 gallons of fuel has been used in flight?

A—Yes, the CG will remain within limits.
B—No, the CG will be located aft of the aft CG limit.
C—Yes, but the CG will be located in the shaded area of the CG envelope.

1. Construct a table with the Item, Weight, and Moment. The moment of each item is found by using the Loading Graph in FAA Figure 38. Locate the weight of each item on the left side of the chart, and proceed to the right to intersect the correct item line. At point of intersection, proceed downward to find the moment for that item at the specified weight.

Item	Weight	Moment
Aircraft empty weight	1,271	102.04
Pilot and copilot	360	32.80
Cargo	340	39.50
Fuel (7.0 gal x 6 lbs/gal)	+ 42	+ 4.00
Totals	2,013	178.34

2. Compare to the limits in FAA Figure 38. Using the CG Envelope Graph, move up the loaded aircraft weight scale to 2,013, then across to 178.34 on the moment/1,000 scale. The CG will remain within limits.

(H15) — FAA-H-8083-1, Chapter 3

AIR
5646. GIVEN:

Total weight .... 4,137 lb
CG location station 67.8
Fuel consumption .... 13.7 GPH
Fuel CG station 68.0

After 1 hour 30 minutes of flight time, the CG would be located at station

A—67.79.
B—68.79.
C—70.78.

1. Find the weight change: 13.7 GPH for 1.5 hours = 20.55 gal

    20.55 gal x 6 lbs/gal = 123.3 lbs

2. New total weight is 4,137 − 123.3 = 4,013.7

3. The distance between the CG and the fuel arm is 68.0 − 67.8 = .2

4. Place the values in the formula:

$$\frac{123.3}{4013.7} = \frac{CG\ change}{.2} = .00614\ inches$$

5. Calculate the new CG:

    Original CG   67.80000
    CG change   − .00614
    New CG   67.79386

(H14) — FAA-H-8083-1, Chapter 3

Answers

5651 [A]   5652 [A]   5646 [A]

AIR
5649. GIVEN:

Total weight .... 3,037 lb
CG location station 68.8
Fuel consumption .... 12.7 GPH
Fuel CG station 68.0

After 1 hour 45 minutes of flight time, the CG would be located at station

A—68.77.
B—68.83.
C—69.77.

1. Find the weight change: 12.7 GPH for 1.75 hours = 22.23 gal

    22.23 gal x 6 lbs/gal = 133.38 lbs

2. New total weight is 3,037 – 133.38 = 2,903.62

3. The distance between the CG and the fuel arm is 68.8 – 68.0 = .8

4. Place the values in the formula:

    $$\frac{133.38}{2903.62} = \frac{CG\ change}{.8} = .03675\ inches$$

5. Calculate the new CG:

    Original CG   68.80000
    CG change   + .03675
    New CG   68.83675

(H14) — FAA-H-8083-1, Chapter 3

AIR
5647. An aircraft is loaded with a ramp weight of 3,650 pounds and having a CG of 94.0, approximately how much baggage would have to be moved from the rear baggage area at station 180 to the forward baggage area at station 40 in order to move the CG to 92.0?

A—52.14 pounds.
B—62.24 pounds.
C—78.14 pounds.

Determine the amount of weight to be moved:

$$\frac{Weight\ shifted}{Total\ weight} = \frac{\Delta\ CG\ (94-92)}{Dist.\ Wt.\ shifted\ (180-40)}$$

$$\frac{Weight\ to\ be\ shifted}{3,650\ lbs} = \frac{2.0}{140}$$

$$\frac{3,650 \times 2}{140} = \frac{7,300}{140} = 52.143$$

(H14) — FAA-H-8083-1, Chapter 3

AIR
5648. An airplane is loaded to a gross weight of 4,800 pounds, with three pieces of luggage in the rear baggage compartment. The CG is located 98 inches aft of datum, which is 1 inch aft of limits. If luggage which weighs 90 pounds is moved from the rear baggage compartment (145 inches aft of datum) to the front compartment (45 inches aft of datum), what is the new CG?

A—96.13 inches aft of datum.
B—95.50 inches aft of datum.
C—99.87 inches aft of datum.

1. Change in CG = Weight shifted x Distance shifted ÷ Total weight

    90 x (145 – 45) ÷ 4800 = 1.875 inches

2. Since the weight shifted forward, the CG also moves forward. The 1.875-inch change is subtracted from the original CG.

    New CG = 98.0 – 1.875 = 96.13 inches

(H14) — FAA-H-8083-1, Chapter 3

Answers

5649 [B]   5647 [A]   5648 [A]

Rotorcraft Weight and Balance

RTC
5682. With respect to using the weight information given in a typical aircraft owner's manual for computing gross weight, it is important to know that if items have been installed in the aircraft in addition to the original equipment, the

A—allowable useful load is decreased.
B—allowable useful load remains unchanged.
C—maximum allowable gross weight is increased.

The empty weight and moment given in most manufacturers' handbooks are for the basic aircraft prior to the installation of additional optional equipment. When the owner later adds items such as radio navigation equipment, autopilot, deicers, etc., the empty weight and the moment are changed. These changes must be recorded in the aircraft's weight and balance data and used in all computations. (H76) — FAA-H-8083-21

RTC
5677. (Refer to Figure 39.)
GIVEN:

	WEIGHT	ARM (IN)	MOMENT (IN.-LBS)
Empty weight	1,700	+6.0	+10,200
Pilot weight	200	-31.0	?
Oil (8 qt, all usable)	?	+1.0	?
Fuel (50 gal, all usable)	?	+2.0	?
Baggage	30	-31.0	?
TOTALS	?		?

If the datum line is located at station 0, the CG is located approximately

A—1.64 inches aft of datum.
B—1.64 inches forward of datum.
C—1.66 inches forward of datum.

1. Calculate the total weight and moment:

Item	Weight (lbs)	Arm (in)	Moment (lbs-in)
Empty weight	1,700	+6.0	+10,200
Pilot	200	-31.0	-6,200
Oil (8 qt.) x 7.5 lbs	15	+1.0	+15
Fuel (50 gal) x 6 lbs	300	+2.0	+600
Baggage	+ 30	-31.0	-930
Totals	2,245		3,685

2. Calculate CG by dividing total moment by total weight:

$$CG = \frac{Total\ Moment}{Total\ Weight} = \frac{3,685}{2,245} = 1.64" \text{ aft of datum}$$

(H76) — FAA-H-8083-21

RTC
5678. (Refer to Figure 40.)
GIVEN:

Basic weight (oil is included) .... 830 lb
Basic weight moment (1,000/in.-lb) 104.8
Pilot weight .... 175 lb
Passenger weight.. 160 lb
Fuel .... 19.2 gal

The CG is located

A—well aft of the aft CG limit.
B—within the CG envelope.
C—forward of the forward CG limit.

1. Calculate the total weight and moments:

Item	Weight (lbs)	Arm (in)	Moment /1,000
Aircraft	830	(given in problem)	104.8
Pilot	175	79.0	13.8
Passenger	160	79.0	12.6
Fuel (19.2 gal x 6)	+ 115.2	108.6	+ 12.5
Totals	1,280.2		143.7

2. Locate the weight and moment in the center of gravity chart. The values fall aft of the CG envelope.

(H76) — FAA-H-8083-21

Answers
5682 [A]   5677 [A]   5678 [A]

Chapter 8 **Aircraft Performance**

RTC
5679. GIVEN:

	WT	LNG. ARM	LNG. MOM.	LAT. ARM.	LAT. MOM.
Empty weight	1700	116.1	?	+0.2	—
Fuel (75 gal at 6.8 ppg)	?	110.0	?	—	—
Oil	12	179.0	?	—	—
Pilot (right seat)	175	65.0	?	+12.5	?
Passenger (left seat)	195	104.0	?	-13.3	?
TOTALS	?	?	?	?	?

Determine the longitudinal and lateral CG respectively.

A—109.35" and -.04".
B—110.43" and +.02".
C—110.83" and -.02".

1. Determine weight and moments for both the longitudinal and the lateral weight distributions:

   a. **Longitudinal**

Item	Weight (lbs)	Arm (in)	Moment (lbs-in)
Empty weight	1,700	116.1	197,370
Fuel (75 x 6.8)	510	110.0	56,100
Oil	12	179.0	2,148
Pilot	175	65.0	11,375
Passenger	+ 195	104.0	+20,280
Totals	2,592		287,273

   b. **Lateral**

Item	Weight (lbs)	Arm (in)	Moment (lbs-in)
Empty weight	1,700	+0.2	+340
Pilot	175	+12.5	+2,188
Passenger	+ 195	-13.3	-2,594
Totals	2,070		-66

2. Calculate the longitudinal and lateral CGs by dividing total moment by total weight:

   $$CG = \frac{Total\ Moment}{Total\ Weight}$$

   a. **Longitudinal**

   $$CG = \frac{287,273}{2,592} = 110.83\ in$$

   b. **Lateral**

   $$CG = \frac{-66}{2,070} = -0.03188\ in$$

(H76) — FAA-H-8083-21

RTC
5644. (Refer to Figure 37.)
GIVEN:

	WEIGHT	MOMENT
Gyroplane basic weight (oil included)	1,315	150.1
Pilot weight	140	?
Passenger weight	150	?
27 gal fuel	162	?

The CG is located

A—outside the CG envelope; the maximum gross weight is exceeded.
B—outside the CG envelope; the maximum gross weight and the gross-weight moment are exceeded.
C—within the CG envelope; neither maximum gross weight nor gross-weight moment is exceeded.

Total the weights and moments to determine if the CG is within limits.

Item	Weight	Moment
Gyroplane	1,315	150.1
Pilot	140	7.2
Passenger	150	12.6
27 gallons gas	+ 162	+ 17.8
Total	1,767	187.7

These figures fall within permitted weight and CG.

(H12) — FAA-H-8083-1, Chapter 3

RTC
5645. (Refer to Figure 37.)
GIVEN:

	WEIGHT	MOMENT
Gyroplane basic weight (oil included)	1,315	154.0
Pilot weight	145	?
Passenger weight	153	?
27 gal fuel	162	?

The CG is located

A—outside the CG envelope; the maximum gross weight is exceeded.
B—outside the CG envelope; but the maximum gross weight is not exceeded.
C—within the CG envelope; neither maximum gross weight nor gross-weight moment is exceeded.

Continued

Answers

5679 [C]   5644 [C]   5645 [B]

Total the weights and moments to determine if the CG is within limits.

Item	Weight	Moment
Gyroplane	1,315	154.0
Pilot	145	7.4
Passenger	153	12.8
27 gallons gas	+ 162	+ 17.6
Total	1,775	191.8

These figures fall within permitted weight but outside CG.

(H12) — FAA-H-8083-1, Chapter 3

RTC
5680. A helicopter is loaded in such a manner that the CG is located aft of the aft allowable CG limit. Which is true about this situation?

A— In case of an autorotation, sufficient aft cyclic control may not be available to flare properly.
B— This condition would become more hazardous as fuel is consumed, if the main fuel tank is located aft of the rotor mast.
C— If the helicopter should pitchup due to gusty winds during high-speed flight, there may not be sufficient forward cyclic control available to lower the nose.

If flight is continued in this condition, the pilot may find it impossible to fly in the upper allowable airspeed range due to insufficient forward cyclic displacement to maintain a nose-low attitude. This particular condition may become quite dangerous if gusty or rough air accelerates the helicopter to a higher airspeed than forward cyclic control will allow. The nose will start to rise and full forward cyclic stick may not be sufficient to hold it down or to lower it once it rises. (H76) — FAA-H-8083-21

RTC
5681. A helicopter is loaded in such a manner that the CG is located forward of the allowable CG limit. Which is true about this situation?

A— This condition would become less hazardous as fuel is consumed if the fuel tank is located aft of the rotor mast.
B— In case of engine failure and the resulting autorotation, sufficient cyclic control may not be available to flare properly to land.
C— Should the aircraft pitchup during cruise flight due to gusty winds, there may not be enough forward cyclic control available to lower the nose.

Flight under this condition should not be continued, since the possibility of running out of rearward cyclic control will increase rapidly as fuel is consumed, and the pilot may find it impossible to decelerate sufficiently to bring the helicopter to a stop. Also, in case of engine failure and the resulting autorotation, sufficient cyclic control may not be available to flare properly for the landing. (H76) — FAA-H-8083-21

Glider Weight and Balance

GLI
5287. In regard to the location of the glider's CG and its effect on glider spin characteristics, which is true? If the CG is too far

A—aft, a flat spin may develop.
B—forward, spin entry will be impossible.
C—aft, spins will degenerate into CG high-speed spirals.

The location of the center of gravity of the glider critically affects its spin characteristics. If the CG is too far back a flat spin will develop and recovery may be impossible. (N21) — Soaring Flight Manual

GLI
5288. The CG of most gliders is located

A—ahead of the aerodynamic center of the wing to increase lateral stability.
B—ahead of the aerodynamic center of the wing to increase longitudinal stability.
C—behind the aerodynamic center of the wing to increase longitudinal stability.

Sailplanes are designed to achieve longitudinal stability by making the aircraft slightly nose heavy with the center of gravity ahead of the center of lift. (N21) — Soaring Flight Manual

Answers

5680 [C]   5681 [B]   5287 [A]   5288 [B]

GLI
5289. Loading a glider so that the CG exceeds the aft limits results in

A— excessive load factor in turns.
B— excessive upward force on the tail, and causes the nose to pitch down.
C— loss of longitudinal stability, and causes the nose to pitch up at slow speeds.

If the CG is displaced too far aft on the longitudinal axis, a tail-heavy condition will result. As the CG moves aft, a less stable condition occurs. (N21) — Soaring Flight Manual

---

GLI
5640. (Refer to Figure 36.)
GIVEN:

	WEIGHT	ARM	MOMENT
Empty weight	610	96.47	?
Pilot (fwd seat)	150	?	?
Passenger (aft seat)	180	?	?
Radio and batteries	10	23.20	?
TOTALS	?	?	?

The CG is located at station

A— 33.20.
B— 59.55.
C— 83.26.

1. Determine the center of gravity under these conditions:

Item	Weight	x Arm	= Moment
Empty weight	610	96.47	58,846.7
Pilot (fwd seat)	150	43.80	6,570.0
Passenger (aft seat)	180	74.70	13,446.0
Radio & batteries	+ 10	23.20	+ 232.0
Totals	950		79,094.7

2. Calculate the CG location by dividing the total moment by the total weight:

$$CG = \frac{Total\ Moment}{Total\ Weight} = \frac{79{,}094.7}{950} = 83.26\ in$$

(H12) — FAA-H-8083-1, Chapter 3

GLI
5641. (Refer to Figure 36.)
GIVEN:

	WEIGHT	ARM	MOMENT
Empty weight	612	96.47	?
Pilot (fwd seat)	170	?	?
Passenger (aft seat)	160	?	?
Radio and batteries	10	23.20	?
Ballast	20	14.75	?
TOTALS	?	?	?

The CG is located at station

A— 81.23.
B— 82.63.
C— 83.26.

1. Determine the center of gravity under these conditions:

Item	Weight	x Arm	= Moment
Empty weight	612	96.47	59,040
Pilot (fwd seat)	170	43.80	7,446
Passenger (aft seat)	160	74.70	11,952
Radio & batteries	10	23.20	232
Ballast	+ 20	14.75	+ 295
Totals	972		78,965

2. Calculate the CG location by dividing total moment by total weight:

$$CG = \frac{Total\ Moment}{Total\ Weight} = \frac{78{,}965}{972} = 81.24\ in$$

(H12) — FAA-H-8083-1, Chapter 3

GLI
5642. (Refer to Figure 36.)
GIVEN:

	WEIGHT	ARM	MOMENT
Empty weight	605	96.47	?
Pilot (fwd seat)	120	?	?
Passenger (aft seat)	160	?	?
Radio and batteries	20	23.20	?
Ballast	40	14.75	?
TOTALS	?	?	?

The CG is located at station

A— 79.77.
B— 80.32.
C— 81.09.

Continued

Answers

5289 [C]   5640 [C]   5641 [A]   5642 [C]

Chapter 8 **Aircraft Performance**

1. Determine the center of gravity under these conditions:

Item	Weight	x Arm	= Moment
Empty weight	605	96.47	58,364.4
Pilot (fwd seat)	120	43.80	5,256.0
Passenger (aft seat)	160	74.70	11,952.0
Radio & batteries	20	23.20	464.0
Ballast	+ 40	14.75	+ 590.0
Totals	945		76,626.4

2. Calculate the CG location by dividing the total moment by the total weight:

   $$CG = \frac{\text{Total Moment}}{\text{Total Weight}} = \frac{76{,}626.4}{945} = 81.09 \text{ in}$$

(H12) — FAA-H-8083-1, Chapter 3

GLI
5643. GIVEN:

	WEIGHT	ARM	MOMENT
Empty weight	957	29.07	?
Pilot (fwd seat)	140	-45.30	?
Passenger (aft seat)	170	+1.60	?
Ballast	15	-45.30	?
TOTALS	?	?	?

The CG is located at station

A— -6.43.
B— +16.43.
C— +27.38.

1. Calculate the moments, and the total weight and moment.

Item	Weight	x Arm	= Moment
Empty weight	957	29.07	27,820.0
Pilot (fwd seat)	140	- 45.30	- 6,342.0
Passenger (aft seat)	170	1.60	272.0
Ballast	+ 15	- 45.30	- 679.5
Totals	1,282		21,070.5

2. CG station = Moment ÷ Weight

   21,070.5 ÷ 1,282 = 16.4356

(H12) — FAA-H-8083-1, Chapter 3

Ground Operations During Windy Conditions

When taxiing in moderate to strong wind conditions, the airplane's control surfaces must be used to counteract the effects of wind. In airplanes equipped with a nose wheel (tricycle-gear), use the appropriate taxi procedure:

1. The elevator should be in the neutral position when taxiing into a headwind.

2. The upwind aileron should be held in the up position when taxiing in a crosswind, or the upwind wing will tend to be lifted.

3. The elevator should be held in the down position and the upwind aileron down when taxiing with a quartering tailwind (the most critical condition for a nosewheel-type airplane).

See Figure 8-3.

When an airplane equipped with a tail wheel is taxied into a headwind, the elevator should be held in the up position to hold the tail down. In a quartering tailwind, both the upwind aileron and the elevator should be in the down position.

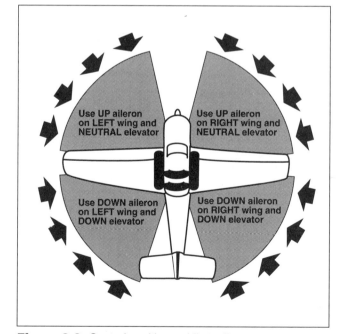

Figure 8-3. Control positions while taxiing

Answers

5643 [B]

Chapter 8 **Aircraft Performance**

AIR
5655. When taxiing during strong quartering tailwinds, which aileron positions should be used?

A—Neutral.
B—Aileron up on the side from which the wind is blowing.
C—Aileron down on the side from which the wind is blowing.

When taxiing with a quartering tailwind, the yoke/stick should be pointed away from the wind. This causes the aileron on the upwind side to go down so the wind will hold down the wing. The elevator will also be down, causing the wind to blow over the top of it and preventing lifting of the tail. (H54) — FAA-H-8083-3, Chapter 2

AIR
5656. While taxiing a light, high-wing airplane during strong quartering tailwinds, the aileron control should be positioned

A—neutral at all times.
B—toward the direction from which the wind is blowing.
C—opposite the direction from which the wind is blowing.

When taxiing with a quartering tailwind, the yoke/stick should be pointed away from the wind. This causes the aileron on the upwind side to go down so the wind will hold down the wing. The elevator will also be down, causing the wind to blow over the top of it and preventing lifting of the tail. (H54) — FAA-H-8083-3, Chapter 2

Headwind and Crosswind Components

In general, taking off into a wind improves aircraft performance, and reduces the runway distance required to become airborne. The stronger the wind, the better the aircraft performs. Crosswinds, however, may make the aircraft difficult or impossible to control. The magnitudes of headwind and crosswind components produced by a wind of a given direction and speed can be determined by using the chart as shown in FAA Figure 31.

Problem:

The wind is reported to be from 190° at 15 knots, and you plan to land on runway 13. What will the headwind and crosswind components be?

Solution:

1. Compute the angle between the wind and the runway:

    190° wind direction
    −130° runway heading
    ─────
    60° wind angle across runway

2. Find the intersection of the 60° angle radial line and the 15-knot wind velocity arc on the graph. From the intersection move downward and read the crosswind component of 13 knots. From the point of intersection move to the left and read the headwind component of 7 knots. *See* Figure 8-4.

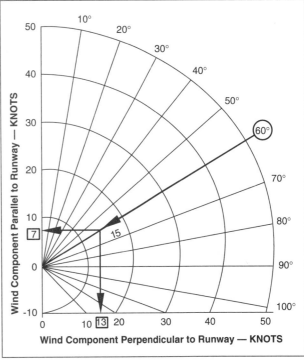

Figure 8-4. Wind components

Answers

5655 [C]   5656 [C]

Chapter 8 Aircraft Performance

AIR

5615. (Refer to Figure 31.) Rwy 30 is being used for landing. Which surface wind would exceed the airplane's crosswind capability of 0.2 V_{SO}, if V_{SO} is 60 knots?

A—260° at 20 knots.
B—275° at 25 knots.
C—315° at 35 knots.

1. Compute 0.2 V_{SO}:

    60 x 0.2 = 12 KIAS crosswind component

2. Compute the angle between the wind and the runway:

    300°   runway direction
    – 260°   wind direction
    040°   wind angle across runway

3. Determine the crosswind component of 20 knots at 40° crossing angle.

(H317) — AC 61-23C, Chapter 4

AIR, RTC, GLI

5616. (Refer to Figure 31.) If the tower-reported surface wind is 010° at 18 knots, what is the crosswind component for a Rwy 08 landing?

A—7 knots.
B—15 knots.
C—17 knots.

1. Compute the angle between the wind and each runway:

    080°   runway direction
    – 010°   wind direction
    070°   wind angle across runway

2. Determine the crosswind component of 18 knots at 070°, 17 KIAS.

(H317) — AC 61-23C, Chapter 4

AIR, RTC, GLI

5617. (Refer to Figure 31.) The surface wind is 180° at 25 knots. What is the crosswind component for a Rwy 13 landing?

A—19 knots.
B—21 knots.
C—23 knots.

1. Compute the angle between the wind and each runway:

    180°   wind direction
    – 130°   runway direction
    050°   wind angle across runway

2. Determine the crosswind component of 25 knots at 050°, 19 KIAS.

(H317) — AC 61-23C, Chapter 4

AIR, RTC, GLI

5618. (Refer to Figure 31.) What is the headwind component for a Rwy 13 takeoff if the surface wind is 190° at 15 knots?

A—7 knots.
B—13 knots.
C—15 knots.

1. Compute the angle between the wind and each runway:

    190°   wind direction
    – 130°   runway direction
    060°   wind angle across runway

2. Determine the headwind component for 15 knots at 060°, 7 KIAS.

(H317) — AC 61-23C, Chapter 4

Answers

5615 [A]   5616 [C]   5617 [A]   5618 [A]

Chapter 8 **Aircraft Performance**

Density Altitude

Aircraft performance charts indicate what performance (rate of climb, takeoff roll, etc.) can be expected of an aircraft under stipulated conditions. Prediction of performance is based upon standard atmospheric conditions, or the International Standard Atmosphere, which at sea level is a temperature of +15°C (+59°F) and an atmospheric pressure of 29.92" Hg (1013.2 hectopascals).

In a standard atmosphere, temperature changes at a rate of 2°C (3.5°F) per 1,000 feet, and pressure changes approximately 1" Hg per 1,000 feet. Temperature and/or pressure deviations from standard will change the air density. The result is a value for density altitude, which affects aircraft performance. Performance charts based on density altitude allow the pilot to predict how an aircraft will perform.

Relative humidity also affects density altitude but is not considered when the performance charts are formulated. A combination of high temperature, high humidity, and high altitude will result in a density altitude higher than the pressure altitude, which results in reduced aircraft performance.

AIR, RTC
5234. The performance tables of an aircraft for takeoff and climb are based on

A—pressure/density altitude.
B—cabin altitude.
C—true altitude.

Pressure and density altitude are the factors used in the performance tables of the aircraft for takeoff and climb. (H66) — AC 61-23C, Chapter 4

Answer (B) is incorrect because the cabin altitude is the altitude that is in the pressurized cabin. Answer (C) is incorrect because true altitude is the height above sea level. Aeronautical charts depict obstacle elevations in true altitude.

AIR, RTC
5300. What effect, if any, would a change in ambient temperature or air density have on gas turbine engine performance?

A—As air density decreases, thrust increases.
B—As temperature increases, thrust increases.
C—As temperature increases, thrust decreases.

An increase in altitude causes the engine air flow to decrease in a manner nearly identical to the altitude density ratio. This causes a significant decrease in thrust at altitude versus thrust at sea level as altitude increases. An increase in inlet air temperature will provide a lower combustion gas energy and cause a lower jet velocity. (H66) — AC 61-23C, Chapter 4

ALL
5302. What is the standard temperature at 10,000 feet?

A— -5°C.
B— -15°C.
C— +5°C.

Standard temperature at sea level is 15°C. The standard lapse rate is 2°C/1,000 feet.

10 x 2 = -20°C temperature decrease to 10,000 feet
+15° – 20° = -5°C standard temperature at 10,000 feet

(I21) — AC 00-6A, Chapter 2

ALL
5303. What is the standard temperature at 20,000 feet?

A— -15°C.
B— -20°C.
C— -25°C.

Standard temperature at sea level is 15°C. The average lapse rate is 2°C/1,000 feet.

20 x -2 = -40°C temperature decrease to 20,000 feet
15°C – 40°C = -25°C standard temperature at 20,000 feet

(I21) — AC 00-6A, Chapter 2

ALL
5305. What are the standard temperature and pressure values for sea level?

A—15°C and 29.92" Hg.
B—59°F and 1013.2" Hg.
C—15°C and 29.92 Mb.

In the standard atmosphere, sea level pressure is 29.92 inches of mercury or 1013.2 hectopascals. The temperature at mean sea level is 15°C (59°F). (I22) — AC 00-6A, Chapter 3

Answers

5234 [A]   5300 [C]   5302 [A]   5303 [C]   5305 [A]

Takeoff and Landing Considerations

During takeoff with crosswind, rudder is used to maintain directional control, and aileron pressure to counter the wind. For both conventional and nosewheel-type airplanes, a higher than normal lift-off airspeed is used. Landing under crosswind conditions requires the direction of motion of the airplane and its longitudinal axis to be parallel to the runway at the moment of touchdown. During gusty wind conditions both the approach and landing should be with power on.

Operating with an uphill runway slope has no effect on the takeoff speed, but will increase the takeoff distance.

In the event of an actual engine failure immediately after takeoff and before a safe maneuvering altitude is attained, it is not recommended to attempt to turn back to the airport. Instead, it is generally safer to maintain a safe airspeed, and select a field directly ahead or slightly to either side of the takeoff path.

While diverting to an alternate airport due to an emergency, the heading should be changed to establish the new course immediately. Apply rule-of-thumb computations, estimates and appropriate shortcuts to calculate wind correction, actual distance and estimated time and fuel required.

ALL
5503. When diverting to an alternate airport because of an emergency, pilots should

A—rely upon radio as the primary method of navigation.
B—climb to a higher altitude because it will be easier to identify checkpoints.
C—apply rule-of-thumb computations, estimates, and other appropriate shortcuts to divert to the new course as soon as possible.

Course, time, speed and distance computations in flight require the same basic procedures as those used in preflight planning. However, because of cockpit space limitations and available equipment, and because the pilot's attention must be divided between solving the problem and operating the aircraft, the pilot must take advantage of all possible shortcuts and rule-of-thumb computations. (H61) — FAA-H-8083-3, Chapter 3

Answer (A) is incorrect because any form of navigation is appropriate. Answer (B) is incorrect because climbs may consume time and fuel.

AIR, GLI
5614. What effect does an uphill runway slope have on takeoff performance?

A—Increases takeoff speed.
B—Increases takeoff distance.
C—Decreases takeoff distance.

The effect of runway slope on takeoff distance is due to the component of weight along the inclined path of the aircraft. An upslope would contribute a retarding force component while a downslope would contribute an accelerating force component. In the case of an upslope, the retarding force component adds to drag and rolling friction to reduce the net accelerating force. (H317) — AC 61-23C, Chapter 4

AIR
5661. With regard to the technique required for a crosswind correction on takeoff, a pilot should use

A—aileron pressure into the wind and initiate the lift-off at a normal airspeed in both tailwheel- and nosewheel-type airplanes.
B—right rudder pressure, aileron pressure into the wind, and higher than normal lift-off airspeed in both tricycle- and conventional-gear airplanes.
C—rudder as required to maintain directional control, aileron pressure into the wind, and higher than normal lift-off airspeed in both conventional- and nosewheel-type airplanes.

Rudder is required at all times to maintain directional control. Aileron pressure into the wind will keep the wing down and prevent side skipping. A slightly higher than normal takeoff speed is desirable to ensure a very definite liftoff with no settling. (H57) — FAA-H-8083-3, Chapter 3

Answer (A) is incorrect because the lift-off speed should be increased slightly and rudder pressure should be applied to maintain directional control. Answer (B) is incorrect because rudder pressure should be applied to maintain directional control, this can be both right or left deflection.

Answers

5503 [C]   5614 [B]   5661 [C]

AIR
5662. When turbulence is encountered during the approach to a landing, what action is recommended and for what primary reason?

A—Increase the airspeed slightly above normal approach speed to attain more positive control.
B—Decrease the airspeed slightly below normal approach speed to avoid overstressing the airplane.
C—Increase the airspeed slightly above normal approach speed to penetrate the turbulence as quickly as possible.

Power-on approaches at an airspeed slightly above the normal approach speed should be used for landing in significantly turbulent air. This provides for more positive control of the airplane when strong horizontal wind gusts, or up- and downdrafts, are experienced. (H58) — FAA-H-8083-3, Chapter 3

Answer (B) is incorrect because, since approach speed is well below V_A, no structural damage will occur. Answer (C) is incorrect because increased approach speed is to increase control effectiveness, not to fly through turbulence faster.

AIR
5663. A pilot's most immediate and vital concern in the event of complete engine failure after becoming airborne on takeoff is

A—maintaining a safe airspeed.
B—landing directly into the wind.
C—turning back to the takeoff field.

The altitude available is, in many ways, the controlling factor in the successful accomplishment of a forced landing. If an actual engine failure should occur immediately after takeoff and before a safe maneuvering altitude is attained, it is usually inadvisable to attempt to turn back to the field from which the takeoff was made. Instead, it is generally safer to immediately establish the proper glide attitude and select a field directly ahead, or slightly to either side, of the takeoff path. (H58) — FAA-H-8083-3, Chapter 3

Answer (B) is incorrect because landing into the wind may not always be the best choice, depending on obstructions, runway lengths, etc. Answer (C) is incorrect because turning back to the airport without sufficient altitude will not be possible.

AIR
5664. Which type of approach and landing is recommended during gusty wind conditions?

A—A power-on approach and power-on landing.
B—A power-off approach and power-on landing.
C—A power-on approach and power-off landing.

Power-on approaches at an airspeed slightly above the normal approach speed should be used for landing in significantly turbulent air. These landing approaches are usually performed at the normal approach speed plus one-half of the wind gust factor. An adequate amount of power should be used to maintain the proper airspeed throughout. (H58) — FAA-H-8083-3, Chapter 3

AIR
5665. A proper crosswind landing on a runway requires that, at the moment of touchdown, the

A—direction of motion of the airplane and its lateral axis be perpendicular to the runway.
B—direction of motion of the airplane and its longitudinal axis be parallel to the runway.
C—downwind wing be lowered sufficiently to eliminate the tendency for the airplane to drift.

Direction of motion and the longitudinal axis must be parallel to the runway or skidding will occur, with possible damage to the aircraft. (H58) — FAA-H-8083-3, Chapter 3

Answer (A) is incorrect because the direction of motion of the airplane must be parallel to the runway. Answer (C) is incorrect because the upwind wing should be lowered to eliminate drift.

Answers

5662 [A]   5663 [A]   5664 [A]   5665 [B]

Chapter 8 **Aircraft Performance**

Takeoff and Landing Distance

The Takeoff Distance Graph such as the graph shown in FAA Figure 32, allows the pilot to determine the ground roll and takeoff distance over a 50-foot obstacle.

Problem:

Using the Obstacle Takeoff Graph (FAA Figure 32), determine the total takeoff distance over a 50-foot obstacle under the following conditions:

Temperature .... 30°F
Pressure Altitude .... 6,000 feet
Weight .... 3,300 lbs
Headwind component .... 20 knots

Solution:

1. Enter the chart at 30°F. Proceed upward until intersecting the 6,000-foot pressure altitude. From this point, proceed to the right until intersecting the first reference line.

2. From this point on the reference line, proceed diagonally upward to the right (on a line spaced proportionally between the existing guide lines) to the vertical line representing 3,300 pounds. From there, proceed to the right until intersecting the second reference line.

3. From this point, proceed diagonally downward and to the right (remaining proportionally between the existing guide lines) to the vertical line representing the 20-knot headwind component. From there, proceed to the far right scale and read the total takeoff distance over a 50-foot obstacle of 1,500 feet.

The Takeoff Distance Graph is often used to find the required takeoff distance based on temperature, pressure altitude, weight and wind. The graph may also be used however, to find the maximum weight allowable under certain conditions for a particular takeoff distance.

Problem:

Using the information shown in FAA Figure 32, determine the maximum weight allowable to take off over a 50-foot obstacle in 1,500 feet under the following conditions:

Temperature .... 30°F
Pressure Altitude .... 6,000 feet
Headwind component .... 20 knots

Solution:

1. Enter the chart at 30°F. Proceed upward until intersecting the 6,000-foot pressure altitude line. From this point, proceed to the right until intersecting the first reference line.

2. From this point, draw a line diagonally upward and to the right (remaining proportionally spaced between the existing guide lines) until intersecting the second reference line.

3. Re-enter the graph from the right at the 1,500-foot takeoff distance over a 50-foot obstacle. Proceed to the left until intersecting the vertical line representing a 20-knot headwind component. From there, proceed diagonally upward and to the left (remaining proportionally spaced between the existing guide lines) until intersecting the second reference line.

Chapter 8 Aircraft Performance

4. From this point, proceed to the left until intersecting the previously drawn diagonal line. From there, proceed vertically downward and read the maximum allowable takeoff weight of 3,300 pounds.

Use the same procedure for Landing Distance Graphs as for the Takeoff Distance Graph.

ALL
5208. At higher elevation airports the pilot should know that indicated airspeed

A—will be unchanged, but groundspeed will be faster.
B—will be higher, but groundspeed will be unchanged.
C—should be increased to compensate for the thinner air.

An airplane at altitude will land at the same indicated airspeed as at sea level but, because of the reduced air density, will have a greater true airspeed. If the true airspeed is greater, the ground speed will be greater. (H66) — FAA-H-8083-3, Chapter 3

AIR
5619. (Refer to Figure 32.)
GIVEN:

Temperature .... 75°F
Pressure altitude .... 6,000 ft
Weight .... 2,900 lb
Headwind .... 20 kts

To safely take off over a 50-foot obstacle in 1,000 feet, what weight reduction is necessary?

A—50 pounds.
B—100 pounds.
C—300 pounds.

1. Start at the right side of the chart and find the 1,000-foot takeoff distance mark. From there, draw a horizontal line to the 20-knot headwind mark, and following the sloped lines, draw a line to the reference line.

2. Draw a horizontal line all the way across the weight section to the other reference line.

3. Starting from the left side of the chart, find the 75°F mark and draw a line up to the 6,000-foot pressure altitude line, and follow that sloped line across to the reference line. Follow the sloped reference lines to the right, up to where your line intercepts the horizontal line drawn previously across that section.

4. Draw a vertical line down from where they meet and read the maximum weight (2,600 pounds) for the aircraft to be able to clear the 50-foot obstacle.

5. Subtract that weight from 2,900 pounds to get 300 pounds.

(H317) — AC 61-23C, Chapter 4

AIR
5620. (Refer to Figure 32.)
GIVEN:

Temperature .... 50°F
Pressure altitude .... 2,000 feet
Weight .... 2,700 lb
Wind .... Calm

What is the total takeoff distance over a 50-foot obstacle?

A—800 feet.
B—650 feet.
C—1,050 feet.

1. Enter chart at 50°F, proceed up to the 2,000 foot pressure altitude line.

2. From this point, go to the right, to the weight reference line.

3. From this point, proceed up and to the right on the trend lines, until intercepting the 2,700 pound weight line. Then proceed to the right, to the wind reference line.

4. From this point, since the winds are calm, proceed to the right and read 800 feet.

(H317) — AC 61-23C, Chapter 4

Answers

5208 [A]   5619 [C]   5620 [A]

Chapter 8 **Aircraft Performance**

AIR
5621. (Refer to Figure 32.)
GIVEN:

Temperature .... 100°F
Pressure altitude .... 4,000 ft
Weight .... 3,200 lb
Wind .... Calm

What is the ground roll required for takeoff over a 50-foot obstacle?

A—1,180 feet.
B—1,350 feet.
C—1,850 feet.

1. Enter chart at 100°F, proceed up to the 4,000-foot pressure altitude line.
2. From this point, go to the right, to the weight reference line.
3. From this point, proceed up and to the right on a line spaced proportionally between the trend lines, until intercepting the 3,200-pound weight line. Then proceed to the right, to the wind reference line.
4. From this point, since the winds are calm, proceed to the right and read 1,850 feet.
5. Calculate the ground roll using the note that the ground roll is approximately 73% of total take-off distance over a 50-foot obstacle:

   1,850 x .73 = 1,350 feet ground roll

(H317) — AC 61-23C, Chapter 4

AIR
5622. (Refer to Figure 32.)
GIVEN:

Temperature .... 30°F
Pressure altitude .... 6,000 ft
Weight .... 3,300 lb
Headwind .... 20 kts

What is the total takeoff distance over a 50-foot obstacle?

A—1,100 feet.
B—1,300 feet.
C—1,500 feet.

1. Enter chart at 30°F, proceed up to the 6,000-foot pressure altitude line.
2. From this point, go to the right to the weight reference line.
3. From this point, proceed up and to the right on a line spaced proportionally between the trend lines, until intercepting the 3,300-pound weight line. Then proceed to the right, to the wind reference line.
4. From this point, proceed down and to the right (maintain proportional spacing), to the 20-knot headwind line. From this point, proceed to the right and read 1,500 feet.

(H317) — AC 61-23C, Chapter 4

AIR
5628. (Refer to Figure 35.)
GIVEN:

Temperature .... 70°F
Pressure altitude .... Sea level
Weight .... 3,400 lb
Headwind .... 16 kts

Determine the approximate ground roll.

A—689 feet.
B—716 feet.
C—1,275 feet.

1. Enter the chart at the 70°F OAT point and proceed upward to intercept the sea level pressure altitude line. Proceed right to the intersection with the vertical reference line.
2. From the reference line, proceed upward to the right following the trend line, until reaching a vertical line representing the weight of 3,400 pounds.
3. Draw a line to the right from the intercept with the weight line to the intercept with the right vertical reference line.
4. Construct a vertical line to represent the headwind, 16 KIAS. Draw a line from the intercept with the reference line to the intersect with the headwind line. Proceed to the right and read the total distance over a 50-foot obstacle: 1,275 feet.
5. Apply the note to calculate the ground roll (53% of the landing distance).

   1,275 x 0.53 = 676 feet ground roll

(H317) — AC 61-23C, Chapter 4

Answers

5621 [B]   5622 [C]   5628 [A]

AIR
5629. (Refer to Figure 35.)

GIVEN:

Temperature	85°F
Pressure altitude	6,000 ft
Weight	2,800 lb
Headwind	14 kts

Determine the approximate ground roll.

A—742 feet.
B—1,280 feet.
C—1,480 feet.

1. Enter the chart at the 85°F OAT point and proceed upward to intercept the 6,000-foot pressure altitude line. Proceed right to the intersection with the vertical reference line.
2. From the reference line, proceed upward to the right following the trend line, until reaching a vertical line representing the weight of 2,800 pounds.
3. Draw a line to the right from the intercept with the weight line, to the intercept with the right vertical reference line.
4. Construct a vertical line to represent the headwind, 14 KIAS. Draw a line from the intercept with the reference line to the intersect with the headwind line. Proceed to the right and read the total distance over a 50-foot obstacle: 1,400 feet.
5. Apply the note to calculate the ground roll (53% of the landing distance).

    1,400 x 0.53 = 742 feet ground roll

(H317) — AC 61-23C, Chapter 4

AIR
5630. (Refer to Figure 35.)

GIVEN:

Temperature	50°F
Pressure altitude	Sea level
Weight	3,000 lb
Headwind	10 kts

Determine the approximate ground roll.

A—425 feet.
B—636 feet.
C—836 feet.

1. Enter the chart at the 50°F OAT point and proceed upward to intercept the sea level pressure altitude line. Proceed right to the intersection with the vertical reference line.
2. From the reference line, proceed upward to the right following the trend line, until reaching a vertical line representing the weight of 3,000 pounds.
3. Draw a line to the right from the intercept with the weight line, to the intercept with the right vertical reference line.
4. Construct a vertical line to represent the headwind, 10 knots. Draw a line from the intercept with the reference line to the intersect with the headwind line. Proceed to the right and read the total distance over a 50-foot obstacle: 1,200 feet.
5. Apply the note to calculate the ground roll (53% of the landing distance).

    1,200 x 0.53 = 636 feet ground roll

(H317) — AC 61-23C, Chapter 4

AIR
5631. (Refer to Figure 35.)

GIVEN:

Temperature	80°F
Pressure altitude	4,000 ft
Weight	2,800 lb
Headwind	24 kts

What is the total landing distance over a 50-foot obstacle?

A—1,125 feet.
B—1,250 feet.
C—1,325 feet.

1. Enter the chart at the 80°F OAT point and proceed upward to intercept the 4,000-foot pressure altitude line. Proceed right to the intersection with the vertical reference line.
2. From the reference line, proceed upward to the right following the trend line, until reaching a vertical line representing the weight of 2,800 pounds.
3. Draw a line to the right from the intercept with the weight line, to the intercept with the right vertical reference line.
4. Construct a vertical line to represent the headwind, 24 knots. Draw a line from the intercept with the reference line to the intercept with the headwind line. Proceed to the right and read the total distance over a 50-foot obstacle: 1,125 feet.

(H317) — AC 61-23C, Chapter 4

Answers

5629 [A]   5630 [B]   5631 [A]

Chapter 8 **Aircraft Performance**

Fuel Consumption vs. Brake Horsepower

The Fuel Consumption vs. Brake Horsepower Graph allows the pilot to determine the gallons per hour used, based on percent of power and mixture settings. *See* FAA Figure 8.

Problem:

How much flight time is available with a 45-minute fuel reserve under the following conditions?

Fuel on board .... 38 gallons
Mixture .... Cruise (lean)
Percent power 55%

Solution:

1. Determine the fuel consumption (GPH) based on a cruise (lean) setting of 55% power. Locate the point where the diagonal cruise (lean) line and the vertical 55% power line intersect. From the point of intersection, proceed to the left and read the fuel consumption of 11.4 GHP.

2. Find the total endurance possible with 38 gallons at 11.4 GHP.

    38 ÷ 11.4 = 3.33 hr (no reserve)

3. Determine the flight time available with a 45-minute (0.75 hours) fuel reserve remaining.

    ```
      3.33 hr   no reserve
    – 0.75 hr   45-minute reserve
      2.58 hr   with 45-minute reserve
    ```

    2.58 hr = 2 hours, 34 minutes

AIR
5451. (Refer to Figure 8.)

GIVEN:

Fuel quantity .... 47 gal
Power-cruise (lean) .... 55 percent

Approximately how much flight time would be available with a night VFR fuel reserve remaining?

A—3 hours 8 minutes.
B—3 hours 22 minutes.
C—3 hours 43 minutes.

1. Enter chart at the point where the 55% maximum continuous power line intersects the curved cruise (lean) line.

2. From that point, proceed to the left and read fuel flow: 11.4 GPH

3. Compute time of flight:

    47 ÷ 11.4 = 4.12 hours

4. Compute available flight time with a 45-minute (0.75 hour) reserve:

    4.12 – 0.75 = 3.37 = 3 hours, 22 minutes

(H342) — AC 61-23C, Chapter 8

AIR
5452. (Refer to Figure 8.)

GIVEN:

Fuel quantity .... 65 gal
Best power (level flight) .... 55 percent

Approximately how much flight time would be available with a day VFR fuel reserve remaining?

A—4 hours 17 minutes.
B—4 hours 30 minutes.
C—5 hours 4 minutes.

1. Enter chart at the point where the 55% maximum continuous power line intersects the curved best power level flight line.

2. From that point, proceed to the left and read fuel flow: 13.0 GPH

3. Compute time of flight:

    65 gal ÷ 13.0 = 5.0 hours

4. Compute available flight time with a 30-minute (0.5 hour) reserve:

    5.0 – 0.5 = 4.5 = 4 hour, 30 minutes

(H342) — AC 61-23C, Chapter 8

Answers

5451 [B]   5452 [B]

AIR
5453. (Refer to Figure 8.) Approximately how much fuel would be consumed when climbing at 75 percent power for 7 minutes?
A—1.82 gallons.
B—1.97 gallons.
C—2.15 gallons.

1. Enter chart at point where the 75% maximum continuous power line intersects the curved take off and climb line.
2. From that point proceed to the left and read fuel flow 18.4 GPH.
3. Compute fuel burn:

   $(18.4 \times 7) \div 60 = 2.1$ gallons

(H342) — AC 61-23C, Chapter 8

AIR
5454. (Refer to Figure 8.) Determine the amount of fuel consumed during takeoff and climb at 70 percent power for 10 minutes.
A—2.66 gallons.
B—2.88 gallons.
C—3.2 gallons.

1. Interpolate a vertical line for 70% power vertically between the 65% and 75% power lines.
2. From the point of intersection of the 70% line with the takeoff and climb curve, draw a line to the left and read the fuel flow, 17.3 GPH.
3. Compute the fuel burned:

   $(17.3 \times 10) \div 60 = 2.88$ gallons

(H342) — AC 61-23C, Chapter 8

AIR
5455. (Refer to Figure 8.) With 38 gallons of fuel aboard at cruise power (55 percent), how much flight time is available with night VFR fuel reserve still remaining?
A—2 hours 34 minutes.
B—2 hours 49 minutes.
C—3 hours 18 minutes.

1. Enter chart at the point where the 55% maximum continuous power line intersects the curved cruise (lean) line.
2. From that point, proceed to the left and read fuel flow 11.4 GPH.
3. Compute time of flight:

   38 gal ÷ 11.4 GPH = 3.33 hours
4. Compute available flight time with a 45-minute (0.75 hour) reserve:

   3.33 − 0.75 = 2.58 = 2 hours, 34 minutes

(H342) — AC 61-23C, Chapter 8

Time, Fuel and Distance to Climb

The Time, Fuel and Distance to Climb tables and graphs allow the pilot to calculate the time required, fuel used and distance covered during a climb to a specified altitude.

Table Method

Problem:

Referring to the Time, Fuel and Distance to Climb Table (FAA Figure 9), determine the amount of fuel used from engine start to a **pressure altitude (PA)** of 12,000 feet using a normal rate of climb under the following conditions:

Weight	3,800 lbs
Airport Pressure altitude	4,000 feet
Temperature at 4,000 feet	26°C

Answers
5453 [C]   5454 [B]   5455 [A]

Chapter 8 **Aircraft Performance**

Solution:

1. Locate the section appropriate to 3,800 pounds weight. Read across the 4,000-foot pressure altitude line to the entry under fuel used, 12 pounds.
2. Read across the 12,000-foot pressure altitude line to the entry under fuel used, 51 pounds.
3. Calculate the fuel required to climb:

    51 − 12 = 39 lbs

4. Apply Note #2. (A temperature of 26°C is +19°C, with respect to the standard atmosphere at 4,000 feet.)

    39 x 1.19 = 46.4 lbs

5. Apply Note #1:

    46.4 lbs + 12.0 lbs start and taxi = 58.4 lbs total

Graph Method

Problem:

Using the Fuel, Time and Distance to Climb Graph shown in FAA Figure 15, determine the fuel, time and distance required to climb to cruise altitude under the following conditions:

Airport pressure altitude .... 2,000 feet
Airport temperature .... 20°C
Cruise pressure altitude .... 10,000 feet
Cruise temperature .... 0°C

Solution:

1. Determine the fuel, time and distance to climb to 10,000 feet at 0°C. Locate the temperature at 0°C on the bottom left side of the graph. Proceed upward to the diagonal line representing the pressure altitude of 10,000 feet.
2. From this point proceed to the right and stop at each of the lines, forming three points of intersection.
3. From each of the three points, draw a line downward to read the fuel used (6 gals), time (11 min) and distance (16 NM). These are the values for a climb from sea level to 10,000 feet.
4. Repeat the steps to find the values for a climb from sea level to 2,000 feet. The fuel used is 1 gallon, time is 2 minutes, and distance is 3 NM.
5. Find the fuel, time and distance values for a climb from 2,000 feet to 10,000 feet, by taking the differences for the two altitudes.

    Fuel   6 (for 10,000 feet) − 1 (for 2,000 feet) = 5 gallons
    Time   11 (for 10,000 feet) − 2 (for 2,000 feet) = 9 minutes
    Distance 16 (for 2,000 feet) − 3 (for 1,000 feet) = 13 NM

AIR
5456. (Refer to Figure 9.) Using a normal climb, how much fuel would be used from engine start to 12,000 feet pressure altitude?

Aircraft weight .... 3,800 lb
Airport pressure altitude .... 4,000 ft
Temperature .... 26°C

A—46 pounds.
B—51 pounds.
C—58 pounds.

1. Locate the section for 3,800 pounds weight. Read across the 4,000-foot PA line to the entry under fuel used, 12 pounds.
2. Read across the 12,000-foot PA line to the entry under fuel used, 51 pounds.
3. Calculate the fuel required to climb:

    51 – 12 = 39 lbs

4. Apply Note #2. (A temperature of 26°C is +19°C, relative to the standard atmosphere at 4,000 feet.)

    39 x 1.19 = 46.4 lbs

5. Apply Note #1:

    46.4 lbs + 12.0 lbs start and taxi = 58.4 lbs total

(H317) — AC 61-23C, Chapter 4

AIR
5457. (Refer to Figure 9.) Using a normal climb, how much fuel would be used from engine start to 10,000 feet pressure altitude?

Aircraft weight .... 3,500 lb
Airport pressure altitude .... 4,000 ft
Temperature .... 21°C

A—23 pounds.
B—31 pounds.
C—35 pounds.

1. Locate the section for 3,500 pounds weight. Read across the 4,000-foot PA line to the entry under fuel used, 11 pounds.
2. Read across the 10,000-foot PA line to the entry under fuel used, 31 pounds.
3. Calculate the fuel required to climb:

    31 – 11 = 20 lbs

4. Apply Note #2. (A temperature of 21°C is +14°C, with respect to the standard atmosphere at 4,000 feet.)

    20 x 1.14 = 22.8 lbs

5. Apply Note #1:

    22.8 lbs + 12.0 lbs start and taxi = 34.8 lbs total

(H317) — AC 61-23C, Chapter 4

AIR
5458. (Refer to Figure 10.) Using a maximum rate of climb, how much fuel would be used from engine start to 6,000 feet pressure altitude?

Aircraft weight .... 3,200 lb
Airport pressure altitude .... 2,000 ft
Temperature .... 27°C

A—10 pounds.
B—14 pounds.
C—24 pounds.

1. Locate the section for 3,200 pounds weight. Read across the 2,000-foot PA line to the entry under fuel used, 4 pounds.
2. Read across the 6,000-foot PA line to the entry under fuel used, 14 pounds.
3. Calculate the fuel required to climb:

    14 – 4 = 10 lbs

4. Apply Note #2. (A temperature of 27°C is +16°C, with respect to the standard atmosphere at 2,000 feet.)

    10 x 1.16 = 11.6 lbs

5. Apply Note #1:

    11.6 lbs + 12.0 lbs start and taxi = 23.6 lbs total

(H317) — AC 61-23C, Chapter 4

Answers

5456 [C]   5457 [C]   5458 [C]

Chapter 8 **Aircraft Performance**

AIR
5459. (Refer to Figure 10.) Using a maximum rate of climb, how much fuel would be used from engine start to 10,000 feet pressure altitude?

Aircraft weight .... 3,800 lb
Airport pressure altitude .... 4,000 ft
Temperature .... 30°C

A—28 pounds.
B—35 pounds.
C—40 pounds.

1. Locate the section for 3,800 pounds weight. Read across the 4,000-foot PA line to the entry under fuel used, 12 pounds.
2. Read across the 10,000-foot PA line to the entry under fuel used, 35 pounds.
3. Calculate the fuel required to climb:

    35 – 12 = 23 lbs

4. Apply Note #2. (A temperature of 30°C is +23°C, with respect to the standard atmosphere at 4,000 feet.)

    23 x 1.23 = 28.3 lbs

5. Apply Note #1:

    28.3 lbs +12.0 lbs start and taxi = 40.3 lbs total

(H317) — AC 61-23C, Chapter 4

AIR
5482. (Refer to Figure 13.)
GIVEN:

Aircraft weight .... 3,400 lb
Airport pressure altitude .... 6,000 ft
Temperature at 6,000 feet .... 10°C

Using a maximum rate of climb under the given conditions, how much fuel would be used from engine start to a pressure altitude of 16,000 feet?

A—43 pounds.
B—45 pounds.
C—49 pounds.

1. Determine the weight of the fuel required to climb from sea level to the PA of the airport (6,000 feet) by interpolating between the 4,000- and 8,000-foot values (for an aircraft weight of 3,400 pounds):

    (9 + 19) ÷ 2 = 14 lbs

2. Determine the weight of fuel required to climb from sea level to 16,000 feet PA: 39 lbs

3. Determine the weight of fuel required to climb from 6,000 feet to 16,000 feet, uncorrected for nonstandard temperature:

    39 – 14 = 25 lbs

4. Correct for nonstandard temperature, Note 2, using 7% increase (+10°C is 7°C relative to the standard atmosphere at 6,000 feet):

    1.07 x 25 = 26.7 lbs

5. Apply Note #1 to determine the total fuel requirement:

    26.7 lbs + 16.0 lbs start and taxi = 42.7
    = 43 lbs total

(H342) — AC 61-23C, Chapter 8

AIR
5483. (Refer to Figure 13.)
GIVEN:

Aircraft weight .... 4,000 lb
Airport pressure altitude .... 2,000 ft
Temperature at 2,000 feet .... 32°C

Using a maximum rate of climb under the given conditions, how much time would be required to climb to a pressure altitude of 8,000 feet?

A—7 minutes.
B—8.4 minutes.
C—11.2 minutes.

1. Determine the time required to climb from sea level to the PA of the airport (2,000 feet) by interpolating between the sea level and the 4,000-foot values (for an aircraft weight of 4,000 lbs):

    (0 + 4) ÷ 2 = 2 minutes

2. Determine the time required to climb from sea level to 8,000-foot pressure altitude: 9 minutes

3. Determine the time required to climb from 2,000 feet to 8,000-foot pressure altitude (uncorrected, for nonstandard temperature):

    9 – 2 = 7 minutes

4. Correct the time required for nonstandard temperature by applying Note #2, using 21% increase (+32°C is +21°C relative to the standard atmosphere at 2,000 feet):

    1.21 x 7 = 8.4 min

(H342) — AC 61-23C, Chapter 8

Answers

5459 [C]   5482 [A]   5483 [B]

AIR
5484. (Refer to Figure 14.)
GIVEN:

Aircraft weight .... 3,700 lb
Airport pressure altitude .... 4,000 ft
Temperature at 4,000 feet................................... 21°C

Using a normal climb under the given conditions, how much fuel would be used from engine start to a pressure altitude of 12,000 feet?

A—30 pounds.
B—37 pounds.
C—46 pounds.

1. Determine the weight of the fuel required to climb from sea level to the PA of the airport (4,000 feet) for an aircraft weight of 3,700 lbs: 12 lbs

2. Determine the weight of the fuel required to climb from sea level to 12,000-foot PA: 37 lbs

3. Determine the weight of fuel required to climb from 4,000 to 12,000 feet, uncorrected for non-standard temperature:

    37 – 12 = 25 lbs

4. Correct the fuel requirement for nonstandard temperature by applying Note 2, using 20% increase (+21°C is +14°C, relative to the standard atmosphere at 4,000 feet):

    1.20 x 25 = 30 lbs

5. Apply Note #1 to determine the total fuel requirement:

    30 lbs + 16 lbs start and taxi = 46 lbs total

(H342) — AC 61-23C, Chapter 8

AIR
5485. (Refer to Figure 14.)
GIVEN:

Aircraft weight .... 3,400 lb
Airport pressure altitude .... 4,000 ft
Temperature at 4,000 feet................................... 14°C

Using a normal climb under the given conditions, how much time would be required to climb to a pressure altitude of 8,000 feet?

A—4.8 minutes.
B—5 minutes.
C—5.5 minutes.

---

1. Determine the time which would have been required, using a 3,400 pounds weight, to climb from sea level to the airport pressure altitude of 4,000 feet, 5 minutes.

2. Determine the time to climb to 8,000 feet from sea level, 10 minutes.

3. The time required for climb, uncorrected for temperature, is 10 – 5 = 5 minutes.

4. Correct for non-standard temperature (+14°C is +7°C, relative to the standard atmosphere.).

5. Apply Note #2 using a 10% increase.

    1.1 x 5 = 5.5 minutes.

(H342) — AC 61-23C, Chapter 8

AIR
5486. (Refer to Figure 15.)
GIVEN:

Airport pressure altitude .... 4,000 ft
Airport temperature .... 12°C
Cruise pressure altitude .... 9,000 ft
Cruise temperature .... -4°C

What will be the distance required to climb to cruise altitude under the given conditions?

A—6 miles.
B—8.5 miles.
C—11 miles.

1. Enter the chart at -4°C and proceed upward until intersecting the 9,000-foot PA line. From this point, proceed to the right until intersecting the Dist-Nautical Miles curved line. From this point proceed downward and read the distance: 14.5 NM.

2. Enter chart at +12°C and proceed upward until intersecting the 4,000-foot PA line. From this point, proceed to the right until intersecting the Dist-Nautical Miles curved line. From this point, proceed downward and read the distance: 6 NM.

3. The difference is the distance required to climb:

    14.5 – 6 = 8.5 NM.

(H342) — AC 61-23C, Chapter 8

Answers

5484 [C]   5485 [C]   5486 [B]

Chapter 8 Aircraft Performance

AIR
5487. (Refer to Figure 15.)
GIVEN:

Airport pressure altitude	2,000 ft
Airport temperature	20°C
Cruise pressure altitude	10,000 ft
Cruise temperature	0°C

What will be the fuel, time, and distance required to climb to cruise altitude under the given conditions?

A—5 gallons, 9 minutes, 13 NM.
B—6 gallons, 11 minutes, 16 NM.
C—7 gallons, 12 minutes, 18 NM.

1. Enter the chart at 0°C and proceed to the 10,000-foot PA curve. From this point proceed to the right. Note the intersections with the fuel, time and distance curves. From these intersections, proceed downward and read:

    Fuel = 6 gal, Time = 11 min, Dist = 16 NM

2. Enter the chart at +20°C and proceed upward to intercept the 2,000-foot PA line, then right to the intersections with the fuel, time and distance curves. From these intersections proceed downward and read:

    Fuel = 1 gal, Time + 2 min, Dist = 3 NM

3. The differences are the fuel, time and distance to climb to cruise altitude:

    Fuel = 6 – 1 = 5 gallons
    Time = 11 – 2 = 9 minutes
    Distance = 16 – 3 = 13 NM

(H342) — AC 61-23C, Chapter 8

Cruise Performance Table

The Cruise Performance Table may be used to determine the expected percent power, true airspeed and fuel flow for a particular altitude and power setting. Based on this information, the pilot can find the estimated time en route and fuel required.

Problem:

Using the Cruise Performance Table (FAA Figure 12), find the approximate flight time available under the following conditions allowing for VFR day fuel reserve:

Pressure Altitude	18,000 feet
Temperature	-1°C
Power (Best fuel economy)	22 RPM — 20 in. manifold pressure
Usable fuel	344 lbs

Solution:

1. Determine the flight time available by finding the rate of fuel consumption. Enter the table on the left column (RPM) and read down to 2,200 RPM.

2. Find line for 20 in. MP under the second column (MP). Follow to the right, to the column that represents a temperature of -1°C (20°C above standard temperature).

3. Continue to the right, noting 43% BHP, and in the next column, read the true airspeed of 124 knots, and a fuel flow of 59 pounds per hour (pph).

4. Find the fuel rate for best fuel economy. The chart notes in the upper right-hand corner state for best fuel economy at 70% power or less, operate at 6 pph leaner than shown in chart. For the given conditions, the engine is operating at 43% BHP, so the note does apply. The fuel burn for best fuel economy is brought down to:

    59 – 6 = 53 pph

Answers
5487 [A]

5. Determine the approximate flight time based on 344 lbs of usable fuel and a fuel rate of 53 pph.

   344 ÷ 53 = 6.50 hours (no reserve)

6. Find the flight time with VFR day fuel reserve (30 minutes).

   ```
     6.50   no reserve
   − 0.50   VFR day reserve
     6.00   with reserve
   ```

AIR
5463. (Refer to Figure 12.)
GIVEN:

Pressure altitude .... 18,000 ft
Temperature .... -21°C
Power .... 2,400 RPM — 28" MP
Recommended lean mixture usable fuel .... 425 lb

What is the approximate flight time available under the given conditions? (Allow for VFR day fuel reserve.)

A—3 hours 46 minutes.
B—4 hours 1 minute.
C—4 hours 31 minutes.

1. Find the 2,400 RPM section and 28" MP row.
2. Follow to the right in the 28" MP row and read 94 pph under the -21°C columns.
3. Compute the time available (note does not apply in this case because BHP = 72%):

   Time = 425 ÷ 94 = 4.52 hr

4. Allowing 30 minutes for day reserve (30 minutes):

   4.52 − 0.50 = 4.02 hr = 4 hours, 1 minute

(H317) — AC 61-23C, Chapter 4

AIR
5464. (Refer to Figure 12.)
GIVEN:

Pressure altitude .... 18,000 ft
Temperature .... -41°C
Power .... 2,500 RPM — 26" MP
Recommended lean mixture usable fuel .... 318 lb

What is the approximate flight time available under the given conditions? (Allow for VFR night fuel reserve.)

A—2 hours 27 minutes.
B—3 hours 12 minutes.
C—3 hours 42 minutes.

1. Find the 2,500 RPM section and 26" MP row.
2. Follow to the right in 26" MP row, and read 99 pph under the -41°C columns.
3. Compute the time available (note does not apply in this case because BHP = 75%):

   Time = 318 ÷ 99 = 3.21 hours

4. Allowing 45 minutes for night reserve (0.75 hour):

   3.21 − 0.75 = 2.46 = 2 hours, 27 minutes

(H317) — AC 61-23C, Chapter 4

AIR
5465. (Refer to Figure 12.)
GIVEN:

Pressure altitude .... 18,000 ft
Temperature .... -1°C
Power .... 2,200 RPM — 20" MP
Best fuel economy usable fuel .... 344 lb

What is the approximate flight time available under the given conditions? (Allow for VFR day fuel reserve.)

A—4 hours 50 minutes.
B—5 hours 20 minutes.
C—5 hours 59 minutes.

1. Find the 2,200 RPM section and 20" MP row.
2. Follow to the right in the 20" MP row, and read 59 under the -1°C columns.
3. Since best fuel economy is required, apply the note:

   59 − 6 = 53 lbs/hr best economy

4. Compute the time available:

   Time = 344 ÷ 53 = 6.49 hours

5. Allowing 30 minutes for day reserve (0.5 hour):

   6.49 − 0.50 = 5.99 hours

(H317) — AC 61-23C, Chapter 4

Answers

5463 [B]   5464 [A]   5465 [C]

Chapter 8 **Aircraft Performance**

AIR
5625. (Refer to Figure 34.)
GIVEN:

Pressure altitude .... 6,000 ft
Temperature .... +3°C
Power .... 2,200 RPM — 22" MP
Usable fuel available .... 465 lb

What is the maximum available flight time under the conditions stated?

A—6 hours 27 minutes.
B—6 hours 39 minutes.
C—6 hours 56 minutes.

1. Enter the Cruise Performance Chart at the 2,200 RPM section (the left-hand column). In that section, move right and find the 22" MP row. Follow that row to the right and read 70 pph under the 3°C columns.

2. Calculate the available flight endurance using:

    Time = Fuel ÷ Fuel Burn Rate = 465 ÷ 70
    = 6.643 hours

3. Convert .643 hours into minutes:

    0.643 x 60 = 38.6 minutes

4. Therefore, the flight time is 6 hours, 39 minutes.

(H317) — AC 61-23C, Chapter 4

AIR
5626. (Refer to Figure 34.)
GIVEN:

Pressure altitude .... 6,000 ft
Temperature .... -17°C
Power .... 2,300 RPM — 23" MP
Usable fuel available .... 370 lb

What is the maximum available flight time under the conditions stated?

A—4 hours 20 minutes.
B—4 hours 30 minutes.
C—4 hours 50 minutes.

1. Enter the Cruise Performance Chart at the 2,300 RPM section (the left-hand column). In that section move right and find the 23" MP row. Follow that row right and read 82 pph under the -17°C columns.

2. Calculate the available flight endurance using:

    Time = Fuel ÷ Fuel Burn Rate = 370 ÷ 82
    = 4.51 hours

3. Convert 0.51 hour to minutes.

    .51 x 60 = 30.6 minutes

4. Therefore, the flight time is 4 hours, 30 minutes.

(H317) — AC 61-23C, Chapter 4

AIR
5627. (Refer to Figure 34.)
GIVEN:

Pressure altitude .... 6,000 ft
Temperature .... +13°C
Power .... 2,500 RPM — 23" MP
Usable fuel available .... 460 lb

What is the maximum available flight time under the conditions stated?

A—4 hours 58 minutes.
B—5 hours 7 minutes.
C—5 hours 12 minutes.

1. Enter the Cruise Performance Chart at the 2,500 RPM section (the left-hand column). In that section, move right and find the 23" MP row. Follow that row to the right until intersecting the 3°C conditions and read 90 pph.

2. Determine the fuel flow at 2,500 RPM, 23" MP and +23°C, 87 pph.

3. Average to determine fuel flow at +13°C, 2,500 RPM and 23" MP:

    (90 + 87) ÷ 2 = 88.5 pph

4. Calculate the available flight endurance using:

    Time = Fuel ÷ Fuel Burn Rate = 460 ÷ 88.5
    = 5.2 hours

5. Convert 0.2 hour into minutes:

    0.2 x 60 = 12 minutes

6. Therefore, the flight time is 5 hours, 12 minutes.

(H317) — AC 61-23C, Chapter 4

Answers

5625 [B]   5626 [B]   5627 [C]

Cruise and Range Performance Table

Referring to the Cruise and Range Performance Table, a pilot can determine true airspeed, fuel consumption, endurance and range based on altitude and power setting.

Problem:

Using FAA Figure 11, find the approximate true airspeed and fuel consumption per hour at an altitude of 7,500 feet, 52% power.

Solution:

1. Enter the chart at the altitude to be flown (7,500 feet) on the left.
2. Under the percent brake horsepower (BHP) column, refer to the line representing 52% power. Read to the right and find the true airspeed (105 MPH) and fuel consumption (6.2 GPH).

AIR
5460. (Refer to Figure 11.) If the cruise altitude is 7,500 feet, using 64 percent power at 2,500 RPM, what would be the range with 48 gallons of usable fuel?

A—635 miles.
B—645 miles.
C—810 miles.

1. Find 7,500 feet in the first column.
2. Go across to third column to find 64% BHP.
3. Using the proper column according to the amount of fuel specified in the question, go across from 64% to find 810 miles endurance under the range column.

(H317) — AC 61-23C, Chapter 4

AIR
5461. (Refer to Figure 11.) What would be the endurance at an altitude of 7,500 feet, using 52 percent power?

NOTE: (With 48 gallons fuel-no reserve.)

A—6.1 hours.
B—7.7 hours.
C—8.0 hours.

1. Find 7,500 feet in the first column.
2. Go across to third column to find 52% BHP.
3. Using the proper column according to the amount of fuel specified in the question, go across from 52% to find 7.7 hours endurance under the Endr. Hours column.

(H317) — AC 61-23C, Chapter 4

AIR
5462. (Refer to Figure 11.) What would be the approximate true airspeed and fuel consumption per hour at an altitude of 7,500 feet, using 52 percent power?

A—103 MPH TAS, 6.3 GPH.
B—105 MPH TAS, 6.1 GPH.
C—105 MPH TAS, 6.6 GPH.

1. Find 7,500 feet in first column.
2. Go across to third column to find 52% BHP.
3. Read 105 TAS in TAS column and 6.2 GPH under the Gal/Hour column.

Answer B is the closest answer.

(H317) — AC 61-23C, Chapter 4

Answers

5460 [C]   5461 [B]   5462 [B]

Chapter 8 **Aircraft Performance**

Maximum Rate of Climb

The Maximum Rate of Climb Table illustrates the expected climb performance based on pressure altitude and temperature.

Problem:

Determine the maximum rate of climb under the given conditions:

Weight .... 3,400 lbs
Pressure altitude .... 7,000 feet
Temperature .... +15°C

Solution:

To find the maximum rate of climb for 7,000 feet at 15°C, interpolation is necessary for both altitude and temperature. The interpolation may be accomplished in any order.

1. To find the value for 15°C, interpolation is used between 0° and 20°C. The difference between the two temperatures at 4,000 feet is 155 fpm (1,220 fpm – 1,065 fpm). The temperature of 15°C is 3/4 of the way between the 0°C and 20°C value. At 15°C for 4,000 feet, the rate of climb is 1,104 fpm.

    1,220 – 116-3/4 = 1,104

    or 1,065 + 39-1/4 = 1,104

2. The same method is applied to 8,000 feet to find the value for 15°C. The difference between 0°C and 20°C is 155 fpm (1,110 fpm – 955 fpm). Taking 3/4 of 155 fpm also gives 116 fpm. The rate of climb at 8,000 feet is 994 fpm.

    1,110 – 116 = 994 or 955 + 39 = 994

3. Interpolate for an altitude of 7,000 feet. The altitude of 7,000 feet is 3/4 of the way between 4,000 feet and 8,000 feet. The overall difference is 110 fpm, 3/4 of which would be 82.5 fpm. The rate for 7,000 feet is 1,021 fpm.

    1,114 – 82.5 = 1,031

    or 994 + 27.5 = 1,021

AIR
5623. (Refer to Figure 33.)
GIVEN:

Weight .... 4,000 lb
Pressure altitude .... 5,000 ft
Temperature .... 30°C

What is the maximum rate of climb under the given conditions?

A—655 ft/min.
B—702 ft/min.
C—774 ft/min.

1. Interpolate the rate of climb for 4,000 feet, altitude at 30°C:

    (800 + 655) ÷ 2 = 728 fpm

2. Interpolate the rate of climb for 8,000 feet, altitude at 30°C:

    (695 + 555) ÷ 2 = 625 fpm

3. Use linear interpolation, in ratio form, to determine the rate of climb valid at 5,000 feet and 30°C. The gap between figure altitudes is four 1,000-foot sections.

    (728 ÷ 4) – (625 ÷ 4) = 26 fpm

4. Subtract this increment from the temperature-adjusted base altitude figure (4,000 feet MSL) and get your maximum rate of climb for all the given conditions.

    728 – 26 = 702 fpm

(H317) — AC 61-23C, Chapter 4

Answers

5623 [B]

Chapter 8 **Aircraft Performance**

AIR
5624. (Refer to Figure 33.)
GIVEN:

Weight .... 3,700 lb
Pressure altitude .... 22,000 ft
Temperature .... -10°C

What is the maximum rate of climb under the given conditions?

A—305 ft/min.
B—320 ft/min.
C—384 ft/min.

1. Interpolate or average to determine the rate of climb at 20,000 feet, -10°C.

    (600 + 470) ÷ 2 = 535 fpm

2. Interpolate or average to determine the rate of climb at 24,000 feet, -10°C.

    (170 + 295) ÷ 2 = 233 fpm

3. Interpolate or average to determine the rate of climb at 22,000 feet and -10°C; incorporate the results for 20,000 and 24,000 feet obtained in Steps 1 and 2.

    (535 + 233) ÷ 2 = 384 fpm

(H317) — AC 61-23C, Chapter 4

Rotorcraft Performance

RTC
5683. (Refer to Figure 41.)
GIVEN:

Helicopter gross weight .... 1,225 lb
Ambient temperature .... 77°F

Determine the in-ground-effect hover ceiling.

A—6,750 feet.
B—7,250 feet.
C—8,000 feet.

1. Enter the "in ground effect" chart at the 1,225-pound point and proceed upward to the 25°C line (visualize midway between 20 and 30).

2. From that intersection proceed left and read the hovering ceiling, 6,750 feet.

(H77) — FAA-H-8083-21

RTC
5684. (Refer to Figure 41.)
GIVEN:

Helicopter gross weight .... 1,175 lb
Ambient temperature .... 95°F

Determine the out-of-ground effect hover ceiling.

A—5,000 feet.
B—5,250 feet.
C—6,250 feet.

1. Enter the "out of ground effect" chart at the 1,175-pound point and proceed upward to the 35°C line (visualize midway between 30 and 40).

2. From that intersection proceed left and read the hovering ceiling, 5,250 feet.

(H77) — FAA-H-8083-21

RTC
5685. (Refer to Figure 41.)
GIVEN:

Helicopter gross weight .... 1,275 lb
Ambient temperature .... 9°F

Determine the in ground effect hover ceiling.

A—6,600 feet.
B—7,900 feet.
C—8,750 feet.

1. Enter the "in ground effect" chart at the 1,275-pound point and proceed upward to the -12.8°C line (visualize between -10 and -20°C).

2. From that intersection proceed left and read the hovering ceiling, 7,900 feet.

(H77) — FAA-H-8083-21

Answers

5624   [C]   5683   [A]   5684   [B]   5685   [B]

Chapter 8 **Aircraft Performance**

RTC
5687. (Refer to Figure 42.) Departure is planned from a heliport that has a reported pressure altitude of 4,100 feet. What rate of climb could be expected in this helicopter if the ambient temperature is 90°F?

A—210 ft/min.
B—250 ft/min.
C—390 ft/min.

1. Visualize, or trace on the plastic overlay, a line midway between and parallel to the 80° and 100° lines. Enter the chart at the 4,100-foot pressure altitude point and proceed to the right until intersecting the 90°F OAT line.
2. From that point of intersection, proceed downward to the rate of climb axis and read 250 fpm.

(H77) — FAA-H-8083-21

RTC
5688. (Refer to Figure 42.) Departure is planned for a flight from a heliport with a pressure altitude of 3,800 feet. What rate of climb could be expected in this helicopter during departure if the ambient temperature is 70°F?

A—330 ft/min.
B—360 ft/min.
C—400 ft/min.

1. Visualize, or trace on the plastic overlay, a line midway between and parallel to the 60° and 80° lines. Enter the chart at the 3,800-foot pressure altitude point and proceed to the right until intersecting your 70°F OAT line.
2. From that point of intersection, proceed downward to the rate of climb axis and read 338 fpm.

(H77) — FAA-H-8083-21

RTC
5689. (Refer to Figure 43.)
GIVEN:

Ambient temperature.. 60°F
Pressure altitude .... 2,000 ft

What is the rate of climb?

A—480 ft/min.
B—515 ft/min.
C—540 ft/min.

1. Enter the chart at the 2,000-foot pressure altitude point and proceed to the right until intersecting the 60°F OAT curve.
2. From that point of intersection, proceed downward to the scale and read the rate of climb, 515 fpm.

(H77) — FAA-H-8083-21

RTC
5690. (Refer to Figure 43.)
GIVEN:

Ambient temperature.. 80°F
Pressure altitude .... 2,500 ft

What is the rate of climb?

A—350 ft/min.
B—395 ft/min.
C—420 ft/min.

1. Enter the chart at the 2,500-foot pressure altitude point and proceed to the right until intersecting the 80°F OAT curve.
2. From that point of intersection, proceed downward to the scale and read the rate of climb, 395 fpm.

(H77) — FAA-H-8083-21

RTC
5691. (Refer to Figure 44.)
GIVEN:

Ambient temperature.. 40°F
Pressure altitude .... 1,000 ft

What is the rate of climb?

A—810 ft/min.
B—830 ft/min.
C—860 ft/min.

1. Enter the chart at the 1,000-foot pressure altitude point and proceed to the right until intersecting the 40°F OAT curve.
2. From that point of intersection proceed downward to the scale and read the rate of climb, 860 fpm.

(H77) — FAA-H-8083-21

Answers

5687 [B]   5688 [A]   5689 [B]   5690 [B]   5691 [C]

RTC
5692. (Refer to Figure 44.)

GIVEN:

Ambient temperature .... 60°F
Pressure altitude .... 2,000 ft

What is the rate of climb?

A—705 ft/min.
B—630 ft/min.
C—755 ft/min.

1. Enter the chart at the 2,000-foot pressure altitude point and proceed to the right until intersecting the 60°F OAT curve.
2. From that point of intersection, proceed downward to the scale and read the rate of climb, 705 fpm.

(H77) — FAA-H-8083-21

RTC
5693. (Refer to Figures 45 and 46.)

GIVEN:

Pressure altitude .... 4,000 ft
Ambient temperature .... 80°F

To clear a 50-foot obstacle, a jump takeoff would require

A—more distance than a running takeoff.
B—less distance than a running takeoff.
C—the same distance as a running takeoff.

1. Determine the running takeoff distance:
   a. Enter the chart at the 4,000-foot pressure altitude and proceed to the right to intersect the 80°F curve.
   b. From the point of intersection move down to the scale and read the total takeoff distance to clear a 50-foot obstacle, 1,230 feet.
2. In the same way as in Step 1, determine the jump takeoff distance to clear a 50-foot obstacle: 1,440 feet.
3. The jump takeoff distance exceeds the running takeoff distance by 210 feet (1,440 – 1,230).

(H77) — FAA-H-8083-21

RTC
5694. (Refer to Figures 45 and 46.)

GIVEN:

Pressure altitude .... 4,000 ft
Ambient temperature .... 80°F

The takeoff distance to clear a 50-foot obstacle is

A—1,225 feet for a jump takeoff.
B—1,440 feet for a running takeoff.
C—less for a running takeoff than for a jump takeoff.

1. Move right from the 4,000-foot pressure altitude line until meeting the 80°F arc.
2. Go down from this intersection by continuing one and one-half small squares off the heavy line.
3. You arrive at the running takeoff line one and one-half squares to the right of 1,200 feet (12 x 100) with each square representing 20 = 1,230 feet.
4. Using FAA Figure 46, move right from the 4,000-foot pressure altitude line until meeting the 80°F arc.
5. Continue down from this intersection, two small squares away from the heavy line.
6. You arrive at the jump takeoff line at 1,440 feet.

(H77) — FAA-H-8083-21

Answers

5692 [A]   5693 [A]   5694 [C]

Chapter 8 **Aircraft Performance**

Glider Performance

GLI
5773. (Refer to Figure 48.) If a dual glider weighs 1,040 pounds and an indicated airspeed of 55 MPH is maintained, how much altitude will be lost while traveling 1 mile?

A—120 feet.
B—240 feet.
C—310 feet.

Note that a weight of 1,040 pounds is associated with dual flight.

1. Enter the chart at 55 MPH and proceed vertically to intercept the "L/D Dual" curve.
2. From the intersection, proceed to the left and read the L/D 22.
3. Calculate the altitude loss using the relation:

    L/D = Glide Ratio = Horizontal Distance Traveled ÷ Vertical Distance Traveled

$$\frac{22}{1} = \frac{5{,}280}{X}$$

    X = 240 feet

(N21) — Soaring Flight Manual

GLI
5774. (Refer to Figure 48.) If a dual glider weighs 1,040 pounds, what is the minimum sink speed and rate of sink?

A—38 MPH and 2.6 ft/sec.
B—42 MPH and 3.1 ft/sec.
C—38 MPH and 3.6 ft/sec.

The gross weight of the glider is 1,040 pounds, therefore the sink speed "dual" would be used. The table at the top of the graph shows this value to be 3.1 fps at 42 MPH. (N21) — Soaring Flight Manual

GLI
5775. (Refer to Figure 48.) If the airspeed of a glider is increased from 54 MPH to 60 MPH, the L/D ratio would

A—decrease and the rate of sink would increase.
B—increase and the rate of sink would decrease.
C—decrease and the rate of sink would decrease.

1. Draw a vertical line representing a speed of 54 MPH to intersect the L/D curves.
2. Draw two lines to the left from the points of intersection to determine the L/D values.
3. In a fashion similar to Steps 1 and 2, determine the L/D values for a speed of 60 MPH.
4. Calculate the changes in L/D for single and dual operation. Single operation L/D decreases from 20.25 to 18.4. Dual operation L/D decreases from 22.1 to 20.8.
5. Note the rate of sink values, V_S, increase with an increase in airspeed for both single and dual curves.

(N21) — Soaring Flight Manual

GLI
5776. Minimum sink speed is the airspeed which results in the

A—least loss of altitude in a given time.
B—least loss of altitude in a given distance.
C—shallowest glide angle in any convective situation.

Minimum sink speed is the speed (indicated) at which the glider loses altitude most slowly. (N21) — Soaring Flight Manual

GLI
5777. (Refer to Figure 49.) If the airspeed is 70 MPH and the sink rate is 5.5 ft/sec, what is the effective L/D ratio with respect to the ground?

A—19:1.
B—20:1.
C—21:1.

1. Enter the chart at the 70 MPH airspeed point and draw a line vertically to intercept the L/D curve.
2. From that intersection proceed to the left to read the value of L/D 19:1.

(N21) — Soaring Flight Manual

Answers

5773 [B]   5774 [B]   5775 [A]   5776 [A]   5777 [A]

GLI
5778. (Refer to Figure 49.) If the airspeed is 50 MPH and the sink rate is 3.2 ft/sec, what is the effective L/D ratio with respect to the ground?

A—20:1.
B—21:1.
C—23:1.

1. Enter the chart at the 50 MPH airspeed point and draw a line vertically to intercept the L/D curve.
2. From that intersection proceed to the left to read the value of L/D, 23:1.

(N21) — Soaring Flight Manual

GLI
5779. The glider has a normal L/D ratio of 23:1 at an airspeed of 50 MPH. What would be the effective L/D ratio with respect to the ground with a 10 MPH tailwind?

A—23:1.
B—25:1.
C—27.6:1.

Glide Ratio = Horizontal Distance Traveled ÷ Vertical Distance Traveled

A tailwind will increase the horizontal distance traveled, in any given time period, but will not affect the vertical distance (sink). Therefore, the new horizontal distance for a 10 MPH tailwind is:

$$\frac{23 \times (50 + 10)}{50} = 27.6$$

Therefore, the effective glide ratio is about 27.6:1.

(N21) — Soaring Flight Manual

GLI
5780. If the glider has drifted a considerable distance from the airport while soaring, the best speed to use to reach the airport when flying into a headwind is the

A—best glide speed.
B—minimum sink speed.
C—speed-to-fly plus half the estimated windspeed at the glider's altitude.

If maximum distance over the ground is desired, the airspeed for best L/D should be used. When gliding into a headwind, maximum distance will be achieved by adding approximately one-half of the estimated headwind velocity to the best L/D speed. (N21) — Soaring Flight Manual

GLI
5781. The maximum airspeed at which abrupt and full deflection of the controls would not cause structural damage to a glider is called the

A—speed-to-fly.
B—maneuvering speed.
C—never-exceed speed.

The speed at which full abrupt control travel at maximum gross weight may be used without exceeding the load limits is called maneuvering speed. (N21) — Soaring Flight Manual

GLI
5782. Which is true regarding minimum control airspeed while thermalling? Minimum control airspeed

A—may coincide with minimum sink airspeed.
B—is greater than minimum sink airspeed.
C—never coincides with minimum sink airspeed.

Minimum control airspeed is the speed at which an increase of either the angle of attack or load factor would result in an immediate stall. Therefore, minimum control speed is only a few miles per hour above the stall speed, nearly coinciding with minimum sink speed. (N21) — Soaring Flight Manual

GLI
5783. (Refer to Figure 50.) Which is true when the glider is operated in the high-performance category and the dive brakes/spoilers are in the closed position? The

A—design dive speed is 150 MPH.
B—never-exceed speed is 150 MPH.
C—design maneuvering speed is 76 MPH.

1. Locate the key showing high performance category (dark lines).
2. Note the V_{NE} for no brake operation (brakes/spoilers closed).
3. Read the intercept of the V_{NE} line with the speed ordinate, 150 MPH.

(N21) — Soaring Flight Manual

Answers

| 5778 | [C] | 5779 | [C] | 5780 | [C] | 5781 | [B] | 5782 | [A] | 5783 | [B] |

Chapter 8 **Aircraft Performance**

GLI
5784. (Refer to Figure 50.) If the glider's airspeed is 70 MPH and a vertical gust of +30 ft/sec is encountered, which would most likely occur? The

A—glider would momentarily stall.
B—maximum load factor would be exceeded.
C—glider would gain 1,800 feet in 1 minute.

1. Enter the chart at 70 MPH velocity.
2. Proceed vertically to an estimated position for a +30 fps gust. The position will be outside the envelope.
3. Note the maneuvering speed V_A, is greater than 70 MPH. Since the flight speed is less than V_A turbulence will not impose a damaging structural load and the glider will momentarily stall.

(N21) — Soaring Flight Manual

GLI
5785. Regarding the effect of loading on glider performance, a heavily loaded glider would

A—have a lower glide ratio than when lightly loaded.
B—have slower forward speed than when lightly loaded.
C—make better flight time on a cross-country flight between thermals than when lightly loaded.

Increasing the weight of the glider increases the airspeed at which the best glide ratio is obtained. Weight does not affect the maximum glide ratio, only the airspeed at which it is attained. (N21) — Soaring Flight Manual

GLI
5786. When flying into a strong headwind on a long final glide or a long glide back to the airport, the recommended speed to use is the

A—best glide speed.
B—minimum sink speed.
C—speed-to-fly plus half the estimated windspeed at the glider's flight altitude.

If maximum distance over the ground is desired, the airspeed for best L/D should be used. When gliding into a headwind, maximum distance will be achieved by adding approximately one-half of the estimated headwind velocity to the best L/D speed. (N21) — Soaring Flight Manual

GLI
5787. Which procedure can be used to increase forward speed on a cross-country flight?

A—Maintain minimum sink speed plus or minus one-half the estimated wind velocity.
B—Use water ballast while thermals are strong and dump the water when thermals are weak.
C—Use water ballast while thermals are weak and dump the water when thermals are strong.

As the weight of the glider increases, the airspeed must be increased to maintain the same glide ratio. During a contest situation, it is desirable to fly faster. (N21) — Soaring Flight Manual

GLI
5788. The reason for retaining water ballast while thermals are strong, is to

A—decrease forward speed.
B—decrease cruise performance.
C—increase cruise performance.

As the weight of the glider increases, the airspeed must be increased to maintain the same glide ratio. (N21) — Soaring Flight Manual

GLI
5789. When flying into a headwind, penetrating speed is the glider's

A—speed-to-fly.
B—minimum sink speed.
C—speed-to-fly plus half the estimated wind velocity.

If maximum distance over the ground is desired, the airspeed for best L/D should be used. When gliding into a headwind, maximum distance will be achieved by adding approximately one-half of the estimated headwind velocity to the best L/D speed. (N21) — Soaring Flight Manual

GLI
5790. Which is true regarding the effect on a glider's performance by the addition of ballast or weight?

A—The glide ratio at a given airspeed will increase.
B—The heavier the glider is loaded, the less the glide ratio will be at all airspeeds.
C—A higher airspeed is required to obtain the same glide ratio as when lightly loaded.

As the weight of the glider increases, the airspeed must be increased to maintain the same glide ratio. (N21) — Soaring Flight Manual

Answers

5784 [A]	5785 [C]	5786 [C]	5787 [B]	5788 [C]	5789 [C]
5790 [C]					

Chapter 8 **Aircraft Performance**

Balloon Performance

LTA
5480. (Refer to Figure 52, point 1.)
GIVEN:

Departure point .... Georgetown Airport (Q61)
Departure time 0637
Winds aloft forecast (FD) at your altitude 1008

At 0755, the balloon should be

A—over Auburn Airport (AUN).
B—over the town of Auburn.
C—slightly west of the town of Garden Valley.

1. Calculate time en route:
   0735 − 0637 = 1:18 = 1.3 hours
2. Compute distance traveled:
   Distance = Speed x Time
   8 x 1.3 = 10.4 NM
3. Plot course of 100° + 180° = 280° out to a distance of 10.4 NM from Georgetown Airport (Q61), point 1. The balloon should be just south of there.

(H344) — AC 61-23C, Chapter 8

LTA
5571. (Refer to Figure 53, point 3.) If at 1,000 feet MSL and drifting at 10 knots toward Firebaugh Airport (Q49), at what approximate distance from the airport should you begin a 100 ft/min ascent to arrive at the center of the airport at 3,000 feet?

A—3.5 NM.
B—5 NM.
C—8 NM.

3,000 feet − 1,000 feet = 2,000 feet to climb.

2,000 feet ÷ 100 fpm = 20 minutes

(10 KTS ÷ 60 min) x 20 minutes = 3.33 NM

(J37) — Sectional Chart Legend

LTA
5573. (Refer to Figure 54, point 5.) A balloon drifts over the town of Brentwood on a magnetic course of 185° at 10 knots. If wind conditions remain the same, after 1 hour 30 minutes the pilot

A—with no radio aboard, must be above 2,900 feet MSL and must have an operating transponder aboard.
B—must remain above 600 feet MSL for national security reasons.
C—with no radio aboard, must be above 2,900 feet MSL.

The balloon traveled a distance of 15 NM, and is located within controlled airspace. With no radio on board, the balloon must be above 2,900 feet MSL. Balloons are not certified with an electrical system, and therefore are not required to have a transponder. (J37) — Sectional Chart Legend, 14 CFR §91.215(b)(3)

LTA
5578. (Refer to Figure 53, point 4.) A balloon departs Mendota Airport (Q84) and drifts for a period of 1 hour and 30 minutes in a wind of 230° at 10 knots. What maximum elevation figure would assure obstruction clearance during the next 1-1/2 hours of flight?

A—1,600 feet MSL.
B—3,200 feet MSL.
C—9,400 feet MSL.

The Maximum Elevation Figures are larger bold-faced blue numbers. They are shown in quadrangles bounded by ticked lines of latitude and longitude and are in thousands and hundreds of feet MSL. They are based on the highest known feature in each quadrangle, including terrain and obstructions. 3 hours into the flight at 10 knots would put the balloon in the upper right grid, which has a maximum altitude of 9,400 feet. (J37) — Sectional Chart Legend

LTA
5579. (Refer to Figure 52, point 4.) If Lincoln Regional Airport (LHM) is departed at 0630, and at 0730 the town of Newcastle is reached, the wind direction and speed would be approximately

A—082° at 6 knots.
B—262° at 11 knots.
C—082° at 17 knots.

The aircraft traveled 13 NM west-to-east. (J37) — Section Chart Legend

Answers
5480 [A]   5571 [A]   5573 [C]   5578 [C]   5579 [B]

Chapter 8 **Aircraft Performance**

LTA
5580. (Refer to Figure 52, point 4.) If you depart Lincoln Airport (O51) and track a true course of 075° with a groundspeed of 12 knots, your position after 1 hour 20 minutes of flight would be over the town of

A—Foresthill.
B—Clipper Gap.
C—Weimar.

With a ground speed of 12 knots, the distance traveled after 1 hour 20 minutes would be:

12 knots x 1:20 = 16.08 NM

On a course of 075°, and a distance of 16.08 NM, your position would be over the town of Clipper Gap.

(J37) — Sectional Chart Legend

LTA
5582. (Refer to Figure 53, point 4.) While drifting above the Mendota Airport (Q84) with a northwesterly wind of 8 knots, you

A—are required to contact ATC on frequency 122.9 Mhz.
B—should remain higher than 2,000 feet AGL until you are at least 8 NM southeast of that airport.
C—will be over Firebaugh Airport (Q49) in approximately 1 hour.

With a northwesterly wind you will be drifting toward the southeast. Just southeast of Mendota Airport is the Mendota State Wildlife Area, which you should overfly at 2,000 feet AGL or higher. (J37) — Sectional Chart Legend

LTA
5586. (Refer to Figure 54, point 3.)
GIVEN:

Departure point .... Meadowlark Airport
Departure time 0710
Wind .... 180°, 8 kts

At 0917 the balloon should be

A—east of VINCO intersection.
B—over the town of Brentwood.
C—3 miles south of the town of Brentwood.

Wind is from 180° pushing the balloon to the north at 8 NM/hr. The time elapsed from 0710 to 0917, which is 2 hours 7 minutes. Therefore, the balloon will travel 17 NM to the north. Using a plotter, measure 17 north, i.e., over Brentwood. (J37) — Sectional Chart Legend

LTA
5589. (Refer to Figure 52, point 5.) A balloon is launched at University Airport (OO5) and drifts south-southwesterly toward the depicted obstruction. If the altimeter was set to 0 feet upon launch, what should it indicate if the balloon is to clear the obstruction by 500 feet above its top?

A—510 feet.
B—813 feet.
C—881 feet.

The field elevation at University Airport is 68 feet MSL. The altimeter in the balloon is set to zero, and a 381-foot obstacle southwest of the airport must be cleared by 500 feet.

381 – 68 = 313 feet AGL
313 + 500 = 813 feet AGL

Therefore, the altimeter would read 813 feet MSL, since it was set to zero at launch.

(J37) — Sectional Chart Legend

Answers

5580 [B]   5582 [B]   5586 [B]   5589 [B]

Chapter 9
Navigation

The Flight Computer   *9–3*

    Finding True Course, Time, Rate, Distance, and Fuel   *9–3*

    Finding Density Altitude   *9–10*

    Finding Wind Direction and Velocity   *9–12*

Off-Course Correction   *9–13*

VHF Omni-Directional Range (VOR)   *9–15*

Estimating Time and Distance To Station Using VOR   *9–22*

Horizontal Situation Indicator (HSI)   *9–25*

Automatic Direction Finder (ADF)   *9–27*

    Determining Bearings with ADF   *9–27*

    Intercepting Bearings with ADF   *9–28*

    Estimated Time and Distance To Station Using ADF   *9–29*

        Wing-Tip Bearing Change Method   *9–29*

        Isosceles Triangle Technique   *9–29*

Chapter 9 **Navigation**

The Flight Computer

ASA's CX-1a Pathfinder is an electronic flight computer and can be used in place of the E6-B. This aviation computer can solve all flight planning problems, as well as perform standard mathematical calculations.

Finding True Course, Time, Rate, Distance, and Fuel

True course is expressed as an angle between the course line and true (geographic) north. Lines of longitude (meridians) designate the direction of north-south at any point on the surface of the Earth and converge at the poles. Because meridians converge toward the poles, true course measurement should be taken at a meridian near the midpoint of the course rather than at the point of departure.

The flight computer can be used to solve problems of time, rate and distance. When two factors are known, the third can be found using the proportion:

$$\text{Rate (speed)} = \frac{\text{Distance}}{\text{Time}}$$

Problem:

Find the time en route and fuel consumption based on the following information:

Wind	175° at 20 knots
Distance	135 NM
True course	075°
True airspeed	80 knots
Fuel consumption	105.0 pounds/hour

Solution using the E6-B:

1. Using the wind face side of the E6-B computer, set the true wind direction (175°) under the true index.
2. Place a wind dot 20 units (wind speed) directly above the grommet.
3. Rotate the plotting disk to set the true course (075°) under the true index.
4. Adjust the sliding grid so that the TAS arc (80 knots) is under the wind dot. Note that the wind dot is 15° right of centerline, so the Wind Correction Angle (WCA) = 15°R.
5. Read the ground speed under the grommet: 81.0 knots.
6. a. Determine the time en route, using the formula:

    $$\frac{\text{Distance}}{\text{Ground Speed}} = \text{Time}$$

    $135 \div 81 = 1.67$ hour

    b. The E6-B may also be used to find the time en route:

    i. Set the 60 (speed) index under 81 knots (outer scale).

    ii. Under 135 NM (outer scale) read the time of 100 minutes or 1 hour 40 minutes (inner scale).

7. Find the amount of fuel consumed, using the formula:

    Time x Fuel Consumption Rate = Total Consumed

    1.67 hours x 105.0 pounds/hour = 175.35 pounds

Continued

Chapter 9 **Navigation**

The E6-B may also be used to find the amount of fuel consumed:
a. Set the 60 (speed) index under 105.0 pounds/hour.
b. Over 100 minutes or 1 hour 40 minutes (inner scale) read 175 pounds required (outer scale).

Solution using the CX-1a:

1. Enter the Hdg/GS function to find the ground speed.
   Wind direction .... 175°
   Wind speed .... 20 knots
   Course .... 075°
   True airspeed .... 80 knots
   Ground speed .... 81.0 knots
2. Enter the Leg Time function to find the time en route.
   Distance .... 135 NM
   Ground speed .... 81.0 knots
   Time 1:39:59
3. Enter the Fuel Burn function to find the amount of fuel consumed.
   Time 1:39:59
   Fuel burn rate .... 105 pounds/hour
   Fuel consumed .... 175.0 pounds

Problem:

Based on the following information, determine the approximate time, fuel consumed, compass headings, and distance traveled during the descent to the airport.

Cruising Altitude .... 10,500 feet
Airport elevation .... 1,700 feet
Descent to .... 1,000 feet AGL
Rate of descent .... 600 ft/min
Average true airspeed .... 135 knots
True course .... 263°
Average wind velocity .... 330° at 30 knots
Variation .... 7° east
Deviation .... +3°
Average fuel consumption .... 11.5 gal/hr

Solution using the E6-B:

1. Find the time en route. The time to descend is based on the vertical distance from the cruising altitude of 10,500 feet down to the lower altitude of 2,700 feet MSL (1,000 feet AGL + 1,700 feet MSL).

       10,500   feet cruising altitude
      – 2,700   feet lower altitude
        7,800   feet altitude change

$$\frac{\text{Distance}}{\text{Rate}} = \text{Time}$$

7,800 feet ÷ 600 feet/min = 13 minutes

2. Find the amount of fuel consumed using the formula:

    Time x Fuel Consumption Rate = Fuel Consumed

    $$\frac{13 \text{ minutes} \times 11.5 \text{ gal/hr}}{60 \text{ min/hr}} = 2.5 \text{ gallons}$$

3. Use the wind face of the E6-B to find the ground speed (120 knots) and wind correction angle (+ 12°R). Due to wind being present, the TAS is not the ground speed. The speed used to calculate the distance covered must be the ground speed.

4. Calculate the distance.

    Time x Speed = Distance

    $$\frac{13 \text{ min} \times 120 \text{ knots}}{60 \text{ min/hr}} = 26 \text{ NM}$$

5. Determine the compass heading, by applying the wind correction angle that was found on the wind face of E6-B to the true course, which gives the true heading. The true heading is corrected by variation to give the magnetic heading. Finally, the deviation is used to correct magnetic heading to give the compass heading as follows:

263°	TC
+ 12°R	±WCA
275°	TH
− 7° E	±VAR
268°	MH
+ 3°	±DEV
271°	CH

Solution using the CX-1a:

1. The time to descend is based on the vertical distance from the cruising altitude of 10,500 feet down to the lower altitude of 2,700 feet MSL (1,000 feet AGL + 1,700 feet MSL).

10,500	feet cruising altitude
− 2,700	feet lower altitude
7,800	feet altitude change

2. Enter the Leg Time function to find the leg time.

    Distance .... 7,800 feet

    Ground speed .... 600 feet/minute

    Leg time .... 13 minutes

3. Enter the Fuel Burn function to find the amount of fuel consumed.

    Time 00:13:00

    Fuel burn rate .... 11.5 gal/hour

    Fuel burned .... 2.5 gallons

Continued

4. Enter the Hdg/GS function to find ground speed.

    Wind direction .... 330°
    Wind speed .... 30 knots
    Course .... 263°
    True airspeed .... 135 knots
    Ground speed .... 120.4 knots
    Heading .... 275°

5. Enter the Dist Fln function to find the distance covered during the descent.

    Time 00:13:00
    Ground speed .... 120.4 knots
    Distance .... 26.1 NM

6. The true heading was found in Step #4. The true heading is corrected by variation to give the magnetic heading, and the deviation is used to correct magnetic heading to give the compass heading:

    ```
     275°    TH
    − 7° E   ±VAR
    ─────
     268°    MH

    + 3°     ±DEV
    ─────
     271°    CH
    ```

ALL
5479. True course measurements on a Sectional Aeronautical Chart should be made at a meridian near the midpoint of the course because the

A—values of isogonic lines change from point to point.
B—angles formed by isogonic lines and lines of latitude vary from point to point.
C—angles formed by lines of longitude and the course line vary from point to point.

Meridians (lines of longitude) converge toward the poles, so course measurement should be taken at a meridian near the midpoint of the course rather than at the point of departure. (H344) — AC 61-23C, Chapter 8

Answers (A) and (B) are incorrect because isogonic lines are used to find magnetic course.

AIR
5469. If fuel consumption is 80 pounds per hour and groundspeed is 180 knots, how much fuel is required for an airplane to travel 460 NM?

A—205 pounds.
B—212 pounds.
C—460 pounds.

1. Calculate the time en route:

    460 NM ÷ 180 knots = 2.556 hours

2. Determine the fuel burn:

    80 lbs/hr x 2.56 hours = 204.8 pounds

(H342) — AC 61-23C, Chapter 8

AIR
5470. If an airplane is consuming 95 pounds of fuel per hour at a cruising altitude of 6,500 feet and the groundspeed is 173 knots, how much fuel is required to travel 450 NM?

A—248 pounds.
B—265 pounds.
C—284 pounds.

1. Calculate the time en route:

    450 NM ÷ 173 knots = 2.6 hour

2. Determine the fuel burn:

    95 lbs/hr x 2.6 hours = 247 pounds

(H342) — AC 61-23C, Chapter 8

Answers

5479 [C]   5469 [A]   5470 [A]

AIR
5471. If an airplane is consuming 12.5 gallons of fuel per hour at a cruising altitude of 8,500 feet and the groundspeed is 145 knots, how much fuel is required to travel 435 NM?

A—27 gallons.
B—34 gallons.
C—38 gallons.

1. Calculate the time en route:

    435 NM ÷ 145 knots = 3 hours

2. Determine the fuel burn:

    12.5 GPH x 3 hours = 37.5 gallons

(H342) — AC 61-23C, Chapter 8

AIR
5472. If an airplane is consuming 9.5 gallons of fuel per hour at a cruising altitude of 6,000 feet and the groundspeed is 135 knots, how much fuel is required to travel 490 NM?

A—27 gallons.
B—30 gallons.
C—35 gallons.

1. Calculate the time en route:

    490 NM ÷ 135 knots = 3.63 hours

2. Determine the fuel burn:

    9.5 GPH x 3.63 hours = 34.5 gallons

(H342) — AC 61-23C, Chapter 8

AIR
5473. If an airplane is consuming 14.8 gallons of fuel per hour at a cruising altitude of 7,500 feet and the groundspeed is 167 knots, how much fuel is required to travel 560 NM?

A—50 gallons.
B—53 gallons.
C—57 gallons.

1. Calculate the time en route:

    560 NM ÷ 167 knots = 3.35 hours

2. Determine the fuel burn:

    14.8 GPH x 3.35 hours = 49.6 gallons

(H342) — AC 61-23C, Chapter 8

AIR
5474. If fuel consumption is 14.7 gallons per hour and groundspeed is 157 knots, how much fuel is required for an airplane to travel 612 NM?

A—58 gallons.
B—60 gallons.
C—64 gallons.

1. Calculate the time en route:

    612 NM ÷ 157 knots = 3.9 hours

2. Determine the fuel burn:

    14.7 GPH x 3.9 = 57.33 gallons

(H342) — AC 61-23C, Chapter 8

AIR, RTC
5481. GIVEN:

Wind .... 175° at 20 kts
Distance .... 135 NM
True course .... 075°
True airspeed .... 80 kts
Fuel consumption .... 105 lb/hr

Determine the time en route and fuel consumption.

A—1 hour 28 minutes and 73.2 pounds.
B—1 hour 38 minutes and 158 pounds.
C—1 hour 40 minutes and 175 pounds.

1. Compute the ground speed:

    Wind direction .... 175°
    Wind speed... 20 knots
    Course ....075°
    True airspeed .... 80 knots
    Ground speed .... 81.0 knots

2. Determine the time en route:

    135 NM ÷ 81 knots = 1.67 hours
        = 1 hour 40 minutes

3. Calculate fuel consumed:

    105.0 lbs/hr x 1.67 hours = 175.0 pounds

(H344) — AC 61-23C, Chapter 8

Answers

5471 [C]   5472 [C]   5473 [A]   5474 [A]   5481 [C]

Chapter 9 **Navigation**

AIR

5466. An airplane descends to an airport under the following conditions:

Cruising altitude	6,500 ft
Airport elevation	700 ft
Descends to	800 ft AGL
Rate of descent	500 ft/min
Average true airspeed	110 kts
True course	335°
Average wind velocity	060° at 15 kts
Variation	3°W
Deviation	+2°
Average fuel consumption	8.5 gal/hr

Determine the approximate time, compass heading, distance, and fuel consumed during the descent.

A— 10 minutes, 348°, 18 NM, 1.4 gallons.
B— 10 minutes, 355°, 17 NM, 2.4 gallons.
C— 12 minutes, 346°, 18 NM, 1.6 gallons.

1. Calculate the time to descend:

    Time = vertical distance ÷ vertical speed

    Where vertical distance = 6,500 – 1,500
    = 5,000 feet

    Time = 5,000 feet ÷ 500 FPM = 10 minutes
    = 0.1667 hr

2. Compute the fuel requirement:

    8.5 gal/hr x 0.1667 hr = 1.42 gallons

3. Calculate the wind correction angle and ground speed using a wind triangle:

    WCA = 8° right
    Ground speed = 108 knots

4. Calculate true heading (TH = TC + WCA):

    335° + 8° = 343° TH

5. Calculate compass heading (CH = TH + Var + Dev):

    343° + 3° + 2° = 348° CH

6. Compute distance flown:

    Distance = 108 knots x 0.1667 hour = 18 NM

(H317) — AC 61-23C, Chapter 4

AIR

5467. An airplane descends to an airport under the following conditions:

Cruising altitude	7,500 ft
Airport elevation	1,300 ft
Descends to	800 ft AGL
Rate of descent	300 ft/min
Average true airspeed	120 kts
True course	165°
Average wind velocity	240° at 20 kts
Variation	4°E
Deviation	-2°
Average fuel consumption	9.6 gal/hr

Determine the approximate time, compass heading, distance, and fuel consumed during the descent.

A— 16 minutes, 168°, 30 NM, 2.9 gallons.
B— 18 minutes, 164°, 34 NM, 3.2 gallons.
C— 18 minutes, 168°, 34 NM, 2.9 gallons.

1. Calculate the time to descend:

    Time = vertical distance ÷ vertical speed

    Where vertical distance = 7,500 – 2,100
    = 5,400 feet

    Time = 5,400 feet ÷ 300 FPM = 18 minutes
    = 0.3 hour

2. Compute the fuel requirement:

    9.6 gal/hr x 0.3 hour = 2.88 gallons

3. Calculate the wind correction angle and ground speed using a wind triangle:

    WCA = 9° right
    Ground speed = 113 knots

4. Calculate true heading (TH = TC + WCA):

    165° + 9° = 174° TH

5. Calculate compass heading (CH = TH + Var + Dev):

    174° – 4° – 2° = 168° CH

6. Compute distance flown:

    Distance = 113 knots x 0.3 hour = 33.9 NM

(H317) — AC 61-23C, Chapter 4

Answers

5466   [A]   5467   [C]

Chapter 9 Navigation

AIR
5468. An airplane descends to an airport under the following conditions:

Cruising altitude .... 10,500 ft
Airport elevation .... 1,700 ft
Descends to .... 1,000 ft AGL
Rate of descent .... 600 ft/min
Average true airspeed .... 135 kts
True course .... 263°
Average wind velocity .... 330° at 30 kts
Variation .... 7°E
Deviation .... +3°
Average fuel consumption .... 11.5 gal/hr

Determine the approximate time, compass heading, distance, and fuel consumed during the descent.

A—9 minutes, 274°, 26 NM, 2.8 gallons.
B—13 minutes, 274°, 28 NM, 2.5 gallons.
C—13 minutes, 271°, 26 NM, 2.5 gallons.

1. Calculate the time to descend:

    Time = vertical distance ÷ vertical speed

    Where vertical distance = 10,500 – 2,700
     = 7,800 feet

    Time = 7,800 feet ÷ 600 FPM = 13 minutes
     = 0.2167 hour

2. Compute the fuel requirement:

    11.5 gal/hr x 0.2167 hour = 2.49 gallon

3. Calculate the wind correction angle and ground speed using a wind triangle:

    WCA = 12° right
    Ground speed = 120 knots

4. Calculate true heading (TH = TC + WCA):

    263° + 12° = 275° TH

5. Calculate compass heading (CH = TH + Var + Dev):

    275° – 7° + 3° = 271° CH

6. Compute distance flown:

    120 knots x 0.2167 hour = 26 NM

(H317) — AC 61-23C, Chapter 4

AIR
5488. An airplane departs an airport under the following conditions:

Airport elevation .... 1,000 ft
Cruise altitude .... 9,500 ft
Rate of climb .... 500 ft/min
Average true airspeed .... 135 kts
True course .... 215°
Average wind velocity .... 290° at 20 kts
Variation .... 3°W
Deviation .... -2°
Average fuel consumption .... 13 gal/hr

Determine the approximate time, compass heading, distance, and fuel consumed during the climb.

A—14 minutes, 234°, 26 NM, 3.9 gallons.
B—17 minutes, 224°, 36 NM, 3.7 gallons.
C—17 minutes, 242°, 31 NM, 3.5 gallons.

1. Calculate the time to climb:

    Time = vertical distance ÷ vertical speed

    Where vertical distance = 9,500 – 1,000
     = 8,500 feet

    Time = 8,500 feet ÷ 500 FPM = 17 minutes
     = 0.283 hour

2. Compute the fuel requirement:

    13 gal/hr x 0.283 hour = 3.679 gallons

3. Calculate the wind correction angle and ground speed using a wind triangle:

    WCA = 8° right
    Ground speed = 128 knots

4. Calculate true heading (TH = TC + WCA):

    215 + 8° = 223° TH

5. Calculate compass heading (CH = TH + Var + Dev):

    223° + 3° – 2° = 224° CH

6. Compute distance flown:

    Distance = 128 knots x 0.283 hour = 36.2 NM

(H344) — AC 61-23C, Chapter 8

Answers
5468 [C]   5488 [B]

Chapter 9 **Navigation**

AIR
5489. An airplane departs an airport under the following conditions:

Airport elevation .... 1,500 ft
Cruise altitude .... 9,500 ft
Rate of climb .... 500 ft/min
Average true airspeed .... 160 kts
True course .... 145°
Average wind velocity .... 080° at 15 kts
Variation .... 5°E
Deviation .... -3°
Average fuel consumption .... 14 gal/hr

Determine the approximate time, compass heading, distance, and fuel consumed during the climb.

A—14 minutes, 128°, 35 NM, 3.2 gallons.
B—16 minutes, 132°, 41 NM, 3.7 gallons.
C—16 minutes, 128°, 32 NM, 3.8 gallons.

1. Calculate the time to climb:

   Time = vertical distance ÷ vertical speed

   Where vertical distance = 9,500 – 1,500
   = 8,000 feet

   Time = 8,000 feet ÷ 500 FPM = 16 minutes
   = 0.27 hour

2. Compute the fuel requirement:

   14 gal/hr x 0.2666 hour = 3.73 gallons

3. Calculate the wind correction angle and ground speed using a wind triangle:

   WCA = 5° left
   Ground speed = 153 knots

4. Calculate true heading (TH = TC + WCA):

   145° – 5° = 140° TH

5. Calculate compass heading (CH = TH + Var + Dev):

   140° – 5° – 3° = 132° CH

6. Compute distance flown:

   Distance = 153 knots x 0.27 hour = 41.31 NM

(H344) — AC 61-23C, Chapter 8

Finding Density Altitude

To find density altitude, refer to the right-hand window on the computer side of the E6-B.

Problem:

Find the density altitude from the following conditions:

Pressure Altitude .... 5,000 feet
True air temperature .... + 40°C

Solution using the E6-B:

1. Refer to the right-hand "Density Altitude" window. Note that the scale above the window is labeled air temperature (°C). The scale inside the window itself is labeled pressure altitude (in thousands of feet). Rotate the disc and place the pressure altitude of 5,000 feet opposite an air temperature of +40°C.

2. The density altitude shown in the window is 8,800 feet.

Solution using the CX-1a:

1. Enter the Plan TAS function to find density altitude.

   Pressure altitude .... 5,000 feet

   Temperature .... 40°C

   Enter a zero when prompted for calibrated airspeed.

2. Density altitude is 8,846 feet.

Answers
5489 [B]

ALL
5306. GIVEN:

Pressure altitude .... 12,000 ft
True air temperature .... +50°F

From the conditions given, the approximate density altitude is

A—11,900 feet.
B—14,130 feet.
C—18,150 feet.

1. Convert 50°F to °C using the temperature conversion table at the bottom of the E6-B or the CX-1a Pathfinder. The result is 10°C.

2. Refer to the right-hand "Density Altitude" window. Note that the scale above the window is labeled air temperature (°C). The scale inside the window itself is labeled pressure altitude (in thousands of feet). Rotate the disc and place the pressure altitude of 12,000 feet opposite an air temperature of 10°C. The density altitude shown in the density altitude window is 14,130 feet.

(I22) — AC 00-6A, Chapter 3

ALL
5307. GIVEN:

Pressure altitude .... 5,000 ft
True air temperature .... +30°C

From the conditions given, the approximate density altitude is

A—7,200 feet.
B—7,800 feet.
C—9,000 feet.

1. Refer to the right-hand "Density Altitude" window. Note that the scale above the window is labeled air temperature (°C). The scale inside the window itself is labeled pressure altitude (in thousands of feet). Rotate the disc and place the pressure altitude of 5,000 feet opposite an air temperature of 30°C.

2. The density altitude shown in the density altitude window is 7,800 feet.

(I22) — AC 00-6A, Chapter 3

ALL
5308. GIVEN:

Pressure altitude .... 6,000 ft
True air temperature .... +30°F

From the conditions given, the approximate density altitude is

A—9,000 feet.
B—5,500 feet.
C—5,000 feet.

1. Convert 30°F to °C using the temperature conversion table at the bottom of the E6-B or the CX-1a Pathfinder. The result is -1°C.

2. Refer to the right-hand "Density Altitude" window. Note that the scale above the window is labeled air temperature (°C). The scale inside the window itself is labeled pressure altitude (in thousands of feet). Rotate the disc and place the pressure altitude of 6,000 feet opposite an air temperature of -1°C. The density altitude shown in the window is 5,500 feet.

(I22) — AC 00-6A, Chapter 3

ALL
5309. GIVEN:

Pressure altitude .... 7,000 ft
True air temperature .... +15°C

From the conditions given, the approximate density altitude is

A—5,000 feet.
B—8,500 feet.
C—9,500 feet.

1. Refer to the right-hand "Density Altitude" window. Note that the scale above the window is labeled air temperature (°C). The scale inside the window itself is labeled pressure altitude (in thousands of feet). Rotate the disc and place the pressure altitude of 7,000 feet opposite an air temperature of 15°C.

2. The density altitude shown in the window is 8,500 feet.

(I22) — AC 00-6A, Chapter 3

Answers

5306 [B]   5307 [B]   5308 [B]   5309 [B]

Chapter 9 **Navigation**

Finding Wind Direction and Velocity

The E6-B can be used to solve for an unknown wind. To determine wind direction and wind speed, the true course, wind correction angle, true airspeed and ground speed are necessary. The wind correction angle (WCA) may not be given directly in a problem, but can be determined from the true course (TC) and true heading (TH).

Problem:

Determine the wind direction and wind speed under the following conditions:

True course .... 095°
True heading .... 075°
True airspeed .... 90 knots
Ground speed .... 77 knots

Solution using the E6-B:

1. Set the true course (095°) under the true index located at the top of the computer.
2. Move the sliding grid to place the ground speed (77 knots) under the center grommet.
3. Determine the wind correction angle. Above the true heading (075°) read the wind correction angle, 20°L. Draw the wind dot over the true airspeed arc (90 knots) and 20° (wind correction angle) to the left.
4. Finally, rotate the window until the wind dot is lined up directly above the grommet. The wind direction is read under the true index (020°). For convenience, the sliding grid may be moved so that 100 knots is placed under the grommet. The difference between the grommet and the wind dot indicates the wind speed (31 knots).

Solution using the CX-1a:

1. Enter the Wind function.

   Heading .... 075°
   Ground speed .... 77 knots
   True airspeed .... 90 knots
   Course .... 095°
   Wind direction .... 019°
   Wind speed .... 32 knots

AIR, RTC
5475. GIVEN:

True course	105°
True heading	085°
True airspeed	95 kts
Groundspeed	87 kts

Determine the wind direction and speed.

A—020° and 32 knots.
B—030° and 38 knots.
C—200° and 32 knots.

Using an E6-B computer:

1. Turn the compass azimuth so 105° is at the true index.
2. Slide board so grommet is on the ground speed arc, 87 knots.
3. Determine the WCA by comparing course to heading.
    105° − 085° = 20° left
4. Plot WCA at TAS arc with a pencil dot, 95 knots.
5. Rotate compass azimuth so pencil dot is on the centerline.
6. Read wind direction under true index and wind speed as number of units between grommet and pencil dot, 020° and 32 knots.

(H341) — AC 61-23C, Chapter 8

AIR, RTC
5476. GIVEN:

True course	345°
True heading	355°
True airspeed	85 kts
Groundspeed	95 kts

Determine the wind direction and speed.

A—095° and 19 knots.
B—113° and 19 knots.
C—238° and 18 knots.

Using an E6-B computer:

1. Turn compass azimuth so 345° is at the true index.
2. Slide board so grommet is on the ground speed arc, 95 knots.
3. Determine the WCA by comparing course to heading.
    345° − 355° = 10° right
4. Plot WCA at TAS arc with a pencil dot, 85 knots.
5. Rotate compass azimuth so pencil dot is on the centerline.
6. Read wind direction under true index and wind speed as number of units between grommet and pencil dot, 114° and 19 knots.

(H341) — AC 61-23C, Chapter 8

Off-Course Correction

Either the off-course correction equation or the E6-B can be used to find the total correction angle that is required to fly to the original destination.

The correction relation can be calculated by using the formula:

$$\frac{\text{Distance Off} \times 60}{\text{Distance Flown}} + \frac{\text{Distance Off} \times 60}{\text{Distance Remaining}} = \text{Degrees Total Correction}$$

Problem:

Find the total correction angle to converge on the destination under the following circumstances:

Distance flown	48 miles
Distance remaining	120 miles
Miles off course	4 miles

Answers

5475 [A]   5476 [B]

Chapter 9 Navigation

Solution using the off-course equation:

$$\frac{(4 \text{ miles})(60)}{48 \text{ miles}} + \frac{(4 \text{ miles})(60)}{120 \text{ miles}} = \text{Total Correction}$$

$$= \frac{240}{48} + \frac{240}{120}$$

$$= 5° + 2° = 7° \text{ total correction angle to converge on destination.}$$

Solution using the E6-B:

1. Set the following relationships on the computer face of the E6-B:

$$\frac{\text{Miles off Course}}{\text{Miles Flown}} = \frac{\text{Degrees to Parallel}}{\Delta}$$

$$\frac{4}{48} = \frac{5° \text{ to Parallel}}{\Delta}$$

2. $$\frac{\text{Miles off Course}}{\text{Miles Flown}} = \frac{\text{Additional DEG to Converge}}{\Delta}$$

$$\frac{4}{120} = \frac{2° \text{ Additional to Converge}}{\Delta}$$

5° to Parallel + 2° Additional to Converge = 7° Total Correction

---

AIR, RTC
5477. You have flown 52 miles, are 6 miles off course, and have 118 miles yet to fly. To converge on your destination, the total correction angle would be
A—3°.
B—6°.
C—10°.

1. Determine the number of degrees correction required to parallel the desired course:

$$\frac{\text{Miles off Course} \times 60}{\text{Number of Miles Flown}} = \frac{\text{Correction}}{\text{to Parallel}}$$

$$\frac{6 \times 60}{52} = 6.92° \text{ to parallel}$$

2. Determine the number of additional degrees correction required to intercept course:

$$\frac{\text{Miles off Course} \times 60}{\text{Number of Miles Remaining}} = \frac{\text{Correction}}{\text{to Intercept}}$$

$$\frac{6 \times 60}{118} = 3.05° \text{ additional to intercept}$$

3. Add the number of degrees required to parallel the course and the number of plus degrees required to intercept the course in order to find total correction angle required to converge at destination:

6.92° to parallel + 3.05° to intercept
= 9.97° total correction angle

(H341) — AC 61-23C, Chapter 8

Answers
5477 [C]

ALL
5478. GIVEN:

Distance off course .... 9 mi
Distance flown .... 95 mi
Distance to fly .... 125 mi

To converge at the destination, the total correction angle would be

A—4°.
B—6°.
C—10°.

1. Determine the number of degrees correction required to parallel the desired course:

$$\frac{\text{Miles off Course} \times 60}{\text{Number of Miles Flown}} = \text{Correction to Parallel}$$

$$\frac{9 \times 60}{95} = 5.68° \text{ to parallel}$$

2. Determine the number of additional degrees correction required to intercept the course:

$$\frac{\text{Miles off Course} \times 60}{\text{Number of Miles Remaining}} = \text{Correction to Intercept}$$

$$\frac{9 \times 60}{125} = 4.32° \text{ additional to intercept}$$

3. Add the number of degrees required to parallel the course and the number of degrees required to intercept the course to find the total correction angle required to converge at destination:

5.68° to parallel + 4.32° to intercept
= 10.00° total to converge

(H341) — AC 61-23C, Chapter 8

VHF Omni-Directional Range (VOR)

All VOR stations transmit an identifier. It is a three-letter Morse code signal interrupted only by a voice identifier on some stations, or to allow the controlling flight service station to speak on the frequency. Absence of a VOR identifier indicates maintenance is being performed on the station and the signal may not be reliable.

All VOR receivers have at least the essential components shown in Figure 9-1. The pilot may select the desired course or radial by turning the Omni-Bearing Selector (OBS). The Course Deviation Indicator (CDI) centers when the aircraft is on the selected radial or its reciprocal. A full-scale deflection of the CDI from the center represents a deviation of approximately 10° to 12°.

The TO/FROM Indicator (ambiguity indicator) shows whether the selected course will take the aircraft TO or FROM the station. A TO indication shows that the OBS selection is on the other side of the VOR station. A FROM indication shows that the OBS selection and the aircraft are on the same side of the VOR station. When an aircraft flies over a VOR, the TO/FROM indicator will reverse, indicating station passage.

The position of the aircraft can always be determined by rotating the OBS until the CDI centers with a FROM indication. The course displayed indicates the radial FROM the station. The VOR indicator displays information as though the

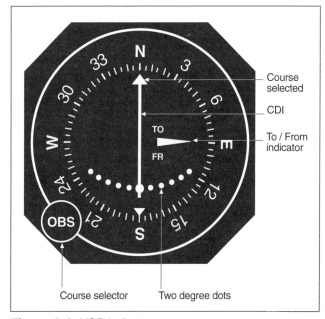

Figure 9-1. VOR indicators

Answers
5478 [C]

Chapter 9 **Navigation**

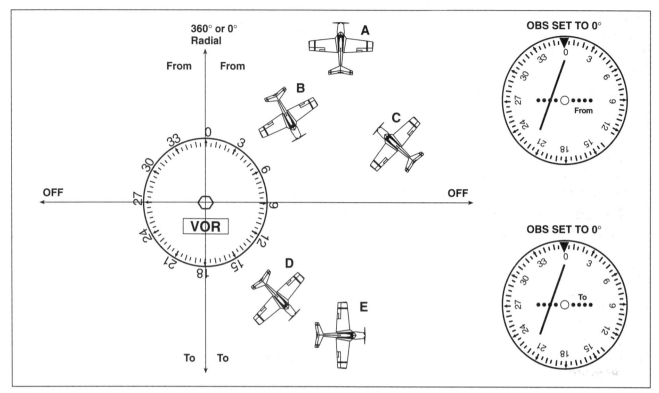

Figure 9-2. VOR display

aircraft were going in the direction of the course selected. However, actual heading does not influence the display. *See* Figure 9-2.

VOR radials, all of which originate at the VOR antenna, diverge as they radiate outward. For example, while the 011° radial and the 012° radial both start at the same point, 1 NM from the antenna, they are 100 feet apart. When they are 2 NM from the antenna, they are 200 feet apart. So at 60 NM, the radials would be 1 NM (6,000 feet) apart. *See* Figure 9-3.

The VOR indicator uses a series of dots to indicate any deviation from the selected course, with each dot equal to approximately 2° of deviation. Thus, a one-dot deviation at a distance of 30 NM from the station would indicate that the aircraft was 1 NM from the selected radial (200 feet x 30 = 6,000 feet).

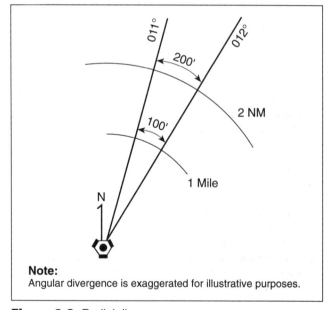

Note:
Angular divergence is exaggerated for illustrative purposes.

Figure 9-3. Radial divergence

To orient where the aircraft is in relation to the VOR, first determine which radial is selected (look at the OBS setting). Next, determine whether the aircraft is flying to or away from the station (look at the TO/FROM indicator), to find which hemisphere the aircraft is in. Last, determine how far off course the aircraft is from the selected course (look at the CDI needle deflection) to find which quadrant the aircraft is in. Remember that aircraft heading does not affect orientation to the VOR.

VOR accuracy may be checked by means of a VOR Test Facility (VOT), ground or airborne checkpoints, or by checking dual VORs against each other. A VOT location and frequency can be found in the Airport/Facility Directory (A/FD) and on the Air-to-Ground Communications Panel of the Low Altitude Enroute Chart.

To use the VOT, tune to the appropriate frequency and center the CDI. The omni-bearing selector should read 0° with a FROM indication, or 180° with a TO indication. The allowable error is ±4°. VOR receiver checkpoints are listed in the A/FD. With the appropriate frequency tuned and the OBS set to the published certified radial, the CDI should center with a FROM indication when the aircraft is over the designated checkpoint. Allowable accuracy is ±4° for a ground check, and ±6° for an airborne check. If the aircraft is equipped with dual VORs, they may be checked against each other. The maximum permissible variation when tuned to the same VOR is 4°.

The pilot must log the results of the VOR accuracy test in the aircraft logbook or other record. The log must include the date, place, bearing error, if any, and a signature.

ALL
5532. When checking the course sensitivity of a VOR receiver, how many degrees should the OBS be rotated to move the CDI from the center to the last dot on either side?

A—5° to 10°.
B—10° to 12°.
C—18° to 20°.

Course sensitivity may be checked by noting the number of degrees of change in course selection as you rotate the OBS to move the CDI from center to the last dot on either side. This should be between 10° and 12°. (I08) — AC 61-27C, Chapter 8

ALL
5533. An aircraft 60 miles from a VOR station has a CDI indication of one-fifth deflection, this represents a course centerline deviation of approximately

A—6 miles.
B—2 miles.
C—1 mile.

Aircraft displacement from course is approximately 200 feet per dot per nautical mile. The CDI deflection indication is one-fifth deflection at 60 miles from the station, and a one-fifth or one-dot deflection indicates a 2-mile displacement of the aircraft from the course centerline. (I08) — AC 61-27C, Chapter 8

ALL
5500. Which situation would result in reverse sensing of a VOR receiver?

A—Flying a heading that is reciprocal to the bearing selected on the OBS.
B—Setting the OBS to a bearing that is 90° from the bearing on which the aircraft is located.
C—Failing to change the OBS from the selected inbound course to the outbound course after passing the station.

With the reciprocal of the inbound course set on the OBS and the indicator showing FROM, the aircraft will be turned away from the needle for direct return to course centerline. This is called reverse sensing. (H348) — AC 61-23C, Chapter 8

AIR, RTC, LTA, MIL
5122-1. When must an operational check on the aircraft VOR equipment be accomplished to operate under IFR? Within the preceding

A—30 days or 30 hours of flight time.
B—10 days or 10 hours of flight time.
C—30 days.

No person may operate a civil aircraft under IFR using the VOR system of radio navigation unless the VOR equipment of that aircraft has been operationally checked within the preceding 30 days, and was found to be within permissible limits. (B10) — 14 CFR §91.171

Answers

5532 [B]	5533 [B]	5500 [A]	5122-1 [C]

Chapter 9 Navigation

AIR, RTC, LTA, MIL

5122-2. Which data must be recorded in the aircraft logbook or other record by a pilot making a VOR operational check for IFR operations?

A—VOR name or identification, place of operational check, amount of bearing error, and date of check.
B—Date of check, place of operational check, bearing error, and signature.
C—VOR name or identification, amount of bearing error, date of check, and signature.

Each person making a VOR operational check shall enter the date, place, bearing error, and sign the aircraft log or other record. (B10) — 14 CFR §91.171

ALL

5501. To track outbound on the 180 radial of a VOR station, the recommended procedure is to set the OBS to

A—360° and make heading corrections toward the CDI needle.
B—180° and make heading corrections away from the CDI needle.
C—180° and make heading corrections toward the CDI needle.

Tracking involves drift correction that is sufficient to maintain a direct course to or from a transmitting station. The course selected for tracking outbound is the course shown under the course index with the TO/FROM indicator showing FROM. Turning toward the needle returns the aircraft to the course centerline and centers the needle. (H348) — AC 61-23C, Chapter 8

ALL

5502. To track inbound on the 215 radial of a VOR station, the recommended procedure is to set the OBS to

A—215° and make heading corrections toward the CDI needle.
B—215° and make heading corrections away from the CDI needle.
C—035° and make heading corrections toward the CDI needle.

Tracking involves drift correction that is sufficient to maintain a direct course to or from a transmitting station. The course selected for tracking outbound is the course shown under the course index with the TO/FROM indicator showing FROM. Turning toward the needle returns the aircraft to the course centerline and centers the needle. (H348) — AC 61-23C, Chapter 8

ALL

5534. (Refer to Figure 20.) Using instrument group 3, if the aircraft makes a 180° turn to the left and continues straight ahead, it will intercept which radial?

A—135 radial.
B—270 radial.
C—360 radial.

RMI 3 indicates a heading of 300°, VOR 2 has been selected for display on the double-barred arrow, and an ADF station is selected for display on the thin arrow. The tail of the double-barred arrow indicates the radial the aircraft is on. This aircraft is inbound southeast of the station, either tracking with a wind correction on, or just crossing the 135° radial. Assuming no wind, a 180° turn to the left would put it outbound, heading 120°, on an intercept heading that would re-cross the 135° radial. See the figure below. (I08) — AC 61-27C, Chapter 8

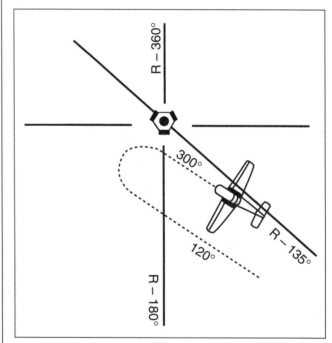

Question 5534

Answers

5122-2 [B]　　　5501 [C]　　　5502 [C]　　　5534 [A]

ALL
5535. (Refer to Figure 20.) Which instrument shows the aircraft in a position where a 180° turn would result in the aircraft intercepting the 150 radial at a 30° angle?

A—2.
B—3.
C—4.

RMI 4 indicates a heading of 360°, VOR 2 has been selected for display on the double-barred arrow, and an ADF station is selected for display on the thin arrow. The tail of the double-barred arrow indicates the radial that the aircraft is on. This aircraft is outbound northeast of the station, either tracking with a wind correction on, or just crossing the 015° radial. A 180° turn in either direction would result in a heading of 180°. Assuming no wind and sufficient distance from the station, this would eventually result in the aircraft being outbound, southeast of the station, and crossing the 150° radial. See the figure below. (I08) — AC 61-27C, Chapter 8

ALL
5536. (Refer to Figure 20.) Which instrument shows the aircraft in a position where a straight course after a 90° left turn would result in intercepting the 180 radial?

A—2.
B—3.
C—4.

RMI 3 indicates a heading of 300°, VOR 2 has been selected for display on the double-barred arrow, and an ADF station is selected for display on the thin arrow. The tail of the double-barred arrow indicates the radial that the aircraft is on. This aircraft is inbound southeast of the station, either tracking with a wind correction on, or just crossing the 135° radial. A 90° left turn would produce a new heading of 210° which, assuming moderate or no wind, would result in intercepting the 180° radial. See the figure below. (I08) — AC 61-27C, Chapter 8

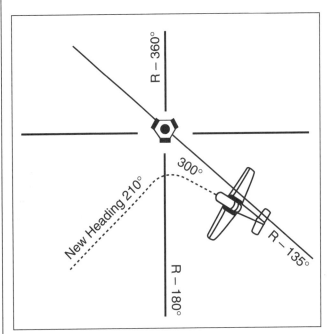

Question 5536

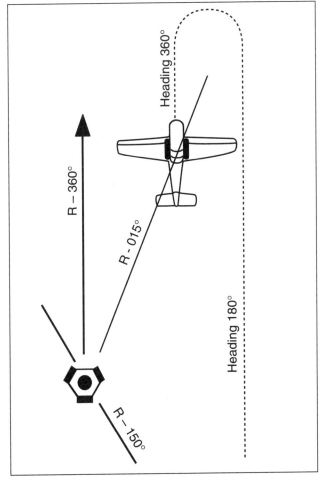

Question 5535

Answers

5535 [C]   5536 [B]

Chapter 9 Navigation

ALL
5537. (Refer to Figure 20.) Which instrument shows the aircraft to be northwest of the VORTAC?

A—1.
B—2.
C—3.

RMI 2 indicates a heading of 125°, VOR 2 has been selected for display on the double-barred arrow, and an ADF station is selected for display on the thin arrow. The tail of the double-barred arrow indicates that the aircraft is on the 310° radial. This aircraft is inbound, northwest of the station, either tracking with a wind correction on, or just crossing the 310° radial. See the figure below. (I08) — AC 61-27C, Chapter 8

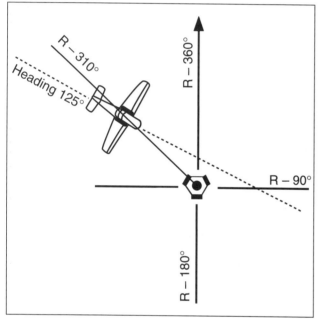

Question 5537

ALL
5538. (Refer to Figure 20.) Which instrument(s) show(s) that the aircraft is getting further from the selected VORTAC?

A—4.
B—1 and 4.
C—2 and 3.

The double-barred bearing indicator gives the magnetic bearing to the tuned VOR. The tail points to the radial the aircraft is on. The RMI rotating compass indicates the aircraft magnetic heading at the top. RMIs 1, 2, and 3 all indicate that the aircraft is heading in the general direction of the VORTAC. Only RMI 4 indicates that the aircraft is moving away from the VORTAC. See the figure on the next page. (I08) — AC 61-27C, Chapter 8

AIR, RTC, LTA, MIL
5062. What is the maximum tolerance (+ or -) allowed for an operational VOR equipment check when using an FAA-approved ground test signal?

A—4 degrees.
B—6 degrees.
C—8 degrees.

The maximum permissible bearing error is ±4° when using a VOT. (B10) — 14 CFR §91.171(b)(1)

ALL, MIL
5551. How should the pilot make a VOR receiver check when the aircraft is located on the designated checkpoint on the airport surface?

A—Set the OBS on 180° plus or minus 4°; the CDI should center with a FROM indication.
B—Set the OBS on the designated radial. The CDI must center within plus or minus 4° of that radial with a FROM indication.
C—With the aircraft headed directly toward the VOR and the OBS set to 000°, the CDI should center within plus or minus 4° of that radial with a TO indication.

Airborne and ground checkpoints consist of certified radials that should be received at specific points on the airport surface or over specific landmarks while airborne in the immediate vicinity of the airport. Should an error in excess of ±4° be indicated through use of a ground check, or ±6° using the airborne check, IFR flight shall not be attempted without first correcting the source of the error. Caution: No correction other than the correction card figures supplied by the manufacturer should be applied in making these VOR receiver checks. (J01) — AIM ¶1-1-4

ALL, MIL
5552. When using VOT to make a VOR receiver check, the CDI should be centered and the OBS should indicate that the aircraft is on the

A—090 radial.
B—180 radial.
C—360 radial.

Answers

5537 [B]   5538 [A]   5062 [A]   5551 [B]   5552 [C]

Chapter 9 Navigation

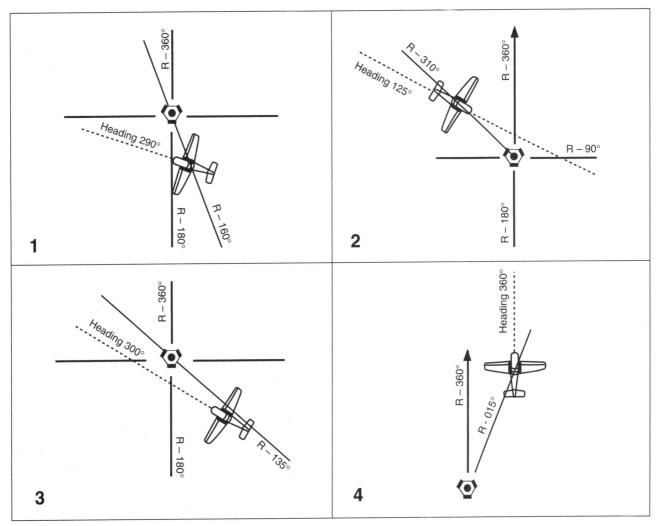

Question 5538

To use the VOT service, tune in the VOT frequency on your VOR receiver. With the Course Deviation Indicator (CDI) centered, the omni-bearing selector (OBS) should read 0° with the TO/FROM indication showing FROM, or the OBS should read 180° with the TO/FROM indication showing TO. Since VOR radials are always expressed as FROM the station, the OBS should indicate the aircraft is on the 360° radial FROM. (J01) — AIM ¶1-1-4

ALL, MIL

5553. When the CDI needle is centered during an airborne VOR check, the omnibearing selector and the TO/FROM indicator should read

A—within 4° of the selected radial.
B—within 6° of the selected radial.
C—0° TO, only if you are due south of the VOR.

If neither a test signal nor a designated checkpoint on the surface is available, use an airborne checkpoint designated by the Administrator or, outside the United States, by appropriate authority. The maximum permissible bearing error is ±6°. (J01) — 14 CFR §91.171(b)(3)

Answers

5553 [B]

Estimating Time and Distance To Station Using VOR

If the aircraft is flown across radials, the approximate time and distance to a station can be determined by using the elapsed time, change of radials (angle) during that time, and the true airspeed (TAS).

$$\text{Time To Station} = \frac{60 \times \text{Elapsed Time}}{\text{Angular Change}}$$

and

$$\text{Distance To Station} = \text{TAS} \times \text{Time To Station}$$

Problem:

Determine the approximate time and distance to the station, given the following:

Heading .... 270°
Crossing .... 360° radial at 1237Z
Crossing .... 350° radial at 1243Z
TAS .... 130 knots

Solution:

1. Calculate elapsed time and angular change:

   ```
   1243 Z          360°
  -1237 Z         -350°
   ─────          ─────
   6 minutes       10°
   ```

2. Calculate time to station using the relation:

   $$\text{Time To Station} = \frac{60 \times \text{Elapsed Time}}{\text{Angular Change}}$$

   $$\frac{(60)(6 \text{ min})}{10°} = 36 \text{ minutes} = 0.6 \text{ hour}$$

3. Calculate the distance to the station using the relation:

   Distance = TAS x Time

   130 knots x 0.6 hours = 78 NM

When the angle to the station is small, that is, the aircraft flies nearly along radials, another method of approximating time to a station can be used. The **Isosceles Triangle Principle** states that if two angles of a triangle are equal, the two sides opposite those angles are also equal. For example, see the problem following, and Figure 9-4.

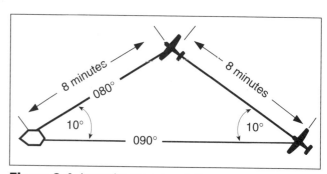

Figure 9-4. Isosceles triangle solution

Chapter 9 **Navigation**

Problem:

While inbound on the 090° radial, a pilot rotates the OBS 10° to the left (080° radial) and turns the aircraft 10° to the right noting the time. Maintaining a constant heading, the pilot determines that the elapsed time for the CDI to center is 8 minutes.

Solution:

Based on this information the ETE (estimated time en route) can be found with the isosceles triangle method since the angle the OBS was changed (90° − 80° = 10°) is equal to the angle the aircraft was turned (010°). The time elapsed for 10° was 8 minutes, and due to the equal angles, the time en route is also determined to be 8 minutes. *See* Figure 9-4.

AIR, RTC, LTA
5539. While maintaining a magnetic heading of 270° and a true airspeed of 120 knots, the 360 radial of a VOR is crossed at 1237 and the 350 radial is crossed at 1244. The approximate time and distance to this station are

A—42 minutes and 84 NM.
B—42 minutes and 91 NM.
C—44 minutes and 96 NM.

1. Calculate the Time to Station using:

$$\text{Time} = \frac{60 \times \text{Time of Bearing Change}}{\text{Bearing Change}}$$

$$= \frac{60 \times (1244 - 1237)}{(360 - 350)} = \frac{60 \times 7}{10} = 42 \text{ minutes}$$

2. Calculate the Distance to Station using:

Distance = Time × Speed

$$\frac{42 \text{ min}}{60 \text{ min/hr}} \times 120 \text{ knots} = 84 \text{ NM}$$

(I08) — AC 61-27C, Chapter 8

AIR, RTC, LTA
5540. (Refer to Figure 21.) If the time flown between aircraft positions 2 and 3 is 13 minutes, what is the estimated time to the station?

A—13 minutes.
B—17 minutes.
C—26 minutes.

Time to Station = Time Required to Complete a Bearing Change

Time to Station = 13 minutes

(I08) — AC 61-27C, Chapter 8

ALL
5541. (Refer to Figure 22.) If the time flown between aircraft positions 2 and 3 is 8 minutes, what is the estimated time to the station?

A—8 minutes.
B—16 minutes.
C—48 minutes.

Time to Station = Time Required to Complete a Bearing Change

Time to Station = 8 minutes

(I08) — AC 61-27C, Chapter 8

AIR, RTC, LTA
5542. (Refer to Figure 23.) If the time flown between aircraft positions 2 and 3 is 13 minutes, what is the estimated time to the station?

A—7.8 minutes.
B—13 minutes.
C—26 minutes.

Time to Station = Time Required to Complete a Bearing Change

Time to Station = 13 minutes

(I08) — AC 61-27C, Chapter 8

AIR, RTC, LTA
5543. (Refer to Figure 24.) If the time flown between aircraft positions 2 and 3 is 15 minutes, what is the estimated time to the station?

A—15 minutes.
B—30 minutes.
C—60 minutes.

Time to Station = Time Required to Complete a Bearing Change

Time to Station = 15 minutes

(I08) — AC 61-27C, Chapter 8

Answers

| 5539 [A] | 5540 [A] | 5541 [A] | 5542 [B] | 5543 [A] |

Chapter 9 **Navigation**

AIR, RTC, LTA
5544. Inbound on the 040 radial, a pilot selects the 055 radial, turns 15° to the left, and notes the time. While maintaining a constant heading, the pilot notes the time for the CDI to center is 15 minutes. Based on this information, the ETE to the station is

A—8 minutes.
B—15 minutes.
C—30 minutes.

Time to Station = Time Required to Complete a Bearing Change

Time to Station = 15 minutes

(I08) — AC 61-27C, Chapter 8

AIR, RTC, LTA
5545. Inbound on the 090 radial, a pilot rotates the OBS 010° to the left, turns 010° to the right, and notes the time. While maintaining a constant heading, the pilot determines that the elapsed time for the CDI to center is 8 minutes. Based on this information, the ETE to the station is

A—8 minutes.
B—16 minutes.
C—24 minutes.

Time to Station = Time Required to Complete a Bearing Change

Time to Station = 8 minutes

(I08) — AC 61-27C, Chapter 8

AIR, RTC, LTA
5546. Inbound on the 315 radial, a pilot selects the 320 radial, turns 5° to the left, and notes the time. While maintaining a constant heading, the pilot notes the time for the CDI to center is 12 minutes. The ETE to the station is

A—10 minutes.
B—12 minutes.
C—24 minutes.

Time to Station = Time Required to Complete a Bearing Change

Time to Station = 12 minutes

(I08) — AC 61-27C, Chapter 8

AIR, RTC, LTA
5547. Inbound on the 190 radial, a pilot selects the 195 radial, turns 5° to the left, and notes the time. While maintaining a constant heading, the pilot notes the time for the CDI to center is 10 minutes. The ETE to the station is

A—10 minutes.
B—15 minutes.
C—20 minutes.

Time to Station = Time Required to Complete a Bearing Change

Time to Station = 10 minutes

(I08) — AC 61-27C, Chapter 8

Answers

5544 [B]   5545 [A]   5546 [B]   5547 [A]

Horizontal Situation Indicator (HSI)

The Horizontal Situation Indicator (HSI) is a combination of two instruments: the heading indicator and the VOR. *See* Figure 9-5.

The aircraft heading displayed on the rotating azimuth card under the upper lubber line in Figure 9-5 is 330°. The course indicating arrowhead that is shown is set to 300°. The tail of the course indicating arrow indicates the reciprocal, or 120°.

The course deviation bar operates with a VOR/LOC navigation receiver to indicate either left or right deviations from the course that is selected with the course indicating arrow. It moves left or right to indicate deviation from the centerline in the same manner that the angular movement of a conventional VOR/LOC needle indicates deviation from course.

The desired course is selected by rotating the course indicating arrow in relation to the azimuth card by means of the course set knob. This gives the pilot a pictorial presentation. The fixed aircraft symbol and the course deviation bar display the aircraft relative to the selected course as though the pilot was above the aircraft looking down.

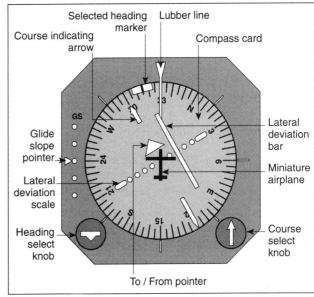

Figure 9-5. Horizontal Situation Indicator (HSI)

The TO/FROM indicator is a triangular-shaped pointer. When this indicator points to the head of the course arrow, it indicates that the course selected, if properly intercepted and flown, will take the aircraft TO the selected facility, and vice versa.

The glide slope deviation pointer indicates the relationship of the aircraft to the glide slope. When the pointer is below the center position, the aircraft is above the glide slope, and an increased rate of descent is required.

To orient where the aircraft is in relation to the facility, first determine which radial is selected (look at the arrowhead). Next, determine whether the aircraft is flying to or away from the station (look at the TO/FROM indicator) to find which hemisphere the aircraft is in. Next, determine how far from the selected course the aircraft is (look at the deviation bar) to find which quadrant the aircraft is in. Finally, consider the aircraft heading (under the lubber line) to determine the aircraft's position within the quadrant.

ALL
5506. (Refer to Figure 17.) Which illustration indicates that the airplane will intercept the 060 radial at a 60° angle inbound, if the present heading is maintained?

A—6.
B—4.
C—5.

To intercept the 060° radial inbound (TO), the reciprocal (240°) would be selected. Only illustrations 4, 5, and 6 have the 240° course selected. The 240° course may be intercepted at a 60° angle with either a heading of 300° (240° + 60°) or 180° (240° – 60°). The only illustration showing either is 6. (I08) — AC 61-27C, Chapter 8

Answers

5506 [A]

AIR, RTC, LTA
5507. (Refer to Figure 17.) Which statement is true regarding illustration 2, if the present heading is maintained? The airplane will

A—cross the 180 radial at a 45° angle outbound.
B—intercept the 225 radial at a 45° angle.
C—intercept the 360 radial at a 45° angle inbound.

Illustration 2 shows the aircraft heading 225°. The bearing pointer indicates a heading of 235° will take the aircraft to the station; therefore, you are on the 055° radial. If the present heading is maintained, the station will remain to the right of the aircraft and you will cross the 180° radial at approximately a 45° angle outbound (225° – 180° = 45°). (I08) — AC 61-27C, Chapter 8

AIR, RTC, LTA
5508. (Refer to Figure 17.) Which illustration indicates that the airplane will intercept the 060 radial at a 75° angle outbound, if the present heading is maintained?

A—4.
B—5.
C—6.

Only two possible headings will intercept the 060° radial at a 75° angle outbound: 135° and 345°. The only one of those shown is illustration 5. The illustration shows that if the aircraft turned to 060°, it would be going FROM the station and would have to fly left to get on the 060 radial. (I08) — AC 61-27C, Chapter 8

AIR, RTC, LTA
5509. (Refer to Figure 17.) Which illustration indicates that the airplane should be turned 150° left to intercept the 360 radial at a 60° angle inbound?

A—1.
B—2.
C—3.

Only two headings intercept the 360° radial inbound at a 60° angle: 120° and 240°. An aircraft heading 270° could turn left 150° to intercept at a 60° angle on a heading of 120°. An aircraft heading 030° could turn left 150° to intercept at a 60° angle on a heading of 240°. Illustration 1 shows the aircraft flying one of the two possible headings. (I08) — AC 61-27C, Chapter 8

AIR, RTC, LTA
5510. (Refer to Figure 17.) Which is true regarding illustration 4, if the present heading is maintained? The airplane will

A—cross the 060 radial at a 15° angle.
B—intercept the 240 radial at a 30° angle.
C—cross the 180 radial at a 75° angle.

Illustration 4 shows the aircraft heading 255° with a magnetic bearing of 275° to the station. The aircraft will pass south of the station and cross the 180° radial at a 75° angle (255° – 180° = 75°). (I08) — AC 61-27C, Chapter 8

Answers
5507 [A]　　　5508 [B]　　　5509 [A]　　　5510 [C]

Automatic Direction Finder (ADF)

ADF equipment consists of a receiver that receives in the low-and-medium frequency bands and an instrument needle that points to the station. The ADF may be used to either home or track to a station. Homing is flying the aircraft on any heading required to keep the azimuth needle on 0° until the station is reached, which results in a curved path that leads to the station (if there is any crosswind at all). Tracking is following a straight geographic path by establishing a heading that will maintain the desired track, regardless of wind effect. When an aircraft is on the desired track outbound from a radio station with the proper drift correction established, the ADF needle will deflect to the windward side.

The azimuth needle, pointing to the selected station, indicates the angular difference between the aircraft heading and the direction to the station, measured clockwise from the nose of the aircraft. This angular difference is the relative bearing to the station, and may be read directly from the fixed-scale indicator.

Determining Bearings with ADF

To determine the magnetic bearing from an aircraft to a selected station, add the magnetic heading of the aircraft to the relative bearing to the station:

Magnetic Heading + Relative Bearing = Magnetic Bearing TO the Station

See Figure 9-6.

Problem:

An aircraft has a magnetic heading of 360° with a relative bearing to a radio beacon of 245°. Calculate the magnetic bearing TO that radio beacon.

Solution:

MH + RB = MB (TO)

360° + 245° = 605° − 360° = 245° (TO)

To find the magnetic bearing being crossed (intercepted) by an aircraft, or the magnetic bearing from a radio beacon (RBN), the reciprocal is taken of the magnetic bearing TO a station.

An aircraft is maintaining a magnetic heading of 275° and the ADF shows a relative bearing of 070°. The magnetic radial FROM that radio beacon can be calculated using the relation:

MH + RB = MB (TO) ± 180° = MB (FROM)

275 + 070° = 345° − 180° = 165°

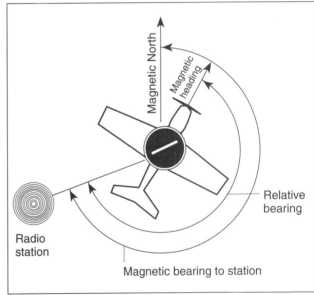

Figure 9-6. Magnetic bearing to the station

Chapter 9 Navigation

Problem:

If the aircraft continues its present heading as shown in Figure 9-7, what will the ADF indicate when the aircraft reaches the magnetic bearing of 030° from the NDB?

Solution:

1. The aircraft's present heading is 330° with a relative bearing of 270°.

    MH + RB = MB

    330° + 270° = 600° − 360° = 240° (TO)

2. The present bearing TO is the reciprocal of the bearing FROM.

    240° − 180° = 060° (FROM)

3. The change from the present MB (060°) to the desired MB (030°) is a decrease of 030° (060° − 030°). As the magnetic bearing decreases by 030°, the relative bearing as indicated on the ADF also decreases. The present relative bearing (270°) is decreased by 030° giving a new relative bearing of 240° (270° − 030°).

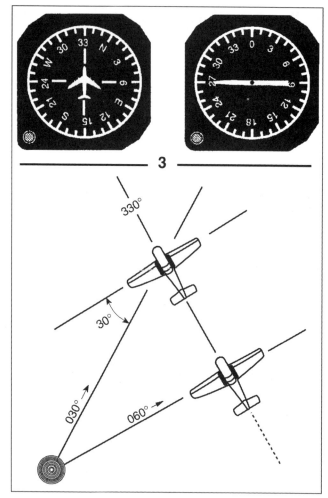

Figure 9-7. Change in relative bearing

Intercepting Bearings with ADF

Problem:

What heading should the airplane turn to, in order to intercept a magnetic bearing of 240° FROM at a 030° angle (while outbound FROM)?

Solution:

1. The magnetic bearing of 240° FROM may be intercepted with a 030° angle at a heading of either 210° (240° − 030°) or 270° (240° + 030°).

2. A heading of 210° would take the aircraft further away from the desired bearing of 240°. Thus, a heading of 270° is the only option.

3. To turn from the current heading of 035° to the desired heading of 270° would require a left turn of 125°.

Chapter 9 **Navigation**

Estimated Time and Distance To Station Using ADF

Relative bearing changes can be used in two ways to estimate flying time, distance and fuel required to fly to a station (or NAVAID): wing-tip bearing change method, and the isosceles triangle technique.

Wing-Tip Bearing Change Method

The first technique is to use the time required for a bearing change when the station is approximately located off the wing tip. The approximate time, distance and fuel to a station can be determined by using the time and amount of wing-tip bearing change, the true airspeed (TAS) and the rate of fuel consumption.

Problem:

If a 15° wing-tip bearing change occurs in 8 minutes at a TAS of 90 knots with 8.6 gallons/hour rate of fuel consumption, what is the time, distance, fuel burn to the station?

Solution:

1. The time-to-station can be calculated using the relation:

   $$\text{Time} = \frac{60 \times \text{minutes}}{\text{degrees}} = \frac{60 \times 8}{15} = 32 \text{ minutes}$$

2. The distance to the station can be calculated using the relation:

   Distance = Time x TAS

   $$\frac{32 \text{ minutes} \times 90 \text{ knots}}{60 \text{ minutes/hour}} = 48 \text{ NM}$$

3. The total fuel burn can be calculated using the relation:

   Fuel = Time x gallons/hour

   $$\frac{32 \text{ minutes} \times 8.6 \text{ gph}}{60 \text{ minutes/hour}} = 4.58 \text{ gallons}$$

Isosceles Triangle Technique

The second method of using ADF to determine time, distance and fuel to station, can be employed when the relative bearing is small. The method is based on the property of an isosceles triangle that if two angles of a triangle are equal, the two sides opposite those angles are themselves equal.

For example, using the isosceles triangle method while cruising at 100 knots and on a constant heading, a relative bearing of 10° doubles in 5 minutes and therefore, the time to the station is equal to the time required for the relative bearing to double. The wing-tip bearing change formula cannot be used since the bearing change was not off the wing tip.

Problem:

Time to Station = 5 minutes

Solution:

The distance to the station can be calculated using the relation:

Distance = Time x Speed

$$\frac{5 \text{ minutes}}{60 \text{ minutes/hour}} \times 100 \text{ knots} = 8.33 \text{ NM}$$

Chapter 9 **Navigation**

ALL
5490. Which is true about homing when using ADF during crosswind conditions? Homing

A—to a radio station results in a curved path that leads to the station.
B—is a practical navigation method for flying both to and from a radio station.
C—to a radio station requires that the ADF have an automatically or manually rotatable azimuth.

If the pilot continues flight toward the station keeping the needle on 0°, the procedure is called homing to the station. If a crosswind exists, the ADF needle will continue to drift away from zero. To keep the needle on zero the pilot will repeatedly turn the aircraft slightly. The result will be a curved flight path to the station. (H348) — AC 61-23C, Chapter 8

Answer (B) is incorrect because homing is not a practical means of navigating both to or from a station. Answer (C) is incorrect because homing can be done with any ADF.

ALL
5491. Which is true regarding tracking on a desired bearing when using ADF during crosswind conditions?

A—To track outbound, heading corrections should be made away from the ADF pointer.
B—When on the desired track outbound with the proper drift correction established, the ADF pointer will be deflected to the windward side of the tail position.
C—When on the desired track inbound with the proper drift correction established, the ADF pointer will be deflected to the windward side of the nose position.

To remain on a desired outbound track from the station when there is a crosswind, the nose of the aircraft must be turned into the wind. The head of the ADF needle, which has been pointing to the NDB behind the aircraft, still points to that station, but the head of the needle is to the windward side of the tail position. (H348) — AC 61-23C, Chapter 8

Answer (A) is incorrect because corrections should be made toward the ADF pointer when tracking outbound. Answer (C) is incorrect because the pointer is deflected to the leeward side when inbound.

ALL
5492. An aircraft is maintaining a magnetic heading of 265° and the ADF shows a relative bearing of 065°. This indicates that the aircraft is crossing the

A—065° magnetic bearing FROM the radio beacon.
B—150° magnetic bearing FROM the radio beacon.
C—330° magnetic bearing FROM the radio beacon.

Use the relation below to determine the bearing FROM:

MB (FROM) = RB + MH ± 180°
MB (FROM) = 065° + 265° − 180°
MB (FROM) = 150°

(H348) — AC 61-23C, Chapter 8

ALL
5493. The magnetic heading is 315° and the ADF shows a relative bearing of 140°. The magnetic bearing FROM the radiobeacon would be

A—095°.
B—175°.
C—275°.

Use the relation below to determine the magnetic bearing FROM:

MB = RB + MH ± 180°
MB = 140° + 315° − 180°
MB = 275°

(H348) — AC 61-23C, Chapter 8

ALL
5494. The magnetic heading is 350° and the relative bearing to a radiobeacon is 240°. What would be the magnetic bearing TO that radiobeacon?

A—050°.
B—230°.
C—295°.

Use the relation below to determine the bearing TO:

MB (TO) = RB + MH
MB (TO) = 240° + 350° − 360°
MB (TO) = 230°

(H348) — AC 61-23C, Chapter 8

ALL
5495. The ADF is tuned to a radiobeacon. If the magnetic heading is 040° and the relative bearing is 290°, the magnetic bearing TO that radiobeacon would be

A—150°.
B—285°.
C—330°.

Use the relation below to determine the bearing TO:

MB = RB + MH
MB = 290° + 040°
MB = 330°

(H348) — AC 61-23C, Chapter 8

Answers

5490 [A]   5491 [B]   5492 [B]   5493 [C]   5494 [B]   5495 [C]

ALL
5496. If the relative bearing to a nondirectional radiobeacon is 045° and the magnetic heading is 355°, the magnetic bearing TO that radiobeacon would be

A—040°.
B—065°.
C—220°.

Use the relation below to determine the bearing TO:

MB = RB + MH
MB = 045° + 355°
MB = 400° – 360° = 040°

(H348) — AC 61-23C, Chapter 8

ALL
5497. (Refer to Figure 16.) If the aircraft continues its present heading as shown in instrument group 3, what will be the relative bearing when the aircraft reaches the magnetic bearing of 030° FROM the NDB?

A—030°.
B—060°.
C—240°.

Use the relation below to determine the relative bearing:

RB = MB – MH
RB = 210° – 330° + 360°
RB = 240°

(H348) — AC 61-23C, Chapter 8

ALL
5498. (Refer to Figure 16.) At the position indicated by instrument group 1, what would be the relative bearing if the aircraft were turned to a magnetic heading of 090°?

A—150°.
B—190°.
C—250°.

Use the relation below to determine the relative bearing:

MB = MH + RB
300° + 040° = 340°

If the aircraft is unchanged relative to the station and its magnetic bearing is reoriented to 090°:

340° = 090° + RB
RB = 250°

(H348) — AC 61-23C, Chapter 8

ALL
5499. (Refer to Figure 16.) At the position indicated by instrument group 1, to intercept the 330° magnetic bearing to the NDB at a 30° angle, the aircraft should be turned

A—left to a heading of 270°.
B—right to a heading of 330°.
C—right to a heading of 360°.

Use the relation below to determine the bearing TO:

MH + RB = MB
300° + 040° = 340°

Only two headings intercept the 330° heading TO the station at a 30° angle: 300° and 360°. On its present 300° MH, the aircraft is diverging from the 330° heading TO and, therefore, should be turned right to a heading of 360°. (H348) — AC 61-23C, Chapter 8

AIR, RTC, LTA
5511. (Refer to Figure 18.) To intercept a magnetic bearing of 240° FROM at a 030° angle (while outbound), the airplane should be turned

A—left 065°.
B—left 125°.
C—right 270°.

Only two headings intercept the 240° magnetic bearing from the station at a 30° intercept angle: 210° and 270°. The aircraft is on the 165° magnetic bearing FROM the station, heading 035°. To intercept the 240° magnetic bearing from the station at a 30° angle while outbound, the aircraft must turn left 125° to a magnetic heading of 270°. See the figure below. (I08) — AC 61-27C, Chapter 8

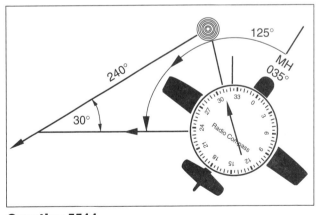

Question 5511

Answers

5496 [A]   5497 [C]   5498 [C]   5499 [C]   5511 [B]

Chapter 9 **Navigation**

AIR, RTC, LTA
5512. (Refer to Figure 18.) If the airplane continues to fly on the heading as shown, what magnetic bearing FROM the station would be intercepted at a 35° angle outbound?

A—035°.
B—070°.
C—215°.

The magnetic bearing from the station which will be intercepted at an angle of 035° outbound while flying a magnetic heading of 035°, is 035° + 035° = 070° FROM. See the figure below. (I08) — AC 61-27C, Chapter 8

The magnetic bearing FROM the station which would be intercepted at an angle of 035° while flying at 340° MH is 340° – 035° = 305° FROM. See the figure below. (I08) — AC 61-27C, Chapter 8

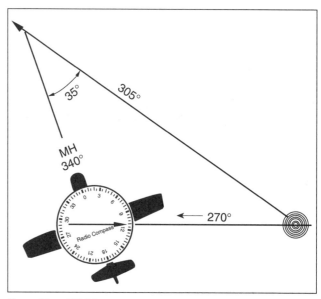

Question 5513

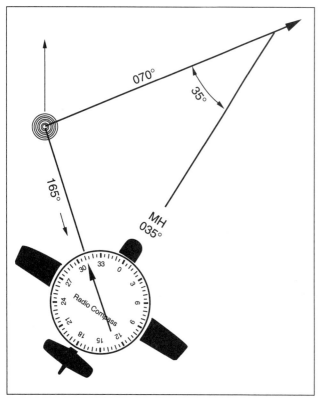

Question 5512

AIR, RTC, LTA
5513. (Refer to Figure 19.) If the airplane continues to fly on the magnetic heading as illustrated, what magnetic bearing FROM the station would be intercepted at a 35° angle?

A—090°.
B—270°.
C—305°.

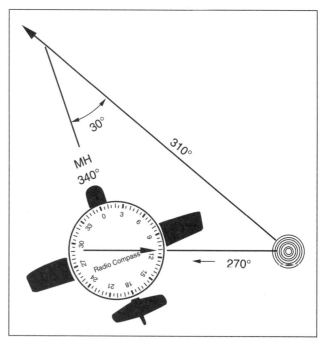

Question 5514

Answers

5512 [B]   5513 [C]

AIR, RTC, LTA
5514. (Refer to Figure 19.) If the airplane continues to fly on the magnetic heading as illustrated, what magnetic bearing FROM the station would be intercepted at a 30° angle?

A—090°.
B—270°.
C—310°.

The magnetic bearing FROM the station which would be intercepted at an angle of 030° while flying at 340° MH is 340°–030°=310° FROM. See the figure for 5514 on the previous page. (I08) — AC 61-27C, Chapter 8

AIR, RTC, LTA
5515. The relative bearing on an ADF changes from 265° to 260° in 2 minutes of elapsed time. If the groundspeed is 145 knots, the distance to that station would be

A—26 NM.
B—37 NM.
C—58 NM.

1. Calculate the Time to station using:

$$\text{Time} = \frac{60 \times \text{Time of Bearing Change}}{\text{Bearing Change (Degrees)}}$$

$$= \frac{60 \times 2}{(265 - 260)} = \frac{120}{5} = 24 \text{ minutes}$$

2. Calculate the Distance to station using:

$$\text{Distance} = \text{Time} \times \text{Speed}$$

$$\frac{24 \text{ min}}{60 \text{ min/hr}} \times 145 = 58 \text{ NM}$$

(I08) — AC 61-27C, Chapter 8

AIR, RTC
5516. The ADF indicates a wingtip bearing change of 10° in 2 minutes of elapsed time, and the TAS is 160 knots. What is the distance to the station?

A—15 NM.
B—32 NM.
C—36 NM.

1. Calculate the Time to station using:

$$\text{Time} = \frac{60 \times \text{Time of Bearing Change}}{\text{Bearing Change (Degrees)}}$$

$$= \frac{60 \times 2}{(10)} = \frac{120}{10} = 12 \text{ minutes}$$

2. Calculate the Distance to station using:

$$\text{Distance} = \text{Time} \times \text{Speed}$$

$$\frac{12 \text{ min}}{60 \text{ min/hr}} \times 160 = 32 \text{ NM}$$

(I08) — AC 61-27C, Chapter 8

AIR, RTC
5517. With a TAS of 115 knots, the relative bearing on an ADF changes from 090° to 095° in 1.5 minutes of elapsed time. The distance to the station would be

A—12.5 NM.
B—24.5 NM.
C—34.5 NM.

1. Calculate the Time to station using:

$$\text{Time} = \frac{60 \times \text{Time of Bearing Change}}{\text{Bearing Change (Degrees)}}$$

$$= \frac{60 \times 1.5}{(095 - 090)} = \frac{90}{5} = 18 \text{ minutes}$$

2. Calculate the Distance to station using:

$$\text{Distance} = \text{Time} \times \text{Speed}$$

$$\frac{18 \text{ min}}{60 \text{ min/hr}} \times 115 = 34.5 \text{ NM}$$

(I08) — AC 61-27C, Chapter 8

AIR, RTC
5518. GIVEN:

Wingtip bearing change .... 5°
Time elapsed between bearing change .... 5 min
True airspeed .... 115 kts

The distance to the station is

A—36 NM.
B—57.5 NM.
C—115 NM.

1. Calculate the Time to station using:

$$\text{Time} = \frac{60 \times \text{Time of Bearing Change}}{\text{Bearing Change (Degrees)}}$$

$$= \frac{60 \times 5}{5} = \frac{300}{5} = 60 \text{ minutes}$$

2. Calculate the Distance to station using:

$$\text{Distance} = \text{Time} \times \text{Speed}$$

$$\frac{60 \text{ min}}{60 \text{ min/hr}} \times 115 \text{ KIAS} = 115 \text{ NM}$$

(I08) — AC 61-27C, Chapter 8

Answers

5514 [C]   5515 [C]   5516 [B]   5517 [C]   5518 [C]

Chapter 9 **Navigation**

AIR, RTC
5519. The ADF is tuned to a nondirectional radiobeacon and the relative bearing changes from 095° to 100° in 1.5 minutes of elapsed time. The time en route to that station would be

A—18 minutes.
B—24 minutes.
C—30 minutes.

Calculate the Time to station using:

$$\text{Time} = \frac{60 \times \text{Time of Bearing Change}}{\text{Bearing Change (Degrees)}}$$

$$= \frac{60 \times 1.5}{(100 - 095)} = \frac{90}{5} = 18 \text{ minutes}$$

(I08) — AC 61-27C, Chapter 8

AIR, RTC
5520. The ADF is tuned to a nondirectional radiobeacon and the relative bearing changes from 270° to 265° in 2.5 minutes of elapsed time. The time en route to that beacon would be

A—9 minutes.
B—18 minutes.
C—30 minutes.

Calculate the Time to station using:

$$\text{Time} = \frac{60 \times \text{Time of Bearing Change}}{\text{Bearing Change (Degrees)}}$$

$$= \frac{60 \times 2.5}{(270 - 265)} = \frac{150}{5} = 30 \text{ minutes}$$

(I08) — AC 61-27C, Chapter 8

AIR, RTC
5521. The ADF is tuned to a nondirectional radiobeacon and the relative bearing changes from 085° to 090° in 2 minutes of elapsed time. The time en route to the station would be

A—15 minutes.
B—18 minutes.
C—24 minutes.

Calculate the Time to station using:

$$\text{Time} = \frac{60 \times \text{Time of Bearing Change}}{\text{Bearing Change (Degrees)}}$$

$$= \frac{60 \times 2}{(090 - 085)} = \frac{120}{5} = 24 \text{ minutes}$$

(I08) — AC 61-27C, Chapter 8

AIR, RTC
5522. If the relative bearing changes from 090° to 100° in 2.5 minutes of elapsed time, the time en route to the station would be

A—12 minutes.
B—15 minutes.
C—18 minutes.

Calculate the Time to station using:

$$\text{Time} = \frac{60 \times \text{Time of Bearing Change}}{\text{Bearing Change (Degrees)}}$$

$$= \frac{60 \times 2.5}{(100 - 090)} = \frac{150}{10} = 15 \text{ minutes}$$

(I08) — AC 61-27C, Chapter 8

AIR, RTC
5523. The ADF is tuned to a nondirectional radiobeacon and the relative bearing changes from 090° to 100° in 2.5 minutes of elapsed time. If the true airspeed is 90 knots, the distance and time en route to that radiobeacon would be

A—15 miles and 22.5 minutes.
B—22.5 miles and 15 minutes.
C—32 miles and 18 minutes.

1. Calculate the Time to station using:

$$\text{Time} = \frac{60 \times \text{Time of Bearing Change}}{\text{Bearing Change (Degrees)}}$$

$$= \frac{60 \times 2.5}{(100 - 090)} = \frac{150}{10} = 15 \text{ minutes}$$

2. Calculate the Distance to station using:

Distance = Time x Speed

$$\frac{15 \text{ min}}{60 \text{ min/hr}} \times 90 = 22.5 \text{ NM}$$

(I08) — AC 61-27C, Chapter 8

AIR
5524. GIVEN:

Wingtip bearing change .... 10°
Elapsed time between bearing change .... 4 min
Rate of fuel consumption .... 11 gal/hr

Calculate the fuel required to fly to the station.

A—4.4 gallons.
B—8.4 gallons.
C—12 gallons.

Answers

5519 [A]   5520 [C]   5521 [C]   5522 [B]   5523 [B]   5524 [A]

1. Calculate the Time to station using:

$$\text{Time} = \frac{60 \times \text{Time of Bearing Change}}{\text{Bearing Change (Degrees)}}$$

$$= \frac{60 \times 4}{10} = \frac{240}{10} = 24 \text{ minutes}$$

2. Calculate the Fuel to station using:

Fuel = Time × Burn Rate

$$\frac{24 \text{ min}}{60 \text{ min/hr}} \times 11 \text{ gal/hr} = 4.4 \text{ gallons}$$

(I08) — AC 61-27C, Chapter 8

AIR
5525. GIVEN:

Wingtip bearing change .... 5°
Elapsed time between bearing change .... 6 min
Rate of fuel consumption .... 12 gal/hr

The fuel required to fly to the station is

A—8.2 gallons.
B—14.4 gallons.
C—18.7 gallons.

1. Calculate the Time to station using:

$$\text{Time} = \frac{60 \times \text{Time of Bearing Change}}{\text{Bearing Change (Degrees)}}$$

$$= \frac{60 \times 6}{5} = \frac{360}{5} = 72 \text{ minutes}$$

2. Calculate the Fuel to station using:

Fuel = Time × Burn Rate

$$\frac{72 \text{ min}}{60 \text{ min/hr}} \times 12 \text{ gal/hr} = 14.4 \text{ gallons}$$

(I08) — AC 61-27C, Chapter 8

AIR
5526. GIVEN:

Wingtip bearing change .... 15°
Elapsed time between bearing change .... 6 min
Rate of fuel consumption .... 8.6 gal/hr

Calculate the approximate fuel required to fly to the station.

A—3.44 gallons.
B—6.88 gallons.
C—17.84 gallons.

---

1. Calculate the Time to station using:

$$\text{Time} = \frac{60 \times \text{Time of Bearing Change}}{\text{Bearing Change (Degrees)}}$$

$$= \frac{60 \times 6}{15} = \frac{360}{15} = 24 \text{ minutes}$$

2. Calculate the Fuel to station using:

Fuel = Time × Burn Rate

$$\frac{24 \text{ min}}{60 \text{ min/hr}} \times 8.6 \text{ gal/hr} = 3.44 \text{ gallons}$$

(I08) — AC 61-27C, Chapter 8

AIR
5527. GIVEN:

Wingtip bearing change .... 15°
Elapsed time between bearing change .... 7.5 min
True airspeed .... 85 kts
Rate of fuel consumption .... 9.6 gal/hr

The time, distance, and fuel required to fly to the station is

A—30 minutes; 42.5 miles; 4.80 gallons.
B—32 minutes; 48 miles; 5.58 gallons.
C—48 minutes; 48 miles; 4.58 gallons.

1. Calculate the Time to station using:

$$\text{Time} = \frac{60 \times \text{Time of Bearing Change}}{\text{Bearing Change (Degrees)}}$$

$$= \frac{60 \times 7.5}{15} = \frac{450}{15} = 30 \text{ minutes}$$

2. Calculate the Distance to station using:

Distance = Time × Speed

$$\frac{30 \text{ min}}{60 \text{ min/hr}} \times 85 = 42.5 \text{ NM}$$

3. Calculate the Fuel to station using:

Fuel = Time × Burn Rate

$$\frac{30 \text{ min}}{60 \text{ min/hr}} \times 9.6 \text{ gal/hr} = 4.8 \text{ gallons}$$

(I08) — AC 61-27C, Chapter 8

Answers

5525 [B]   5526 [A]   5527 [A]

Chapter 9 Navigation

AIR, RTC
5528. While maintaining a constant heading, a relative bearing of 15° doubles in 6 minutes. The time to the station being used is

A—3 minutes.
B—6 minutes.
C—12 minutes.

The bearings to the station are not off the wing tip. Therefore, the isosceles triangle method should be used. Time to Station = Time for Relative Bearing to Double. Thus, Time to Station = 6 minutes. (I08) — AC 61-27C, Chapter 8

AIR, RTC
5529. While maintaining a constant heading, the ADF needle increases from a relative bearing of 045° to 090° in 5 minutes. The time to the station being used is

A—5 minutes.
B—10 minutes.
C—15 minutes.

The bearings to the station are not off the wing tip. Therefore, the isosceles triangle method should be used. Time to Station = Time for Relative Bearing to Double. Thus, Time to Station = 5 minutes (I08) — AC 61-27C, Chapter 8

AIR, RTC
5530. While cruising at 135 knots and on a constant heading, the ADF needle decreases from a relative bearing of 315° to 270° in 7 minutes. The approximate time and distance to the station being used is

A—7 minutes and 16 miles.
B—14 minutes and 28 miles.
C—19 minutes and 38 miles.

The bearings to the station are not off the wing tip. Therefore, the isosceles triangle method should be used. Time to Station = Time for Relative Bearing to Double. Thus, Time to Station = 7 minutes. Calculate the distance to station using:

Distance = Time x Speed

$$\frac{7 \text{ min}}{60 \text{ min/hr}} \times 135 \text{ knots} = 15.75 = 16 \text{ NM}$$

(I08) — AC 61-27C, Chapter 8

AIR, RTC
5531. While maintaining a constant heading, a relative bearing of 10° doubles in 5 minutes. If the true airspeed is 105 knots, the time and distance to the station being used is approximately

A—5 minutes and 8.7 miles.
B—10 minutes and 17 miles.
C—15 minutes and 31.2 miles.

The bearings to the station are not off the wing tip. Therefore, the isosceles triangle method should be used. Time to Station = Time for Relative Bearing to Double. Thus, Time to Station = 5 minutes. Calculate the distance to station using:

Distance = Time x Speed

$$\frac{5 \text{ min}}{60 \text{ min/hr}} \times 105 \text{ knots} = 8.7 \text{ NM}$$

(I08) — AC 61-27C, Chapter 8

Answers

5528 [B]   5529 [A]   5530 [A]   5531 [A]

Chapter 10
IFR Operations

Instrument Approach Procedures   *10–3*

Departure Procedures (DPs)   *10–6*

Enroute Procedures   *10–6*

Standard Terminal Arrivals (STARs)   *10–7*

Chapter 10 IFR Operations

Instrument Approach Procedures

Studies have shown that an untrained pilot, deprived of visual reference to the horizon, quickly becomes spatially disoriented. Influenced by physical sensations, the aviator refuses to believe information displayed by the flight instruments and soon loses control of the aircraft. The key to successful instrument flying is thorough training. In flight, the pilot learns to rely entirely on the indications of the flight instruments, and while on the ground, learns the procedures and regulations that apply under instrument flight rules (IFR).

Federal Aviation Regulations Part 61 stipulates that no person may act as pilot-in-command (PIC) of a civil aircraft under IFR or in weather conditions less than the minimums prescribed for visual flight rules (VFR) unless he/she holds an instrument rating. The rating must be for the category of aircraft to be flown; e.g., airplane or rotorcraft. In addition, any flight in Class A airspace (from 18,000 feet MSL to and including FL600) requires an instrument rating, as VFR flight is not allowed. Commercial airplane pilots who carry passengers for hire at night or on cross-country flights of more than 50 nautical miles (NM) are also required to hold an instrument rating.

Unless otherwise authorized, each person operating an aircraft shall, when an instrument let-down to an airport is necessary, use a standard instrument approach procedure prescribed for that airport. Instrument approach procedures are depicted on Jeppesen and National Ocean Service (NOS) Instrument Approach Procedure Charts.

ATC approach procedures depend upon the facilities available at the terminal area, the type of instrument approach executed, and the existing weather conditions. The ATC facilities, navigation aids, and associated frequencies appropriate to each standard instrument approach are given on the approach chart.

AIR, RTC, LTA
5548. (Refer to Figure 25.) During the ILS RWY 13L procedure at DSM, what altitude minimum applies if the glide slope becomes inoperative?

A—1,420 feet.
B—1,340 feet.
C—1,121 feet.

The approach becomes S-LOC 13L if the glide slope becomes inoperative. Therefore, the altitude minimum becomes 1,340 feet. (I10) — AC 61-27C, Chapter 10

AIR, RTC, LTA
5549. What does the absence of the procedure turn barb on the plan view on an approach chart indicate?

A—A procedure turn is not authorized.
B—Teardrop-type procedure turn is authorized.
C—Racetrack-type procedure turn is authorized.

The absence of the procedure turn barb in the plan view indicates that a procedure turn is not authorized for that approach. (I10) — AC 61-27C, Chapter 10

AIR, RTC, LTA
5550. When making an instrument approach at the selected alternate airport, what landing minimums apply?

A—Standard alternate minimums.
B—The IFR alternate minimums listed for that airport.
C—The landing minimums published for the type of procedure selected.

The alternate airport becomes the new destination, and the pilot uses the landing minimum appropriate to the type of procedure selected. (I10) — AC 61-27C, Chapter 10

Answers (A) and (B) are incorrect because alternate minimums refer to weather minimums that allow a specific airport to be filed as an alternate. Once the alternate becomes the destination, the minimums specified for the approach flown must be used.

Answers

5548 [B]   5549 [A]   5550 [C]

Chapter 10 **IFR Operations**

AIR, RTC, LTA, MIL
5561. (Refer to Figure 26.) The final approach fix for the precision approach is located at

A—DENAY Intersection.
B—Glide slope intercept.
C—ROMEN Intersection/Locator outer marker.

The lightning flash shows a glide slope intercept altitude of 2,200 feet MSL, which is the FAF for the precision approach. (J33) — IAP Legend

Answer (A) is incorrect because DENAY intersection is the initial approach fix. Answer (C) is incorrect because ROMEN Intersection/Locator outer marker has the maltese cross, which shows the non-precision final approach fix.

AIR, RTC, LTA, MIL
5594. (Refer to Figure 27.) In the DEN ILS RWY 35R procedure, the glide slope intercept altitude is

A—7,000 feet MSL.
B—11,000 feet MSL.
C—9,000 feet MSL.

The glide slope intercept altitude is indicated by the lightning bolt. (J42) — IAP Legend

AIR, RTC, LTA, MIL
5595. (Refer to Figure 27.) The symbol [9200] in the MSA circle of the ILS RWY 35R procedure at DEN represents a minimum safe sector altitude within 25 NM of

A—Dymon outer marker.
B—Cruup I-AQD DME fix.
C—Denver VORTAC.

The [9200] is in the MSA circle, in the bottom left corner of the plan view. Minimum safe altitudes (MSA) provide at least 1,000 feet of clearance above the highest obstacle in the defined sector to a distance of 25 NM from the facility. (J42) — IAP Legend

AIR, RTC, LTA, MIL
5596. (Refer to Figure 28.) During the ILS RWY 31R procedure at DSM, the minimum altitude for glide slope interception is

A—2,365 feet MSL.
B—2,400 feet MSL.
C—3,000 feet MSL.

The lightning bolt shows a glide slope intercept altitude of 2,400 feet MSL, which is the FAF for the precision approach. (J42) — IAP Legend

AIR, RTC, LTA, MIL
5597. (Refer to Figure 28.) If the glide slope becomes inoperative during the ILS RWY 31R procedure at DSM, what MDA applies?

A—1,157 feet.
B—1,320 feet.
C—1,360 feet.

The approach becomes S-LOC 31R if the glide slope becomes inoperative. (J42) — IAP Legend

Answer (A) is incorrect because 1,157 feet is the minimum altitude for the ILS approach, which requires the glide slope to be operative. Answer (C) is incorrect because 1,360 feet is the minimum altitude for the circling approach.

AIR, RTC, LTA, MIL
5598. (Refer to Figure 29.) When approaching the ATL ILS RWY 8L, how far from the FAF is the missed approach point?

A—4.8 NM.
B—5.2 NM.
C—12.0 NM.

When flying the S-ILS 8L approach, the FAF is glide slope intercept, which could occur at 5,000 feet at BAHRR intersection, or at 2,800 feet at CATTA LOM if 2,800 feet was authorized by ATC. The FAF is 5.2 NM for CATTA LOM from the MAP. (J42) — IAP Legend

AIR, RTC, LTA, MIL
5599. (Refer to Figure 30.) When approaching the VOR/DME-A, the symbol [2800] in the MSA circle represents a minimum safe sector altitude within 25 NM of

A—DEANI intersection.
B—White Cloud VORTAC.
C—Baldwin Municipal Airport.

The symbol in the center of the MSA circle (bottom right corner of the plan view) is a VORTAC. The facility identifier on the edge of the circle is HIC which represents White Cloud. (J42) — IAP Legend

Answers (A) and (C) are incorrect because an MSA is always based on a VOR or an NDB, not an intersection or airport.

Answers

| 5561 | [B] | 5594 | [A] | 5595 | [C] | 5596 | [B] | 5597 | [B] | 5598 | [B] |
| 5599 | [B] | | | | | | | | | | |

Chapter 10 IFR Operations

AIR, RTC, LTA, MIL

5600. (Refer to Figure 30.) What minimum navigation equipment is required to complete the VOR/DME-A procedure?

A—One VOR receiver.
B—One VOR receiver and DME.
C—Two VOR receivers and DME.

One VOR receiver is required to track the 345° radial from White Cloud VORTAC. DME must be used to identify the step-down at DEANI, the final approach fix at HOPPR, the missed approach point, and the fix for the missed approach holding pattern (also HOPPR). (J42) — IAP Legend

AIR, RTC, LTA, MIL

5123-1. For an airport with an approved instrument approach procedure to be listed as an alternate airport on an IFR flight plan, the forecasted weather conditions at the time of arrival must be at or above the following weather minimums.

A—Ceiling 600 feet and visibility 2 NM for precision.
B—Ceiling 800 feet and visibility 2 SM for nonprecision.
C—Ceiling 800 feet and visibility 2 NM for nonprecision.

No person may include an alternate airport in an IFR flight plan unless current weather forecasts indicate that, at the estimated time of arrival at the alternate airport, the ceiling and visibility at that airport will be at or above 800 feet and 2 statute miles for a nonprecision approach procedure. (B10) — 14 CFR §91.169

AIR, RTC, LTA, MIL

5123-2. For an airport without an approved instrument approach procedure to be listed as an alternate airport on an IFR flight plan, the forecasted weather conditions at the time of arrival must have at least a

A—ceiling of 2,000 feet and visibility 3 SM.
B—ceiling and visibility that allows for a descent, approach, and landing under basic VFR.
C—ceiling of 1,000 feet and visibility 3 NM.

If no instrument approach procedure has been published at an airport to be listed as an alternate, the minimum ceiling and visibility must allow descent from the MEA, approach, and landing under basic VFR. (B10) — 14 CFR §91.169

Answers (A) and (C) are incorrect because to be exempted from listing an alternate airport, the weather reports or forecasts must indicate for at least 1 hour before and 1 hour after the estimated time of arrival that the ceiling will be at least 2,000 feet above the airport elevation and the visibility will be at least 3 statute miles.

AIR, RTC, LTA, MIL

5124-1. On an instrument approach where a DH or MDA is applicable, the pilot may not operate below, or continue the approach unless the

A—aircraft is continuously in a position from which a descent to a normal landing, on the intended runway, can be made.
B—approach and runway lights are distinctly visible to the pilot.
C—flight visibility and ceiling are at, or above, the published minimums for that approach.

Where a DH or MDA is applicable, no pilot may operate an aircraft at any airport below the authorized MDA or continue an approach below the authorized DH unless the aircraft is continuously in a position from which a descent to a landing on the intended runway can be made at a normal rate of descent using normal maneuvers. (B10) — 14 CFR §91.175

AIR, RTC, LTA, MIL

5124-2. Pilots are not authorized to land an aircraft from an instrument approach unless the

A—flight visibility is at, or exceeds the visibility prescribed in the approach procedure being used.
B—flight visibility and ceiling are at, or exceeds the minimums prescribed in the approach being used.
C—visual approach slope indicator and runway references are distinctly visible to the pilot.

No pilot operating an aircraft, except a military aircraft of the United States, may land that aircraft when the flight visibility is less than the visibility prescribed in the standard instrument approach procedure being used. (B10) — 14 CFR §91.175

AIR, RTC, LTA, MIL

5124-3. A pilot performing a published instrument approach is not authorized to perform a procedure turn when

A—receiving a radar vector to a final approach course or fix.
B—maneuvering at minimum safe altitudes.
C—maneuvering at radar vectoring altitudes.

A limitation on procedure turns states that in the case of a radar vector to a final approach course or fix, a timed approach from a holding fix, or an approach for which the procedure specifies "No PT," no pilot may make a procedure turn unless cleared to do so by ATC. (B10) — 14 CFR §91.175

Answers
5600 [B]   5123-1 [B]   5123-2 [B]   5124-1 [A]   5124-2 [A]   5124-3 [A]

Chapter 10 **IFR Operations**

AIR, RTC, LTA, MIL
5125-1. The pilot in command of an aircraft operated under IFR, in controlled airspace, not in radar contact, shall report by radio as soon as possible when

A—passing FL 180.
B—passing each designated reporting point, to include time and altitude.
C—changing control facilities.

When not under radar control, the pilot-in-command of each aircraft operated under IFR in controlled airspace shall have a continuous watch maintained on the appropriate frequency and shall report by radio as soon as possible the time and altitude of passing each designated reporting point, or the reporting points specified by ATC. (B10) — 14 CFR §91.183

AIR, RTC, LTA
5125-2. The pilot in command of an aircraft operated under IFR, in controlled airspace, shall report as soon as practical to ATC when

A—climbing or descending to assigned altitudes.
B—experiencing any malfunctions of navigational, approach, or communications equipment, occurring in flight.
C—requested to contact a new controlling facility.

The pilot-in-command of each aircraft operated in controlled airspace under IFR shall report as soon as practical to ATC any malfunctions of navigational, approach, or communication equipment occurring in flight. (B10) — 14 CFR §91.187

Departure Procedures (DPs)

To simplify air traffic control clearance delivery, coded ATC departure procedures called Instrument Departure Procedures (DPs) have been established at certain airports. Pilots may be issued a DP whenever ATC deems it appropriate. If the PIC does not wish to use a DP he/she is expected to advise ATC. If the pilot does elect to use a DP, he/she must possess at least the textual description.

AIR, RTC, LTA, MIL
5556. Which is true regarding the use of an Instrument Departure Procedure (DP) chart?

A—At airfields where DPs have been established, DP usage is mandatory for IFR departures.
B—To use a DP, the pilot must possess at least the textual description of the approved standard departure.
C—To use a DP, the pilot must possess both the textual and graphic form of the approved standard departure.

Use of a DP requires pilot possession of at least the textual description of the DP. (J16) — AIM ¶5-2-6

Answer (A) is incorrect because DPs are not mandatory. Answer (C) is incorrect because the pilot must possess at least a textual description, but to have both is not mandatory.

Enroute Procedures

Normal procedures en route will vary according to the proposed route, the traffic environment, and the ATC facilities controlling the flight. Some IFR flights are under radar surveillance and control from departure to arrival; others rely entirely on pilot navigation. Flights proceeding from Class E airspace to Class G airspace are outside ATC jurisdiction as soon as the aircraft is outside of Class E airspace.

Where ATC has no jurisdiction, an IFR clearance is not issued. ATC has no control over the flight; nor does the pilot have any assurance of separation from other traffic.

With the increasing use of the national airspace, the amount of Class G airspace is diminishing, and the average pilot will usually file IFR via airways and under ATC control.

Answers

5125-1 [B]   5125-2 [B]   5556 [B]

AIR, RTC, LTA, MIL
5591. (Refer to Figure 55.) En route on V112 from BTG VORTAC to LTJ VORTAC, the minimum altitude crossing GYMME intersection is

A—6,400 feet.
B—6,500 feet.
C—7,000 feet.

The question specifies an eastbound flight. The direction arrow on the altitude from BTG toward indicates the minimum altitude crossing GYMME intersection is 7,000 feet MSL. (J37) — Enroute Low-Altitude Chart

Answer (A) is incorrect because 6,400 feet is the minimum obstruction clearance altitude. Answer (B) is incorrect because 6,500 feet is the minimum altitude crossing GYMME intersection on a westbound flight.

AIR, RTC, LTA, MIL
5592. (Refer to Figure 55.) En route on V448 from YKM VORTAC to BTG VORTAC, what minimum navigation equipment is required to identify ANGOO intersection?

A—One VOR receiver.
B—One VOR receiver and DME.
C—Two VOR receivers.

Only one VOR receiver is necessary to identify ANGOO intersection. (J37) — Sectional Chart Legend

AIR, RTC, LTA, MIL
5593. (Refer to Figure 55.) En route on V468 from BTG VORTAC to YKM VORTAC, the minimum en route altitude at TROTS intersection is

A—7,100 feet.
B—10,000 feet.
C—11,500 feet.

Airway data such as the airway identifications, bearings or radials, mileages and altitude (e.g., MEA, MOCA, MAA) are shown aligned with the airway. The Minimum Crossing Altitude for this direction (northeast) is printed next to the "X" flag, and is 11,500 feet. (J37) — Enroute Low-Altitude Chart

Standard Terminal Arrivals (STARs)

Standard Terminal Arrivals (STARs) are used in much the same way as DPs—to relieve frequency congestion and to expedite the arrival of aircraft into the terminal area. Like DPs, STARs may be issued by air traffic control whenever it is deemed appropriate. As with DPs, the PIC may either accept or decline a STAR, but if the STAR is accepted, the pilot must possess at least the textual description of the procedure. Should the pilot not wish to use a STAR, the pilot should make a notation to that effect in the remarks section of the flight plan.

AIR, RTC, LTA, MIL
5557. Which is true regarding STARs? STARs are

A—used to separate IFR and VFR traffic.
B—established to simplify clearance delivery procedures.
C—used at certain airports to decrease traffic congestion.

A STAR is an ATC-coded IFR arrival route established for the use of arriving IFR aircraft destined for certain airports. Its purpose is to simplify clearance delivery procedures. (J18) — AIM ¶5-4-1

Answer (A) is incorrect because STARs are designed for IFR traffic only. Answer (C) is incorrect because STARs do not decrease traffic congestion, they simplify clearance procedures and frequency congestion.

AIR, RTC, LTA, MIL
5558. While being radar vectored, an approach clearance is received. The last assigned altitude should be maintained until

A—reaching the FAF.
B—advised to begin descent.
C—established on a segment of a published route or instrument approach procedure.

When operating on an unpublished route or while being radar vectored, and when an approach clearance is received, the pilot shall, in addition to complying with the minimum altitudes for IFR operations, maintain the last assigned altitude unless a different altitude is assigned by ATC, or until the aircraft is established on a segment of a published route or IAP. (J18) — AIM ¶5-4-7

Answers

| 5591 | [C] | 5592 | [A] | 5593 | [C] | 5557 | [B] | 5558 | [C] |

Cross-Reference A:
Answer, Subject Matter Knowledge Code, Category & Page Number

The FAA does not publish a category list for any exam; therefore, the category assigned below is based on historical data and the best judgement of our researchers.

ALL All aircraft
AIR Airplane
GLI Glider
LTA Lighter-Than-Air (applies to hot air balloon, gas balloon and airship)
RTC Rotorcraft (applies to both helicopter and gyroplane)
MIL Military Competency

Question	Answer	SMK Code	Category	Page
5001	[C]	(G10)	ALL, MIL	4-34
5002	[C]	(G11)	ALL, MIL	4-34
5003-1	[A]	(G11)	ALL, MIL	4-35
5003-2	[C]	(G11)	ALL, MIL	4-35
5004-1	[A]	(G11)	ALL, MIL	4-35
5004-2	[C]	(G11)	ALL, MIL	4-35
5005	[C]	(G11)	ALL, MIL	4-35
5006	[A]	(G11)	ALL, MIL	4-35
5007	[C]	(G13)	ALL, MIL	4-36
5008	[C]	(G13)	ALL, MIL	4-36
5009	[B]	(J08)	ALL, MIL	5-6
5010	[C]	(A01)	ALL, MIL	4-12
5011	[C]	(A01)	ALL, MIL	4-12
5012	[C]	(A01)	ALL, MIL	4-12
5013	[B]	(A02)	AIR, GLI, MIL	3-4
5014	[A]	(A02)	AIR, GLI, MIL	3-4
5015	[A]	(A02)	AIR, GLI, MIL	3-4
5016-1	[A]	(A02)	AIR, MIL	3-4
5016-2	[B]	(A02)	AIR, GLI, MIL	3-4
5016-3	[C]	(A02)	AIR, GLI, MIL	3-4
5017	[B]	(A150)	AIR, MIL	1-17
5018	[C]	(A20)	ALL, MIL	4-4
5019	[C]	(A20)	AIR, RTC, LTA, MIL	4-3
5020	[A]	(A20)	ALL, MIL	4-4
5021	[A]	(A20)	AIR, RTC, LTA, MIL	4-6
5022	[C]	(A20)	AIR, RTC, LTA, MIL	4-4
5023	[B]	(A20)	AIR, RTC, MIL	4-4
5024	[B]	(A20)	AIR, MIL	4-7
5025	[B]	(A20)	AIR, RTC, MIL	4-7
5026	[A]	(A20)	ALL, MIL	4-7
5027	[A]	(A20)	ALL, MIL	4-9
5028	[C]	(A20)	ALL, MIL	4-9
5029	[C]	(A20)	RTC, MIL	4-36
5030	[A]	(A20)	LTA	4-38
5031	[C]	(A20)	ALL, MIL	4-9
5032	[A]	(A20)	ALL, MIL	4-10
5033	[A]	(A21)	AIR, GLI, MIL	4-10
5034	[B]	(A21)	AIR, MIL	4-11
5035-1	[B]	(A22)	GLI, LTA	4-37
5035-2	[C]	(A22)	LTA	4-38
5036	[C]	(A22)	LTA	4-38
5037	[C]	(A20)	LTA	4-38
5038	[A]	(A20)	GLI	4-37
5039	[C]	(A24)	AIR, MIL	4-4
5040	[C]	(A26)	LTA	4-39
5041	[C]	(A26)	LTA	4-39
5042	[B]	(A26)	LTA	4-39
5043	[B]	(A66)	ALL, MIL	5-5
5044	[C]	(B07)	ALL, MIL	4-12
5045	[B]	(B07)	ALL, MIL	4-14
5046	[B]	(B07)	ALL, MIL	4-14
5047	[A]	(B07)	ALL, MIL	4-12
5048	[C]	(B08)	LTA	4-40
5049-1	[B]	(B08)	ALL, MIL	4-14
5049-2	[B]	(B08)	ALL, MIL	4-14
5050-1	[C]	(B08)	ALL, MIL	4-15
5050-2	[C]	(B08)	ALL, MIL	4-15
5051-1	[B]	(B08)	ALL, MIL	4-15
5051-2	[C]	(B08)	ALL, MIL	4-16
5052	[B]	(B08)	AIR, MIL	4-16
5053	[B]	(B12)	GLI	1-27
5054	[C]	(B12)	GLI	1-27
5055	[B]	(B12)	AIR, MIL	4-11
5056-1	[C]	(B07)	ALL, MIL	4-17
5056-2	[A]	(B07)	ALL, MIL	4-17
5057	[A]	(B07)	LTA	4-40
5058	[A]	(B09)	RTC, MIL	4-36
5059	[B]	(B10)	AIR, LTA, MIL	4-18
5060	[A]	(B11)	ALL, MIL	4-18
5061	[A]	(B11)	AIR, RTC, MIL	4-18
5062	[A]	(B10)	AIR, RTC, LTA, MIL	9-20
5063	[C]	(B11)	ALL, MIL	4-19
5064	[C]	(B11)	ALL, MIL	4-19
5065	[A]	(B11)	AIR, RTC, MIL	4-26
5066	[B]	(B11)	AIR, RTC, MIL	4-20
5067-1	[B]	(B11)	AIR, MIL	4-20
5067-2	[B]	(B11)	RTC, MIL	4-20
5068-1	[C]	(B12)	RTC, MIL	4-37
5068-2	[C]	(B12)	AIR, MIL	4-21
5068-3	[B]	(B12)	AIR, MIL	4-21

Cross-Reference A: Answer, SMK Code, Category & Page Number

Question	Answer	SMK Code	Category	Page
5069	[B]	(B12)	ALL, MIL	4-20
5070	[C]	(B11)	AIR, RTC, MIL	4-21
5071	[A]	(B07)	AIR, MIL	4-22
5072-1	[A]	(B11)	RTC, MIL	4-37
5072-2	[A]	(B11)	ALL, MIL	4-18
5073-1	[C]	(B08)	ALL, MIL	4-23
5073-2	[A]	(B08)	ALL, MIL	4-23
5073-3	[A]	(B08)	ALL, MIL	4-23
5074	[A]	(B08)	AIR, RTC, MIL	4-23
5075	[B]	(B08)	ALL, MIL	4-24
5076-1	[B]	(B08)	AIR, RTC, MIL	4-24
5076-2	[A]	(B08)	AIR, RTC, MIL	4-24
5076-3	[C]	(B08)	AIR, RTC, MIL	4-24
5076-4	[A]	(B08)	AIR, RTC, MIL	4-24
5077	[B]	(B08)	AIR, RTC, MIL	4-25
5078	[B]	(B08)	AIR, RTC, MIL	4-25
5079	[A]	(B08)	AIR, GLI, MIL	4-26
5080-1	[A]	(B11)	ALL, MIL	4-27
5080-2	[A]	(B11)	ALL, MIL	4-27
5080-3	[A]	(B11)	LTA	4-27
5080-4	[A]	(B11)	LTA	4-27
5081	[B]	(B11)	LTA	4-40
5082-1	[B]	(B08)	ALL, MIL	5-6
5082-2	[B]	(B08)	ALL, MIL	5-6
5082-3	[A]	(B08)	ALL, MIL	5-6
5083	[B]	(B08)	ALL, MIL	5-12
5084	[C]	(B08)	GLI	4-38
5085	[B]	(B08)	AIR, LTA, GLI, MIL	5-12
5086	[C]	(B08)	RTC, MIL	4-37
5087	[C]	(B08)	RTC, MIL	4-37
5088	[A]	(B09)	AIR, MIL	5-11
5089	[B]	(B09)	AIR, MIL	5-11
5090	[C]	(B09)	AIR, MIL	5-11
5091	[B]	(B09)	ALL, MIL	5-16
5092	[C]	(B10)	ALL, MIL	4-28
5093	[C]	(B13)	ALL, MIL	4-29
5094	[C]	(B13)	ALL, MIL	4-33
5095	[B]	(B13)	ALL, MIL	4-32
5096	[A]	(B13)	ALL, MIL	4-30
5097	[B]	(B13)	ALL, MIL	4-29
5098	[B]	(B13)	ALL, MIL	4-33
5099	[C]	(B13)	ALL, MIL	4-30
5100	[B]	(B13)	ALL, MIL	4-31
5101	[C]	(B13)	ALL, MIL	4-31
5102	[C]	(B13)	ALL, MIL	4-32
5103	[B]	(B13)	ALL, MIL	4-33
5104	[C]	(B13)	AIR, RTC, MIL	4-32
5105	[A]	(B13)	ALL, MIL	4-31
5106	[C]	(A20)	AIR, MIL	4-8
5107	[C]	(A20)	AIR, MIL	4-8
5108	[C]	(A20)	AIR, MIL	4-8
5109	[B]	(B07)	ALL, MIL	4-12
5110-1	[B]	(B08)	AIR, MIL	4-16
5110-2	[A]	(B08)	RTC, MIL	4-16
5111	[C]	(B08)	ALL, MIL	4-5
5112	[B]	(B08)	ALL, MIL	4-26
5113-1	[B]	(B08)	RTC, MIL	4-28
5113-2	[B]	(B08)	RTC, MIL	4-28
5114	[B]	(B08)	ALL, MIL	3-6
5115	[C]	(B08)	ALL, MIL	5-5
5116-1	[A]	(B08)	AIR, GLI, LTA, MIL	5-8
5116-2	[A]	(B08)	RTC, MIL	5-8
5117	[C]	(B08)	ALL, MIL	5-7
5118	[C]	(B08)	ALL, MIL	5-7
5119-1	[C]	(B08)	ALL, MIL	5-7
5119-2	[A]	(B08)	ALL, MIL	5-7
5120-1	[B]	(B08)	ALL, MIL	5-7
5120-2	[B]	(B08)	ALL, MIL	5-8
5121	[A]	(B08)	ALL, MIL	5-13
5122-1	[C]	(B10)	AIR, RTC, LTA, MIL	9-17
5122-2	[B]	(B10)	AIR, RTC, LTA, MIL	9-18
5123-1	[B]	(B10)	AIR, RTC, LTA, MIL	10-5
5123-2	[B]	(B10)	AIR, RTC, LTA, MIL	10-5
5124-1	[A]	(B10)	AIR, RTC, LTA, MIL	10-5
5124-2	[A]	(B10)	AIR, RTC, LTA, MIL	10-5
5124-3	[A]	(B10)	AIR, RTC, LTA, MIL	10-5
5125-1	[B]	(B10)	AIR, RTC, LTA, MIL	10-6
5125-2	[B]	(B10)	AIR, RTC, LTA	10-6
5126-1	[A]	(A24)	ALL, MIL	4-5
5126-2	[B]	(A24)	ALL, MIL	4-5
5127	[B]	(A20)	GLI	4-11
5128	[C]	(A20)	AIR, MIL	4-8
5129	[C]	(B12)	ALL, MIL	4-20
5130	[B]	(H564)	ALL	5-18
5131	[B]	(A24)	LTA	4-39
5132	[A]	(A24)	LTA	4-39
5133	[A]	(H568)	ALL	5-18
5134	[C]	(H568)	ALL	5-18
5135	[B]	(H572)	ALL	5-18
5136	[B]	(H574)	ALL	5-18
5137	[A]	(H574)	ALL	5-18
5138	[C]	(J13)	ALL	5-14
5139	[A]	(J13)	ALL	5-14
5140	[A]	(J13)	ALL	5-15
5141	[B]	(A20)	ALL, MIL	4-13
5142	[C]	(A20)	ALL, MIL	4-13
5143	[C]	(A20)	ALL, MIL	4-13
5144	[C]	(A20)	ALL, MIL	4-13
5145	[Removed by the FAA]			
5146	[Removed by the FAA]			
5147	[Removed by the FAA]			
5148	[Removed by the FAA]			
5149	[Removed by the FAA]			
5150	[Removed by the FAA]			
5151	[A]	(H303)	AIR, GLI	1-18
5152	[A]	(H303)	AIR, GLI	1-18
5153	[A]	(H303)	AIR	1-18
5154	[B]	(H303)	AIR	1-18
5155	[A]	(H303)	AIR, GLI	1-18
5156	[B]	(H303)	AIR, GLI	1-18

Cross-Reference A: Answer, SMK Code, Category & Page Number

Question	Answer	SMK Code	Category	Page
5157	[B]	(H303)	AIR, GLI	1-19
5158	[A]	(H300)	AIR, GLI	1-7
5159	[A]	(H303)	AIR, GLI	1-19
5160	[B]	(H303)	AIR, GLI	1-20
5161	[C]	(H300)	AIR	1-8
5162	[B]	(H300)	AIR	1-9
5163	[C]	(H303)	AIR	1-19
5164	[B]	(H303)	AIR	8-7
5165	[A]	(H300)	AIR	1-10
5166	[B]	(H300)	AIR	1-10
5167	[B]	(H300)	AIR, GLI	1-9
5168	[B]	(H307)	RTC	2-13
5169	[B]	(H307)	AIR, RTC	2-3
5170	[C]	(H307)	AIR, RTC, LTA	2-6
5171	[A]	(H307)	AIR, RTC	2-3
5172	[A]	(H307)	AIR, RTC, LTA	2-4
5173	[C]	(H307)	AIR, RTC	2-3
5174	[C]	(H307)	AIR, RTC	2-4
5175	[B]	(H307)	AIR	2-8
5176	[C]	(H307)	AIR, RTC, LTA	2-4
5177	[C]	(H312)	AIR, GLI, MIL	3-4
5178	[B]	(H314)	ALL	3-6
5179	[C]	(H303)	AIR	1-19
5180	[A]	(H303)	AIR	1-19
5181	[B]	(H305)	AIR, GLI	1-17
5182	[B]	(H305)	AIR, GLI	1-22
5183	[C]	(H51)	AIR	2-10
5184	[B]	(H308)	AIR	2-10
5185-1	[A]	(H307)	AIR, RTC, LTA	2-7
5185-2	[C]	(H307)	AIR, RTC, LTA	2-7
5186	[C]	(H51)	AIR, RTC, LTA	2-8
5187	[B]	(H51)	AIR, RTC, LTA	2-5
5188	[A]	(H307)	AIR, RTC, LTA	2-5
5189	[A]	(H307)	AIR, RTC, LTA	2-6
5190	[C]	(H51)	AIR, RTC, LTA	2-8
5191	[C]	(H55)	AIR	3-8
5192	[C]	(H55)	AIR, GLI	1-16
5193	[A]	(H55)	AIR	1-16
5194	[A]	(H55)	AIR	1-16
5195	[B]	(H55)	AIR	1-17
5196	[A]	(H66)	AIR	1-21
5197	[B]	(H66)	AIR, GLI	1-22
5198	[A]	(H66)	AIR, GLI	1-4
5199	[C]	(H66)	AIR, GLI	1-4
5200	[C]	(H66)	AIR	1-7
5201	[C]	(H66)	AIR, GLI	1-7
5202	[B]	(H66)	AIR, GLI	1-9
5203	[A]	(H66)	AIR	1-8
5204	[A]	(H66)	AIR, GLI	1-21
5205	[B]	(H66)	AIR	1-13
5206	[A]	(H66)	AIR,	1-21
5207	[B]	(H66)	AIR	1-14
5208	[A]	(H66)	ALL	8-23
5209	[B]	(H66)	AIR	1-24
5210	[C]	(H66)	AIR	1-17
5211	[C]	(H66)	AIR	1-21
5212	[A]	(H66)	AIR	1-14
5213	[B]	(H66)	AIR	1-11
5214	[C]	(H66)	AIR	1-11
5215	[C]	(H66)	AIR, GLI	1-11
5216	[A]	(H66)	AIR, GLI	1-24
5217	[B]	(H66)	AIR	1-11
5218	[C]	(H66)	AIR, GLI	1-9
5219	[B]	(H66)	AIR, GLI	1-8
5220	[C]	(H66)	AIR	1-10
5221	[C]	(H66)	AIR	1-20
5222	[C]	(H66)	AIR, GLI	1-20
5223	[C]	(H66)	AIR	1-8
5224	[A]	(H66)	AIR	1-25
5225	[A]	(H66)	AIR, GLI	1-17
5226	[A]	(H66)	AIR, GLI	1-14
5227	[B]	(H66)	AIR, GLI	1-14
5228	[B]	(H66)	AIR	1-14
5229	[B]	(H66)	AIR	1-8
5230	[B]	(H66)	AIR	1-14
5231	[B]	(H66)	AIR, GLI	1-12
5232	[A]	(H66)	AIR, GLI	1-12
5233	[A]	(H66)	AIR, GLI	3-3
5234	[A]	(H66)	AIR, RTC	8-19
5235	[A]	(H66)	AIR	2-10
5236	[B]	(H66)	AIR	2-10
5237	[C]	(H66)	AIR	2-10
5238	[B]	(H66)	AIR	1-24
5239	[A]	(H70)	RTC	1-4
5240	[B]	(H71)	RTC	2-13
5241	[A]	(H71)	RTC	2-13
5242	[B]	(H71)	RTC	2-13
5243	[A]	(H71)	RTC	2-13
5244	[C]	(H71)	RTC	2-13
5245	[C]	(H71)	RTC	2-14
5246	[C]	(H71)	RTC	2-14
5247	[C]	(H71)	RTC	2-14
5248	[A]	(H72)	RTC	2-14
5249	[A]	(H73)	RTC	2-14
5250	[B]	(H73)	RTC	2-14
5251	[C]	(H73)	RTC	2-14
5252	[C]	(H73)	RTC	2-15
5253	[B]	(H74)	RTC	2-15
5254	[B]	(H74)	RTC	2-15
5255	[A]	(H74)	RTC	2-15
5256	[C]	(H74)	RTC	2-15
5257	[A]	(H74)	RTC	2-16
5258	[C]	(H77)	RTC	2-16
5259	[B]	(H77)	RTC	2-16
5260	[A]	(H77)	RTC	2-16
5261	[B]	(H78)	RTC	2-16
5262	[C]	(H78)	RTC	2-16
5263	[B]	(H78)	RTC	2-16
5264	[A]	(H78)	RTC	2-17
5265	[A]	(H78)	RTC	2-17

Cross-Reference A: Answer, SMK Code, Category & Page Number

Question	Answer	SMK Code	Category	Page
5266	[C]	(H78)	RTC	2-17
5267	[C]	(H80)	RTC	2-17
5268	[C]	(I04)	AIR	3-7
5269	[A]	(I04)	AIR, RTC, LTA	3-8
5270	[B]	(I05)	AIR, RTC, LTA	3-8
5271	[A]	(K20)	AIR	2-9
5272	[C]	(L34)	ALL	5-17
5273	[B]	(N22)	GLI	2-27
5274	[C]	(N22)	GLI	2-27
5275	[A]	(N22)	GLI	2-27
5276	[C]	(N20)	GLI	1-22
5277	[B]	(N20)	GLI	1-27
5278	[A]	(N20)	GLI	1-28
5279	[C]	(N20)	GLI	1-28
5280	[C]	(N20)	GLI	1-9
5281	[C]	(N20)	GLI	1-28
5282	[A]	(N20)	GLI	1-22
5283	[B]	(N20)	GLI	1-28
5284	[C]	(N20)	GLI	1-28
5285	[C]	(N20)	GLI	1-28
5286	[C]	(N20)	GLI	1-28
5287	[A]	(N21)	GLI	8-14
5288	[B]	(N21)	GLI	8-14
5289	[C]	(N21)	GLI	8-15
5290	[B]	(N20)	GLI	1-29
5291	[A]	(N20)	GLI	1-29
5292	[B]	(N20)	GLI	1-29
5293	[B]	(N20)	GLI	1-29
5294	[B]	(N20)	GLI	1-29
5295	[C]	(N21)	GLI	1-30
5296	[B]	(N21)	GLI	1-30
5297	[C]	(N22)	GLI	2-28
5298	[B]	(H51)	AIR, RTC, LTA	2-5
5299	[C]	(H51)	AIR, RTC, LTA	2-8
5300	[C]	(H66)	AIR, RTC	8-19
5301	[A]	(I21)	ALL	6-4
5302	[A]	(I21)	ALL	8-19
5303	[C]	(I21)	ALL	8-19
5304	[A]	(I21)	ALL	6-6
5305	[A]	(I22)	ALL	8-19
5306	[B]	(I22)	ALL	9-11
5307	[B]	(I22)	ALL	9-11
5308	[B]	(I22)	ALL	9-11
5309	[B]	(I22)	ALL	9-11
5310	[C]	(I23)	ALL	6-4
5311	[A]	(I23)	ALL	6-4
5312	[A]	(I23)	ALL	6-4
5313	[B]	(I23)	ALL	6-7
5314	[C]	(I23)	ALL	6-8
5315	[A]	(I23)	ALL	6-4
5316	[A]	(I23)	ALL	6-8
5317	[C]	(I23)	ALL	6-8
5318	[B]	(I23)	ALL	6-8
5319	[B]	(I23)	ALL	6-8
5320	[B]	(I24)	ALL	6-10
5321	[A]	(I24)	ALL	6-8
5322	[A]	(I24)	ALL	6-21
5323	[C]	(I24)	ALL	6-10
5324	[B]	(I24)	ALL	6-19
5325	[A]	(I24)	ALL	6-19
5326	[C]	(I24)	ALL	6-19
5327	[C]	(I25)	ALL	6-13
5328	[C]	(I25)	ALL	6-16
5329	[C]	(I25)	ALL	6-13
5330	[B]	(I25)	ALL	6-13
5331	[B]	(I25)	ALL	6-16
5332	[B]	(I25)	ALL	6-14
5333	[A]	(I25)	ALL	6-10
5334	[B]	(I25)	ALL	6-11
5335	[C]	(I25)	ALL	6-14
5336	[B]	(I25)	ALL	6-11
5337	[B]	(I26)	ALL	6-14
5338	[C]	(I26)	ALL	6-14
5339	[B]	(I26)	ALL	6-17
5340	[B]	(I27)	ALL	6-13
5341	[B]	(I27)	ALL	6-14
5342	[A]	(I27)	ALL	6-14
5343	[B]	(I27)	ALL	6-15
5344	[C]	(I27)	ALL	6-15
5345	[C]	(I27)	ALL	6-15
5346	[C]	(I27)	ALL	6-15
5347	[B]	(I27)	ALL	6-17
5348	[B]	(I27)	ALL	6-15
5349	[C]	(I27)	ALL	6-15
5350	[C]	(I27)	ALL	6-24
5351	[C]	(I28)	ALL	6-26
5352	[C]	(I28)	ALL	6-26
5353	[B]	(I28)	ALL	6-26
5354	[A]	(I28)	ALL	6-27
5355	[A]	(I28)	ALL	6-27
5356	[A]	(I28)	ALL	6-5
5357	[B]	(I28)	ALL	6-18
5358	[C]	(I28)	AIR, RTC	6-27
5359	[A]	(I28)	ALL	6-27
5360	[C]	(I29)	ALL	6-19
5361	[C]	(I30)	ALL	6-21
5362	[B]	(I30)	ALL	6-21
5363	[B]	(I30)	ALL	6-21
5364	[C]	(I30)	ALL	6-21
5365	[A]	(I30)	ALL	6-21
5366	[C]	(I30)	ALL	6-21
5367	[C]	(I30)	ALL	6-22
5368	[B]	(I30)	ALL	6-22
5369	[C]	(I30)	ALL	6-22
5370	[A]	(I30)	ALL	6-22
5371	[B]	(I30)	ALL	6-22
5372	[C]	(I30)	ALL	6-22
5373	[C]	(I30)	ALL	6-23
5374	[C]	(I31)	ALL	6-24
5375	[A]	(I31)	ALL	6-23

Cross-Reference A: Answer, SMK Code, Category & Page Number

Question	Answer	SMK Code	Category	Page
5376	[B]	(I31)	ALL	6-24
5377	[C]	(I31)	ALL	6-24
5378	[C]	(I31)	ALL	6-24
5379	[A]	(I31)	ALL	6-24
5380	[C]	(I31)	ALL	6-25
5381	[B]	(I32)	ALL	6-5
5382	[A]	(I32)	ALL	6-5
5383	[B]	(I32)	ALL	6-5
5384	[B]	(I32)	ALL	6-5
5385	[A]	(I32)	ALL	6-5
5386	[B]	(I35)	GLI	6-28
5387	[B]	(I35)	GLI	6-28
5388	[A]	(I35)	GLI	6-28
5389	[B]	(I35)	GLI	6-28
5390	[A]	(I35)	GLI	6-29
5391	[C]	(I35)	GLI	6-29
5392	[B]	(I35)	GLI	6-31
5393	[A]	(I35)	ALL	6-18
5394	[B]	(I35)	GLI	6-29
5395	[C]	(I35)	GLI	6-29
5396	[A]	(I35)	GLI	6-29
5397	[C]	(I35)	GLI	6-30
5398	[A]	(I40)	ALL	7-3
5399	[A]	(I40)	ALL	7-4
5400	[B]	(J25)	ALL, MIL	7-11
5401	[C]	(J25)	ALL, MIL	7-12
5402	[C]	(J25)	ALL, MIL	7-4
5403	[B]	(J25)	ALL, MIL	7-4
5404	[B]	(J25)	ALL, MIL	7-4
5405	[C]	(I41)	ALL	7-4
5406	[B]	(J25)	ALL, MIL	7-6
5407	[A]	(I42)	ALL	7-6
5408	[C]	(I42)	ALL	7-15
5409	[B]	(J25)	ALL, MIL	7-8
5410	[C]	(J25)	ALL, MIL	7-8
5411	[C]	(J25)	ALL, MIL	7-8
5412	[B]	(J25)	ALL, MIL	7-8
5413	[A]	(J25)	ALL, MIL	7-9
5414	[B]	(I43)	ALL	7-9
5415	[C]	(I43)	ALL	7-9
5416	[A]	(I43)	ALL	7-9
5417	[C]	(J25)	ALL	7-11
5418	[A]	(J25)	ALL	7-12
5419	[B]	(I43)	ALL	7-9
5420	[C]	(I43)	ALL	7-13
5421	[A]	(J25)	ALL, MIL	7-12
5422	[A]	(I43)	ALL	7-11
5423	[C]	(I43)	ALL	7-11
5424	[B]	(I43)	ALL	7-10
5425	[A]	(I43)	ALL	7-15
5426	[A]	(I44)	ALL	7-16
5427	[A]	(I44)	ALL	7-16
5428	[B]	(I44)	ALL	7-16
5429	[B]	(I44)	ALL	7-16
5430	[C]	(I45)	ALL	7-17
5431	[B]	(I45)	ALL	7-17
5432	[A]	(I46)	ALL	7-15
5433	[C]	(I47)	ALL	7-20
5434	[B]	(I47)	ALL	7-20
5435	[C]	(I47)	ALL	7-20
5436	[B]	(I47)	ALL	7-20
5437	[B]	(I49)	GLI	6-30
5438	[C]	(I49)	ALL	7-21
5439	[A]	(I49)	ALL	7-21
5440	[B]	(I51)	ALL	7-18
5441	[A]	(I51)	ALL	7-18
5442	[C]	(I51)	ALL	7-18
5443	[B]	(I52)	ALL	7-18
5444	[B]	(I53)	ALL	7-6
5445	[C]	(I53)	ALL	7-7
5446	[B]	(I53)	ALL	7-7
5447	[B]	(K02)	ALL	6-6
5448	[C]	(K02)	ALL	6-28
5449	[B]	(J25)	ALL	6-26
5450	[A]	(N33)	ALL	6-18
5451	[B]	(H342)	AIR	8-26
5452	[B]	(H342)	AIR	8-26
5453	[C]	(H342)	AIR	8-27
5454	[B]	(H342)	AIR	8-27
5455	[A]	(H342)	AIR	8-27
5456	[C]	(H317)	AIR	8-29
5457	[C]	(H317)	AIR	8-29
5458	[C]	(H317)	AIR	8-29
5459	[C]	(H317)	AIR	8-30
5460	[C]	(H317)	AIR	8-35
5461	[B]	(H317)	AIR	8-35
5462	[B]	(H317)	AIR	8-35
5463	[B]	(H317)	AIR	8-33
5464	[A]	(H317)	AIR	8-33
5465	[C]	(H317)	AIR	8-33
5466	[A]	(H317)	AIR	9-8
5467	[C]	(H317)	AIR	9-8
5468	[C]	(H317)	AIR	9-9
5469	[A]	(H342)	AIR	9-6
5470	[A]	(H342)	AIR	9-6
5471	[C]	(H342)	AIR	9-7
5472	[C]	(H342)	AIR	9-7
5473	[A]	(H342)	AIR	9-7
5474	[A]	(H342)	AIR	9-7
5475	[A]	(H341)	AIR, RTC	9-13
5476	[B]	(H341)	AIR, RTC	9-13
5477	[C]	(H341)	AIR, RTC	9-14
5478	[C]	(H341)	ALL	9-15
5479	[C]	(H344)	ALL	9-6
5480	[A]	(H344)	LTA	8-43
5481	[C]	(H344)	AIR, RTC	9-7
5482	[A]	(H342)	AIR	8-30
5483	[B]	(H342)	AIR	8-30
5484	[C]	(H342)	AIR	8-31
5485	[C]	(H342)	AIR	8-31

Cross-Reference A: Answer, SMK Code, Category & Page Number

Question	Answer	SMK Code	Category	Page
5486	[B]	(H342)	AIR	8-31
5487	[A]	(H342)	AIR	8-32
5488	[B]	(H344)	AIR	9-9
5489	[B]	(H344)	AIR	9-10
5490	[A]	(H348)	ALL	9-30
5491	[B]	(H348)	ALL	9-30
5492	[B]	(H348)	ALL	9-30
5493	[C]	(H348)	ALL	9-30
5494	[B]	(H348)	ALL	9-30
5495	[C]	(H348)	ALL	9-30
5496	[A]	(H348)	ALL	9-31
5497	[C]	(H348)	ALL	9-31
5498	[C]	(H348)	ALL	9-31
5499	[C]	(H348)	ALL	9-31
5500	[A]	(H348)	ALL	9-17
5501	[C]	(H348)	ALL	9-18
5502	[C]	(H348)	ALL	9-18
5503	[C]	(H61)	ALL	8-20
5504	[A]	(J22)	ALL	5-13
5505	[B]	(H66)	ALL	1-11
5506	[A]	(I08)	ALL	9-25
5507	[A]	(I08)	AIR, RTC, LTA	9-26
5508	[B]	(I08)	AIR, RTC, LTA	9-26
5509	[A]	(I08)	AIR, RTC, LTA	9-26
5510	[C]	(I08)	AIR, RTC, LTA	9-26
5511	[B]	(I08)	AIR, RTC, LTA	9-31
5512	[B]	(I08)	AIR, RTC, LTA	9-32
5513	[C]	(I08)	AIR, RTC, LTA	9-32
5514	[C]	(I08)	AIR, RTC, LTA	9-33
5515	[C]	(I08)	AIR, RTC, LTA	9-33
5516	[B]	(I08)	AIR, RTC	9-33
5517	[C]	(I08)	AIR, RTC	9-33
5518	[C]	(I08)	AIR, RTC	9-33
5519	[A]	(I08)	AIR, RTC	9-34
5520	[C]	(I08)	AIR, RTC	9-34
5521	[C]	(I08)	AIR, RTC	9-34
5522	[B]	(I08)	AIR, RTC	9-34
5523	[B]	(I08)	AIR, RTC	9-34
5524	[A]	(I08)	AIR	9-34
5525	[B]	(I08)	AIR	9-35
5526	[A]	(I08)	AIR	9-35
5527	[A]	(I08)	AIR	9-35
5528	[B]	(I08)	AIR, RTC	9-36
5529	[A]	(I08)	AIR, RTC	9-36
5530	[A]	(I08)	AIR, RTC	9-36
5531	[A]	(I08)	AIR, RTC	9-36
5532	[B]	(I08)	ALL	9-17
5533	[B]	(I08)	ALL	9-17
5534	[A]	(I08)	ALL	9-18
5535	[C]	(I08)	ALL	9-19
5536	[B]	(I08)	ALL	9-19
5537	[B]	(I08)	ALL	9-20
5538	[A]	(I08)	ALL	9-20
5539	[A]	(I08)	AIR, RTC, LTA	9-23
5540	[A]	(I08)	AIR, RTC, LTA	9-23
5541	[A]	(I08)	ALL	9-23
5542	[B]	(I08)	AIR, RTC, LTA	9-23
5543	[A]	(I08)	AIR, RTC, LTA	9-23
5544	[B]	(I08)	AIR, RTC, LTA	9-24
5545	[A]	(I08)	AIR, RTC, LTA	9-24
5546	[B]	(I08)	AIR, RTC, LTA	9-24
5547	[A]	(I08)	AIR, RTC, LTA	9-24
5548	[B]	(I10)	AIR, RTC, LTA	10-3
5549	[A]	(I10)	AIR, RTC, LTA	10-3
5550	[C]	(I10)	AIR, RTC, LTA	10-3
5551	[B]	(J01)	ALL, MIL	9-20
5552	[C]	(J01)	ALL, MIL	9-20
5553	[B]	(J01)	ALL, MIL	9-21
5554	[C]	(J08)	LTA	4-40
5555	[C]	(J15)	LTA	2-43
5556	[B]	(J16)	AIR, RTC, LTA, MIL	10-6
5557	[B]	(J18)	AIR, RTC, LTA, MIL	10-7
5558	[C]	(J18)	AIR, RTC, LTA, MIL	10-7
5559	[B]	(J25)	ALL, MIL	7-13
5560	[A]	(J25)	ALL, MIL	7-13
5561	[B]	(J33)	AIR, RTC, LTA, MIL	10-4
5562	[C]	(J14)	LTA	2-43
5563	[C]	(J33)	LTA	2-43
5564	[C]	(J37)	ALL, MIL	5-6
5565	[C]	(J37)	ALL, MIL	5-10
5566	[C]	(J37)	ALL, MIL	5-10
5567	[B]	(J37)	ALL, MIL	5-10
5568	[B]	(J37)	ALL, MIL	5-10
5569	[B]	(J37)	ALL, MIL	5-9
5570	[B]	(J37)	ALL, MIL	5-10
5571	[A]	(J37)	LTA	8-43
5572	[B]	(J37)	ALL, MIL	5-8
5573	[C]	(J37)	LTA	8-43
5574-1	[B]	(J37)	LTA	5-8
5574-2	[A]	(J37)	RTC, MIL	5-8
5575	[C]	(J37)	ALL, MIL	5-10
5576	[B]	(J37)	ALL, MIL	5-9
5577	[C]	(J37)	ALL, MIL	5-12
5578	[C]	(J37)	LTA	8-43
5579	[B]	(J37)	LTA	8-43
5580	[B]	(J37)	LTA	8-44
5581	[C]	(J37)	ALL, MIL	5-10
5582	[B]	(J37)	LTA	8-44
5583	[C]	(J37)	ALL, MIL	5-9
5584	[A]	(J37)	ALL, MIL	5-9
5585	[B]	(J37)	ALL, MIL	5-11
5586	[B]	(J37)	LTA	8-44
5587	[A]	(J37)	ALL, MIL	5-9
5588	[B]	(J37)	AIR, LTA, GLI, MIL	5-13
5589	[B]	(J37)	LTA	8-44
5590	[C]	(J37)	LTA	4-40
5591	[C]	(J37)	AIR, RTC, LTA, MIL	10-7
5592	[A]	(J37)	AIR, RTC, LTA, MIL	10-7
5593	[C]	(J37)	AIR, RTC, LTA, MIL	10-7
5594	[A]	(J42)	AIR, RTC, LTA, MIL	10-4

Cross-Reference A: Answer, SMK Code, Category & Page Number

Question	Answer	SMK Code	Category	Page
5595	[C]	(J42)	AIR, RTC, LTA, MIL	10-4
5596	[B]	(J42)	AIR, RTC, LTA, MIL	10-4
5597	[B]	(J42)	AIR, RTC, LTA, MIL	10-4
5598	[B]	(J42)	AIR, RTC, LTA, MIL	10-4
5599	[B]	(J42)	AIR, RTC, LTA, MIL	10-4
5600	[B]	(J42)	AIR, RTC, LTA, MIL	10-5
5601	[A]	(H312)	ALL	3-5
5602	[C]	(H312)	ALL	3-5
5603	[A]	(B08)	LTA	2-43
5604	[B]	(H312)	ALL	3-4
5605	[B]	(H312)	AIR	3-5
5606	[C]	(H307)	AIR, RTC, LTA	2-7
5607	[B]	(H307)	AIR, RTC, LTA	2-9
5608	[C]	(H307)	AIR, RTC, LTA	2-5
5609	[C]	(H307)	AIR, RTC, LTA	2-5
5610	[A]	(H307)	AIR, RTC, LTA	2-5
5611	[B]	(H307)	AIR, RTC, LTA	2-6
5612	[A]	(H314)	GLI	2-28
5613	[B]	(H314)	GLI	2-28
5614	[B]	(H317)	AIR, GLI	8-20
5615	[A]	(H317)	AIR	8-18
5616	[C]	(H317)	AIR, RTC, GLI	8-18
5617	[A]	(H317)	AIR, RTC, GLI	8-18
5618	[A]	(H317)	AIR, RTC, GLI	8-18
5619	[C]	(H317)	AIR	8-23
5620	[A]	(H317)	AIR	8-23
5621	[B]	(H317)	AIR	8-24
5622	[C]	(H317)	AIR	8-24
5623	[B]	(H317)	AIR	8-36
5624	[C]	(H317)	AIR	8-37
5625	[B]	(H317)	AIR	8-34
5626	[B]	(H317)	AIR	8-34
5627	[C]	(H317)	AIR	8-34
5628	[A]	(H317)	AIR	8-24
5629	[A]	(H317)	AIR	8-25
5630	[B]	(H317)	AIR	8-25
5631	[A]	(H317)	AIR	8-25
5632	[A]	(H315)	ALL	8-7
5633	[B]	(H316)	ALL	8-8
5634	[C]	(H12)	ALL	8-8
5635	[B]	(H12)	ALL	8-8
5636	[B]	(H12)	ALL	8-8
5637	[B]	(H12)	ALL	8-8
5638	[A]	(H12)	ALL	8-9
5639	[C]	(H12)	ALL	8-9
5640	[C]	(H12)	GLI	8-15
5641	[A]	(H12)	GLI	8-15
5642	[C]	(H12)	GLI	8-15
5643	[B]	(H12)	GLI	8-16
5644	[C]	(H12)	RTC	8-13
5645	[B]	(H12)	RTC	8-13
5646	[A]	(H14)	AIR	8-10
5647	[A]	(H14)	AIR	8-11
5648	[A]	(H14)	AIR	8-11
5649	[B]	(H14)	AIR	8-11
5650	[A]	(H15)	AIR	8-9
5651	[A]	(H15)	AIR	8-10
5652	[A]	(H15)	AIR	8-10
5653	[A]	(H351)	AIR, RTC	2-11
5654	[C]	(H308)	AIR	2-10
5655	[C]	(H54)	AIR	8-17
5656	[C]	(H54)	AIR	8-17
5657	[C]	(J05)	AIR, RTC	5-15
5658	[B]	(J05)	AIR, RTC	5-15
5659-1	[C]	(J05)	AIR, RTC	5-15
5659-2	[A]	(J05)	AIR, RTC	5-16
5660	[A]	(J05)	AIR, RTC	5-16
5661	[C]	(H57)	AIR	8-20
5662	[A]	(H58)	AIR	8-21
5663	[A]	(H58)	AIR	8-21
5664	[A]	(H58)	AIR	8-21
5665	[B]	(H58)	AIR	8-21
5666	[A]	(H63)	ALL	4-27
5667	[B]	(H66)	AIR	2-11
5668	[A]	(H66)	AIR	2-11
5669	[B]	(H66)	AIR	3-5
5670	[B]	(H66)	AIR	3-5
5671	[B]	(H71)	RTC	2-17
5672	[C]	(H71)	RTC	2-17
5673	[B]	(H73)	RTC	2-18
5674	[C]	(H73)	RTC	2-18
5675	[C]	(H73)	RTC	2-18
5676	[A]	(H75)	RTC	2-18
5677	[A]	(H76)	RTC	8-12
5678	[A]	(H76)	RTC	8-12
5679	[C]	(H76)	RTC	8-13
5680	[C]	(H76)	RTC	8-14
5681	[B]	(H76)	RTC	8-14
5682	[A]	(H76)	RTC	8-12
5683	[A]	(H77)	RTC	8-37
5684	[B]	(H77)	RTC	8-37
5685	[B]	(H77)	RTC	8-37
5686	[B]	(H77)	RTC	2-18
5687	[B]	(H77)	RTC	8-38
5688	[A]	(H77)	RTC	8-38
5689	[B]	(H77)	RTC	8-38
5690	[B]	(H77)	RTC	8-38
5691	[C]	(H77)	RTC	8-38
5692	[A]	(H77)	RTC	8-39
5693	[A]	(H77)	RTC	8-39
5694	[C]	(H77)	RTC	8-39
5695	[B]	(H78)	RTC	2-18
5696	[C]	(H78)	RTC	2-18
5697	[B]	(H78)	RTC	2-19
5698	[C]	(H78)	RTC	2-19
5699	[B]	(H78)	RTC	2-19
5700	[C]	(H78)	RTC	2-19
5701	[B]	(H78)	RTC	2-19
5702	[A]	(H78)	RTC	2-20
5703	[C]	(H78)	RTC	2-20

Cross-Reference A: Answer, SMK Code, Category & Page Number

Question	Answer	SMK Code	Category	Page
5704	[C]	(H78)	RTC	2-20
5705	[A]	(H78)	RTC	2-20
5706	[B]	(H78)	RTC	2-20
5707	[A]	(H79)	RTC	2-21
5708	[B]	(H79)	RTC	2-21
5709	[B]	(H80)	RTC	2-21
5710	[B]	(H80)	RTC	2-21
5711	[C]	(H80)	RTC	2-21
5712	[B]	(H80)	RTC	2-21
5713	[A]	(H80)	RTC	2-22
5714	[C]	(H80)	RTC	2-22
5715	[C]	(H80)	RTC	2-22
5716	[C]	(H80)	RTC	2-22
5717	[B]	(H80)	RTC	2-22
5718	[C]	(H80)	RTC	2-23
5719	[C]	(H80)	RTC	2-23
5720	[A]	(H80)	RTC	2-23
5721	[A]	(H80)	RTC	2-23
5722	[B]	(H80)	RTC	2-23
5723	[B]	(H80)	RTC	2-24
5724	[A]	(H80)	RTC	2-24
5725	[C]	(H80)	RTC	2-24
5726	[B]	(H80)	RTC	2-24
5727	[A]	(H80)	RTC	2-24
5728	[C]	(H80)	RTC	2-24
5729	[B]	(H81)	RTC	2-25
5730	[C]	(H81)	RTC	2-25
5731	[B]	(H81)	RTC	2-25
5732	[A]	(H81)	RTC	2-25
5733	[C]	(H767)	RTC	2-25
5734	[C]	(H762)	RTC	2-26
5735	[B]	(H765)	RTC	2-26
5736	[C]	(H765)	RTC	2-26
5737	[B]	(H766)	RTC	2-26
5738	[A]	(H766)	RTC	2-26
5739	[B]	(I29)	AIR	6-19
5740	[B]	(I04)	ALL	3-6
5741	[C]	(I30)	AIR	3-5
5742	[B]	(N23)	GLI	6-30
5743	[C]	(N23)	GLI	6-30
5744	[B]	(N23)	GLI	6-30
5745	[C]	(N23)	GLI	6-31
5746	[C]	(N23)	GLI	6-31
5747	[C]	(I35)	GLI	6-31
5748	[C]	(J13)	AIR, RTC, MIL	5-15
5749	[B]	(J14)	ALL, MIL	5-17
5750	[B]	(J27)	ALL, MIL	1-26
5751	[A]	(J27)	ALL, MIL	1-26
5752	[A]	(J27)	ALL, MIL	1-26
5753	[A]	(J27)	ALL, MIL	1-27
5754	[A]	(J27)	ALL, MIL	1-27
5755	[C]	(J27)	RTC	2-27
5756	[B]	(J27)	RTC	2-27
5757	[C]	(J31)	ALL, MIL	5-20
5758	[C]	(J31)	ALL, MIL	5-17
5759	[A]	(J31)	ALL, MIL	5-20
5760	[C]	(J31)	ALL, MIL	5-20
5761	[B]	(J31)	ALL, MIL	5-20
5762	[B]	(J31)	ALL, MIL	5-20
5763	[C]	(J31)	ALL, MIL	5-20
5764	[B]	(J31)	ALL, MIL	5-21
5765	[C]	(H351)	AIR, RTC, LTA, MIL	5-21
5766	[C]	(L52)	AIR, RTC	2-12
5767	[A]	(L52)	AIR, RTC	2-12
5768	[A]	(L52)	AIR	5-14
5769	[B]	(N32)	GLI	2-28
5770	[B]	(N30)	GLI	2-29
5771	[C]	(N20)	GLI	2-29
5772	[A]	(N20)	GLI	2-29
5773	[B]	(N21)	GLI	8-40
5774	[B]	(N21)	GLI	8-40
5775	[A]	(N21)	GLI	8-40
5776	[A]	(N21)	GLI	8-40
5777	[A]	(N21)	GLI	8-40
5778	[C]	(N21)	GLI	8-41
5779	[C]	(N21)	GLI	8-41
5780	[C]	(N21)	GLI	8-41
5781	[B]	(N21)	GLI	8-41
5782	[A]	(N21)	GLI	8-41
5783	[B]	(N21)	GLI	8-41
5784	[A]	(N21)	GLI	8-42
5785	[C]	(N21)	GLI	8-42
5786	[C]	(N21)	GLI	8-42
5787	[B]	(N21)	GLI	8-42
5788	[C]	(N21)	GLI	8-42
5789	[C]	(N21)	GLI	8-42
5790	[C]	(N21)	GLI	8-42
5791	[B]	(N22)	GLI	2-29
5792	[C]	(N28)	GLI	2-28
5793	[A]	(N29)	GLI	2-29
5794	[A]	(N29)	GLI	2-28
5795	[A]	(N29)	GLI	2-29
5796	[B]	(N30)	GLI	2-30
5797	[B]	(N30)	GLI	2-30
5798	[A]	(N30)	GLI	2-30
5799	[C]	(N30)	GLI	2-30
5800	[B]	(N30)	GLI	2-30
5801	[C]	(N30)	GLI	2-30
5802	[C]	(N30)	GLI	2-31
5803	[C]	(N30)	GLI	2-31
5804	[A]	(N30)	GLI	2-31
5805	[C]	(N31)	GLI	2-31
5806	[C]	(N31)	GLI	2-31
5807	[A]	(N31)	GLI	2-31
5808	[A]	(N31)	GLI	2-32
5809	[C]	(N31)	GLI	2-32
5810	[A]	(N31)	GLI	2-32
5811	[B]	(N31)	GLI	2-32
5812	[B]	(N31)	GLI	2-32
5813	[B]	(N32)	GLI	2-33

Cross-Reference A: Answer, SMK Code, Category & Page Number

Question	Answer	SMK Code	Category	Page
5814	[B]	(N32)	GLI	2-33
5815	[C]	(N32)	GLI	2-33
5816	[C]	(N32)	GLI	2-33
5817	[A]	(N32)	GLI	2-33
5818	[C]	(N33)	GLI	2-33
5819	[B]	(N33)	GLI	2-33
5820	[C]	(N33)	GLI	2-34
5821	[C]	(N34)	GLI	2-34
5822	[B]	(N34)	GLI	2-34
5823	[C]	(N34)	GLI	2-34
5824	[A]	(N34)	GLI	2-34
5825	[C]	(O277)	LTA	2-35
5826	[B]	(O171)	LTA	2-35
5827	[C]	(O220)	LTA	2-35
5828	[B]	(O263)	LTA	2-35
5829	[B]	(O252)	LTA	2-35
5830	[C]	(O220)	LTA	2-35
5831	[C]	(O220)	LTA	2-36
5832	[C]	(O257)	LTA	2-36
5833	[C]	(O270)	LTA	2-36
5834	[C]	(O265)	LTA	2-36
5835	[B]	(O261)	LTA	2-36
5836	[B]	(O150)	LTA	2-36
5837	[B]	(O150)	LTA	2-36
5838	[B]	(O220)	LTA	2-36
5839	[B]	(O220)	LTA	2-37
5840	[B]	(O220)	LTA	2-37
5841	[A]	(O220)	LTA	2-37
5842	[B]	(O220)	LTA	2-37
5843	[C]	(O220)	LTA	2-37
5844	[B]	(O220)	LTA	2-37
5845	[A]	(O170)	LTA	2-37
5846	[B]	(O170)	LTA	2-38
5847	[B]	(O220)	LTA	2-38
5848	[C]	(O155)	LTA	2-38
5849	[C]	(O170)	LTA	2-38
5850	[B]	(O30)	LTA	2-38
5851	[C]	(O30)	LTA	2-38
5852	[C]	(O30)	LTA	2-39
5853	[A]	(O30)	LTA	2-39
5854	[C]	(O30)	LTA	2-39
5855	[C]	(O30)	LTA	2-39
5856	[C]	(O30)	LTA	2-39
5857	[A]	(O30)	LTA	2-39
5858	[C]	(O30)	LTA	2-39
5859	[B]	(O30)	LTA	2-40
5860	[C]	(O30)	LTA	2-40
5861	[C]	(O30)	LTA	2-40
5862	[A]	(O30)	LTA	2-40
5863	[B]	(O30)	LTA	2-40
5864	[C]	(P01)	LTA	2-40
5865	[C]	(P01)	LTA	2-40
5866	[B]	(P01)	LTA	2-41
5867	[A]	(P01)	LTA	2-41
5868	[A]	(P03)	LTA	2-41
5869	[B]	(P04)	LTA	2-41
5870	[C]	(P04)	LTA	2-41
5871	[A]	(P04)	LTA	2-41
5872	[C]	(P05)	LTA	2-41
5873	[B]	(P05)	LTA	2-41
5874	[C]	(P11)	LTA	2-42
5875	[C]	(P11)	LTA	2-42
5876	[B]	(P11)	LTA	2-42
5877	[A]	(P11)	LTA	2-42
5878	[C]	(P11)	LTA	2-42
5879	[A]	(P11)	LTA	2-42
5880	[A]	(P11)	LTA	2-42
5881	[A]	(P12)	LTA	2-43
5882	[A]	(H20)	LTA	4-41
5883	[B]	(H20)	LTA	4-41
5884	[B]	(H20)	LTA	4-41
5885	[B]	(H20)	LTA	4-41
5886	[C]	(H20)	LTA	4-41
5887	[B]	(H20)	LTA	4-41
5888	[A]	(H20)	LTA	4-41
5889	[C]	(H20)	LTA	4-42
5890	[A]	(H20)	LTA	4-42
5891	[C]	(H20)	LTA	4-42
5892	[A]	(H20)	LTA	4-42
5893	[C]	(H20)	LTA	4-42
5894	[B]	(H20)	LTA	4-42
5895	[C]	(H21)	LTA	4-42
5896	[C]	(H21)	LTA	4-43
5897	[A]	(H21)	LTA	4-43
5898	[A]	(H21)	LTA	4-43
5899	[C]	(H21)	LTA	4-43
5900	[B]	(H21)	LTA	4-43
5901	[A]	(H21)	LTA	4-43
5902	[C]	(H21)	LTA	4-43
5903	[B]	(H22)	LTA	4-43
5904	[C]	(H22)	LTA	4-44
5905	[B]	(H22)	LTA	4-44
5906	[C]	(H23)	LTA	4-44
5907	[B]	(H23)	LTA	4-44
5908	[A]	(H23)	LTA	4-44
5909	[C]	(H24)	LTA	4-44
5910	[B]	(H24)	LTA	4-44
5911	[B]	(H24)	LTA	4-45
5912	[C]	(H24)	LTA	4-45
5913	[B]	(H24)	LTA	4-45
5914	[A]	(H24)	LTA	4-45
5915	[C]	(H24)	LTA	4-45
5916	[C]	(H25)	LTA	4-45
5917	[C]	(H25)	LTA	4-45
5918	[C]	(H25)	LTA	4-45
5919	[A]	(H26)	LTA	4-46
5920	[A]	(H26)	LTA	4-46
5921	[C]	(H26)	LTA	4-46
5922	[A]	(H26)	LTA	4-46
5923	[C]	(H26)	LTA	4-46

Cross-Reference A: Answer, SMK Code, Category & Page Number

Question	Answer	SMK Code	Category	Page
5924	[A]	(H27)	LTA	4-46
5925	[B]	(H27)	LTA	4-46
5926	[C]	(H30)	LTA	4-46
5927	[A]	(H30)	LTA	4-47
5928	[A]	(H30)	LTA	4-47
5929	[B]	(H30)	LTA	4-47
5930	[C]	(H30)	LTA	4-47
5931	[A]	(H30)	LTA	4-47
5932	[B]	(H30)	LTA	4-47
5933	[A]	(H30)	LTA	4-48
5934	[A]	(H30)	LTA	4-48
5935	[C]	(H30)	LTA	4-48
5936	[A]	(H30)	LTA	4-48
5937	[B]	(H30)	LTA	4-48
5938-1	[A]	(H31)	LTA	4-48
5938-2	[A]	(H31)	LTA	4-48
5939	[C]	(H31)	LTA	4-49
5940	[A]	(H32)	LTA	4-49
5941	[C]	(L05)	ALL, MIL	5-24
5942	[A]	(L05)	ALL, MIL	5-24
5943	[C]	(L05)	ALL, MIL	5-24
5944	[C]	(L05)	ALL, MIL	5-24
5945	[B]	(L05)	ALL, MIL	5-24
5946	[C]	(L05)	ALL, MIL	5-24
5947	[A]	(L05)	ALL, MIL	5-25
5948	[C]	(L05)	ALL, MIL	5-25
5949	[A]	(L05)	ALL, MIL	5-25
5950	[C]	(L05)	ALL, MIL	5-25
5951	[C]	(L05)	ALL, MIL	5-25
5952	[C]	(L05)	ALL, MIL	5-26
5953	[B]	(L05)	ALL, MIL	5-26
5954	[C]	(L05)	ALL, MIL	5-26
5955	[A]	(L05)	ALL, MIL	5-26
5956	[A]	(L05)	ALL, MIL	5-26
5957	[A]	(L05)	ALL, MIL	5-26
5958	[B]	(L05)	ALL, MIL	5-27
5959	[C]	(L05)	ALL, MIL	5-27
5960	[A]	(L05)	ALL, MIL	5-27
5961	[B]	(L05)	ALL, MIL	5-27
5962	[A]	(L05)	ALL, MIL	5-27
5963	[A]	(L05)	ALL, MIL	5-27

Cross-Reference B:
Subject Matter Knowledge Code & Question Number

The subject matter knowledge codes establish the specific reference for the knowledge standard. When reviewing results of your knowledge test, you should compare the subject matter knowledge code(s) on your test report to the ones found below. All the questions on the Commercial and Military Competency tests have been broken down into their subject matter knowledge codes and listed under the appropriate reference. This will be helpful for both review and preparation for the practical test.

14 CFR Part 1—Definitions and Abbreviations

A01 .... General Definitions
 5010, 5011, 5012

A02 .... Abbreviations and Symbols
 5013, 5014, 5015, 5016-1, 5016-2, 5016-3

14 CFR Part 23—Airworthiness Standards: Normal, Utility, and Acrobatic Category Aircraft

A150 .... General
 5017

14 CFR Part 61—Certification: Pilots, Flight Instructors, and Ground Instructors

A20 .... General
 5018, 5019, 5020, 5021, 5022, 5023, 5024, 5025, 5026, 5027, 5028, 5029, 5030, 5031, 5032, 5037, 5038, 5106, 5107, 5108, 5127, 5128, 5141, 5142, 5143, 5144

A21 .... Aircraft Ratings and Pilot Authorizations
 5033, 5034

A22 .... Student Pilots
 5035-1, 5035-2, 5036

A24 .... Commercial Pilots
 5039, 5126-1, 5126-2, 5131, 5132

A26 .... Flight Instructors
 5040, 5041, 5042

14 CFR Part 71—Designation of Class A, Class B, Class C, Class D, and Class E Airspace Areas; Airways; Routes; and Reporting Points

A66......... Class E Airspace
5043

14 CFR Part 91—General Operating and Flight Rules

B07......... General
5044, 5045, 5046, 5047, 5056-1, 5056-2, 5057, 5071, 5109

B08......... Flight Rules: General
5048, 5049-1, 5049-2, 5050-1, 5050-2, 5051-1, 5051-2, 5052, 5073-1, 5073-2, 5073-3, 5074, 5075, 5076-1, 5076-2, 5076-3, 5076-4, 5077, 5078, 5079, 5082-1, 5082-2, 5082-3, 5083, 5084, 5085, 5086, 5087, 5110-1, 5110-2, 5111, 5112, 5113-1, 5113-2, 5114, 5115, 5116-1, 5116-2, 5117, 5118, 5119-1, 5119-2, 5120-1, 5120-2, 5121, 5603

B09......... Visual Flight Rules
5058, 5088, 5089, 5090, 5091

B10......... Instrument Flight Rules
5059, 5062, 5092, 5122-1, 5122-2, 5123-1, 5123-2, 5124-1, 5124-2, 5124-3, 5125-1, 5125-2

B11......... Equipment, Instrument, and Certificate Requirements
5060, 5061, 5063, 5064, 5065, 5066, 5067-1, 5067-2, 5070, 5072-1, 5072-2, 5080-1, 5080-2, 5080-3, 5080-4, 5081

B12......... Special Flight Operations
5053, 5054, 5055, 5068-1, 5068-2, 5068-3, 5069, 5129

B13......... Maintenance, Preventive Maintenance, and Alterations
5093, 5094, 5095, 5096, 5097, 5098, 5099, 5100, 5101, 5102, 5103, 5104, 5105

NTSB 830—Rules Pertaining to the Notification and Reporting of Aircraft Accidents or Incidents and Overdue Aircraft, and Preservation of Aircraft Wreckage, Mail, Cargo, and Records

G10 .... General
5001

G11 .... Initial Notification of Aircraft Accidents, Incidents, and Overdue Aircraft
5002, 5003-1, 5003-2, 5004-1, 5004-2, 5005, 5006

G13 .... Reporting of Aircraft Accidents, Incidents, and Overdue Aircraft
5007, 5008

AC 91-23—Pilot's Weight and Balance Handbook

H12......... Empty Weight Center of Gravity
5634, 5635, 5636, 5637, 5638, 5639, 5640, 5641, 5642, 5643, 5644, 5645

H14......... Change of Weight
5646, 5647, 5648, 5649

H15......... Control of Loading: General Aviation
5650, 5651, 5652

AC 60-14—Aviation Instructor's Handbook

H20......... The Learning Process
5882, 5883, 5884, 5885, 5886, 5887, 5888, 5889, 5890, 5891, 5892, 5893, 5894

H21......... Human Behavior
5895, 5896, 5897, 5898, 5899, 5900, 5901, 5902

H22......... Effective Communication
5903, 5904, 5905

H23......... The Teaching Process
5906, 5907, 5908

H24......... Teaching Methods
5909, 5910, 5911, 5912, 5913, 5914, 5915

H25......... The Instructor as a Critic
5916, 5917, 5918

H26......... Evaluation
5919, 5920, 5921, 5922, 5923

H27......... Instructional Aids
5924, 5925

H30......... Flight Instructor Characteristics and Responsibilities
5926, 5927, 5928, 5929, 5930, 5931, 5932, 5933, 5934, 5935, 5936, 5937

H31......... Techniques of Flight Instruction
5938-1, 5938-2, 5939

H32......... Planning Instructional Activity
5940

AC 61-21—Flight Training Handbook

H51 .... Introduction to Airplanes and Engines
5183, 5186, 5187, 5190, 5298, 5299

H54 .... Ground Operations
5655, 5656

H55 .... Basic Flight Maneuvers
5191, 5192, 5193, 5194, 5195

H57 .... Takeoffs and Departure Climbs
5661

H58 .... Landing Approaches and Landings
5662, 5663, 5664, 5665

H61 .... Cross-Country Flying
5503

H63 .... Night Flying
5666

H66 .... Principles of Flight and Performance Characteristics
5196, 5197, 5198, 5199, 5200, 5201, 5202, 5203, 5204, 5205, 5206, 5207, 5208, 5209, 5210, 5211, 5212, 5213, 5214, 5215, 5216, 5217, 5218, 5219, 5220, 5221, 5222, 5223, 5224, 5225, 5226, 5227, 5228, 5229, 5230, 5231, 5232, 5233, 5234, 5235, 5236, 5237, 5238, 5300, 5505, 5667, 5668, 5669, 5670

AC 61-13—Basic Helicopter Handbook

H70 .... General Aerodynamics
5239

H71 .... Aerodynamics of Flight
5240, 5241, 5242, 5243, 5244, 5245, 5246, 5247, 5671, 5672

H72 .... Loads and Load Factors
5248

H73 .... Function of the Controls
5249, 5250, 5251, 5252, 5673, 5674, 5675

H74 .... Other Helicopter Components and Their Functions
5253, 5254, 5255, 5256, 5257

H75 .... Introduction to the Helicopter Flight Manual
5676

H76 .... Weight and Balance
5677, 5678, 5679, 5680, 5681, 5682

H77 .... Helicopter Performance
5258, 5259, 5260, 5683, 5684, 5685, 5686, 5687, 5688, 5689, 5690, 5691, 5692, 5693, 5694

H78......... Some Hazards of Helicopter Flight
5261, 5262, 5263, 5264, 5265, 5266, 5695, 5696, 5697, 5698, 5699, 5700, 5701, 5702, 5703, 5704, 5705, 5706

H79......... Precautionary Measures and Critical Conditions
5707, 5708

H80......... Helicopter Flight Maneuvers
5267, 5709, 5710, 5711, 5712, 5713, 5714, 5715, 5716, 5717, 5718, 5719, 5720, 5721, 5722, 5723, 5724, 5725, 5726, 5727, 5728

H81......... Confined Area, Pinnacle, and Ridgeline Operations
5729, 5730, 5731, 5732

AC 61-23—Pilot's Handbook of Aeronautical Knowledge

H300....... Forces Acting on the Airplane in Flight
5158, 5161, 5162, 5165, 5166, 5167

H303....... Loads and Load Factors
5151, 5152, 5153, 5154, 5155, 5156, 5157, 5159, 5160, 5163, 5164, 5179, 5180

H305....... Flight Control Systems
5181, 5182

H307....... Engine Operation
5168, 5169, 5170, 5171, 5172, 5173, 5174, 5175, 5176, 5185-1, 5185-2, 5188, 5189, 5606, 5607, 5608, 5609, 5610, 5611

H308....... Propeller
5184, 5654

H312....... The Pitot-Static System and Associated Instruments
5177, 5601, 5602, 5604, 5605

H314....... Magnetic Compass
5178, 5612, 5613

H315....... Weight Control
5632

H316....... Balance, Stability, and Center of Gravity
5633

H317....... Airplane Performance
5456, 5457, 5458, 5459, 5460, 5461, 5462, 5463, 5464, 5465, 5466, 5467, 5468, 5614, 5615, 5616, 5617, 5618, 5619, 5620, 5621, 5622, 5623, 5624, 5625, 5626, 5627, 5628, 5629, 5630, 5631

H341....... Effect of Wind
5475, 5476, 5477, 5478

H342....... Basic Calculations
5451, 5452, 5453, 5454, 5455, 5469, 5470, 5471, 5472, 5473, 5474, 5482, 5483, 5484, 5485, 5486, 5487

Cross-Reference B: SMK Code & Question Number

H344 .... Dead Reckoning
5479, 5480, 5481, 5488, 5489

H348 .... Radio Navigation
5490, 5491, 5492, 5493, 5494, 5495, 5496, 5497, 5498, 5499, 5500, 5501, 5502

H351 .... Environmental Factors Which Affect Pilot Performance
5653, 5765

FAA-H-8083-3—Airplane Flying Handbook

H564 .... Night Vision
5130

H568 .... Airport and Navigation Lighting Aids
5133, 5134

H572 .... Orientation and Navigation
5135

H574 .... Night Emergencies
5136, 5137

FAA-H-8083-21—Rotorcraft Flying Handbook

H762 .... Aerodynamics of the Gyroplane
5734

H765 .... Rotorcraft Flight Manual
5735, 5736

H766 .... Gyroplane Flight Operations
5737, 5738

H767 .... Gyroplane Emergencies
5733

AC 61-27—Instrument Flying Handbook

I04 .... Basic Flight Instruments
5268, 5269, 5740

I05 .... Attitude Instrument Flying: Airplanes
5270

I08 .... Using the Navigation Instruments
5506, 5507, 5508, 5509, 5510, 5511, 5512, 5513, 5514, 5515, 5516, 5517, 5518, 5519, 5520, 5521, 5522, 5523, 5524, 5525, 5526, 5527, 5528, 5529, 5530, 5531, 5532, 5533, 5534, 5535, 5536, 5537, 5538, 5539, 5540, 5541, 5542, 5543, 5544, 5545, 5546, 5547

I10 .... The Federal Airways System and Controlled Airspace
　　　　　5548, 5549, 5550

AC 00-6—Aviation Weather

I21 .... Temperature
　　　　　5301, 5302, 5303, 5304

I22 .... Atmospheric Pressure and Altimetry
　　　　　5305, 5306, 5307, 5308, 5309

I23 .... Wind
　　　　　5310, 5311, 5312, 5313, 5314, 5315, 5316, 5317, 5318, 5319

I24 .... Moisture, Cloud Formation, and Precipitation
　　　　　5320, 5321, 5322, 5323, 5324, 5325, 5326

I25 .... Stable and Unstable Air
　　　　　5327, 5328, 5329, 5330, 5331, 5332, 5333, 5334, 5335, 5336

I26 .... Clouds
　　　　　5337, 5338, 5339

I27 .... Air Masses and Fronts
　　　　　5340, 5341, 5342, 5343, 5344, 5345, 5346, 5347, 5348, 5349, 5350

I28 .... Turbulence
　　　　　5351, 5352, 5353, 5354, 5355, 5356, 5357, 5358, 5359

I29 .... Icing
　　　　　5360, 5739

I30 .... Thunderstorms
　　　　　5361, 5362, 5363, 5364, 5365, 5366, 5367, 5368, 5369, 5370, 5371, 5372, 5373, 5741

I31 .... Common IFR Producers
　　　　　5374, 5375, 5376, 5377, 5378, 5379, 5380

I32 .... High Altitude Weather
　　　　　5381, 5382, 5383, 5384, 5385

I35 .... Soaring Weather
　　　　　5386, 5387, 5388, 5389, 5390, 5391, 5392, 5393, 5394, 5395, 5396, 5397, 5747

AC 00-45—Aviation Weather Services

I40 .... The Aviation Weather Service Program
5398, 5399

I41 .... Surface Aviation Weather Reports
5405

I42 .... Pilot and Radar Reports and Satellite Pictures
5407, 5408

I43 .... Aviation Weather Forecasts
5414, 5415, 5416, 5419, 5420, 5422, 5423, 5424, 5425

I44 .... Surface Analysis Chart
5426, 5427, 5428, 5429

I45 .... Weather Depiction Chart
5430, 5431

I46 .... Radar Summary Chart
5432

I47 .... Significant Weather Prognostics
5433, 5434, 5435, 5436

I49 .... Composite Moisture Stability Chart
5437, 5438, 5439

I51 .... Constant Pressure Charts
5440, 5441, 5442

I52 .... Tropopause Data Chart
5443

I53 .... Tables and Conversion Graphs
5444, 5445, 5446

AIM—Aeronautical Information Manual

J01 .... Air Navigation Radio Aids
5551, 5552, 5553

J05 .... Airport Marking Aids and Signs
5657, 5658, 5659-1, 5659-2, 5660

J08 .... Controlled Airspace
5009, 5554

J13 .... Airport Operations
5138, 5139, 5140, 5748

J14 .... ATC Clearance/Separations
5562, 5749

J15 .... Preflight
5555

J16 .... Departure Procedures
5556

J18 .... Arrival Procedures
5557, 5558

J22 .... Emergency Services Available to Pilots
5504

J25 .... Meteorology
5400, 5401, 5402, 5403, 5404, 5406, 5409, 5410, 5411, 5412, 5413, 5417, 5418, 5421, 5449, 5559, 5560

J27 .... Wake Turbulence
5750, 5751, 5752, 5753, 5754, 5755, 5756

J31 .... Fitness for Flight
5757, 5758, 5759, 5760, 5761, 5762, 5763, 5764

J33 .... Pilot Controller Glossary
5561, 5563

J37 .... Sectional Chart
5564, 5565, 5566, 5567, 5568, 5569, 5570, 5571, 5572, 5573, 5574-1, 5574-2, 5575, 5576, 5577, 5578, 5579, 5580, 5581, 5582, 5583, 5584, 5585, 5586, 5587, 5588, 5589, 5590, 5591, 5592, 5593

J42 .... Instrument Approach Procedures
5594, 5595, 5596, 5597, 5598, 5599, 5600

Cross-Reference B: SMK Code & Question Number

Additional Advisory Circulars

K02......... AC 00-30, Rules of Thumb for Avoiding or Minimizing Encounters with Clear Air Turbulence
5447, 5448

K20......... AC 20-103, Aircraft Engine Crankshaft Failure
5271

L05......... AC 60-22, Aeronautical Decision Making
5941, 5942, 5943, 5944, 5945, 5946, 5947, 5948, 5949, 5950, 5951, 5952, 5953, 5954, 5955, 5956, 5957, 5958, 5959, 5960, 5961, 5962, 5963

L34......... AC 90-48, Pilots' Role in Collision Avoidance
5272

L52......... AC 91-13, Cold Weather Operation of Aircraft
5766, 5767, 5768

Soaring Flight Manual—Jeppesen Sanderson, Inc.

N20......... Sailplane Aerodynamics
5276, 5277, 5278, 5279, 5280, 5281, 5282, 5283, 5284, 5285, 5286, 5290, 5291, 5292, 5293, 5294, 5771, 5772

N21......... Performance Considerations
5287, 5288, 5289, 5295, 5296, 5773, 5774, 5775, 5776, 5777, 5778, 5779, 5780, 5781, 5782, 5783, 5784, 5785, 5786, 5787, 5788, 5789, 5790

N22......... Flight Instruments
5273, 5274, 5275, 5297, 5791

N23......... Weather for Soaring
5742, 5743, 5744, 5745, 5746

N28......... Personal Equipment
5792

N29......... Preflight and Ground Operations
5793, 5794, 5795

N30......... Aerotow Launch Procedures
5770, 5796, 5797, 5798, 5799, 5800, 5801, 5802, 5803, 5804

N31......... Ground Launch Procedures
5805, 5806, 5807, 5808, 5809, 5810, 5811, 5812

N32......... Basic Flight Maneuvers and Traffic
5769, 5813, 5814, 5815, 5816, 5817

N33......... Soaring Techniques
5450, 5818, 5819, 5820

N34......... Cross-Country Soaring
5821, 5822, 5823, 5824

Powerline Excerpts—Balloon Federation of America

O30 .... Excerpts
5850, 5851, 5852, 5853, 5854, 5855, 5856, 5857, 5858, 5859, 5860, 5861, 5862, 5863

Balloon Digest — Balloon Federation of America

O150 ...... Balloon — Theory and Practice
5836, 5837

O155 ...... Structure of the Modern Balloon
5848

O170 ...... Propane and Fuel Management
5845, 5846, 5849

O171 ...... Chemical and Physical Properties
5826

O252 ...... Physics
5829

O257 ...... The Standard Burn
5832

O261 ...... Ascents and Descents
5835

O263 ...... Maneuvering
5828

O265 ...... Landings
5834

O277 ...... Maintenance
5825

Balloon Ground School—Balloon Publishing Co.

O220 ...... Balloon Operations
5827, 5830, 5831, 5838, 5839, 5840, 5841, 5842, 5843, 5844, 5847

How To Fly A Balloon—Balloon Publishing Co.

O270 ...... Propane: Management and Fueling
5833

Cross-Reference B: SMK Code & Question Number

Goodyear Airship Operations Manual

P01 .... Buoyancy
5864, 5865, 5866, 5867

P03 .... Free Ballooning
5868

P04 .... Aerostatics
5869, 5870, 5871

P05 .... Envelope
5872, 5873

P11 .... Operating Instructions
5874, 5875, 5876, 5877, 5878, 5879, 5880

P12 .... History
5881

NOTE: AC 00-2, Advisory Circular Checklist, transmits the status of all FAA advisory circulars (ACs), as well as FAA internal publications and miscellaneous flight information, such as Aeronautical Information Manual, Airport/Facility Directory, knowledge test guides, practical test standards, and other material directly related to a certificate or rating. To obtain a free copy of AC 00-2, send your request to:

U.S. Department of Transportation
Subsequent Distribution Office, SVC-121.23
Ardmore East Business Center
3341 Q 75 Ave.
Landover, MD 20785

Notes

Notes

Notes

Notes

Notes

Notes

More Commercial Pilot Products from ASA

ASA has many other books and supplies for the Commercial Pilot. They are listed below and available from an aviation retailer in your area. Need help locating a retailer? Call ASA at 1-800-ASA-2-FLY. We can also send you the latest ASA Catalog which includes our *complete* line of publications and pilot supplies…for all types of pilots and aviation technicians.

Commercial Pilot Books

	Product Code	Suggested Price
Flight Training by Trevor Thom	ASA-PM-1	$29.95
Private & Commercial by Trevor Thom	ASA-PM-2	29.95
Instrument Flying by Trevor Thom	ASA-PM-3	29.95
Instrument Pilot Syllabus (for Trevor Thom series)	ASA-PM-S-I2	10.95
Commercial Pilot Syllabus (for Trevor Thom series)	ASA-PM-S-C	10.95
The Complete Advanced Pilot by Bob Gardner	ASA-CAP-2	24.95
The Complete Multi-Engine Pilot by Bob Gardner	ASA-MPT-2	19.95
Say Again, Please: Guide to Radio Communications	ASA-SAP	14.95
Guide to the Biennial Flight Review	ASA-OEG-BFR	9.95
Commercial Oral Exam Guide	ASA-OEG-C4	9.95
Commercial Pilot Practical Test Standards	ASA-8081-12A	4.95
Rotorcraft/Helicopter Practical Test Standards	ASA-8081-HD	4.95
Visualized Flight Maneuvers for High-Wing Aircraft	ASA-VFM-HI	16.95
Visualized Flight Maneuvers for Low-Wing Aircraft	ASA-VFM-LO	16.95
Certified Flight Instructor Test Prep	ASA-TP-CFI	19.95
The Savvy Flight Instructor	ASA-SFI	19.95
FAR/AIM	ASA-FR-AM-BK	15.95
FAR for Flight Crew (Parts 1, 25, 63, 65, and 121)	ASA-FAR-FC	16.95
Dictionary of Aeronautical Terms	ASA-DAT-3	19.95

Commercial Pilot Software

Prepware™ for Commercial Pilot	ASA-TW-C	$49.95
Flight Library CD-ROM (over 500 publications)	ASA-CD-FL	29.95
Pro-Flight Library CD-ROM (over 850 publications)	ASA-CD-FL-PRO	99.95

Commercial Pilot Supplies

Pilot Master Log	ASA-SP-6	$24.95
Flightlight™	ASA-FL-1	14.95
Tri-Fold Kneeboard	ASA-KB-3	29.95
E6-B Flight Computer	ASA-E6B	26.95
Micro E6-B Flight Computer	ASA-E6B-1	26.95
Ultimate Rotating Plotter	ASA-CP-RLX	9.95
Accordion-Fold Chart Wallet (holds 10 charts)	ASA-CW-10	9.95
Book-Style Chart Wallet (holds 22 charts)	ASA-CW-22	19.95

All prices based on U.S. currency.
Prices subject to change without notice.

Call **1-800-ASA-2-FLY** for the retailer nearest you.   **Products** from ASA

ASA's On Top™ IFR Proficiency Simulator

An indispensible proficiency tool

Version 6.0 of the popular On Top desktop simulator introduces new features and benefits, including:

- An updated U.S. FAA database. All airports (both VFR and IFR), airways, NAVAIDs, intersections, and fixes are included.
- A new worldwide database, including airports, NAVAIDs, intersections, fixes and airways.
- The Beech 1900 turboprop twin aircraft has been added, bringing the total On Top fleet to nine aerodynamically precise aircraft.
- On Top not only has the performance associated with each aircraft, but now features panels representative of each model.

Each aircraft continues to provide for flexibility in the panel: choose a DG or an HSI, an ADF or RMI, add a moving map or GPS, an attitude indicator or a flight director, and adjust the weight to best simulate your flight. Every aircraft and every airport is shown in photo-realistic detail, from perfectly replicated instruments in your panel to the correct approach lights for that airport.

Version 6.0 includes the following singles, high-performance, twin, and turbine airplanes:

Cessna 172—representing the 1983 172P, with panel updated to 1998 model

Cessna 182—representing the 1985 182R, with the panel updated to 1998 model

Cessna 182RG—representing the 1985 182R, with the panel updated to the 1998 model

Warrior—representing the 1976 panel

Arrow—representing the 1976 panel

Mooney—representing the 1996 MSE panel

Bonanza—representing the 1972 V35B, with the panel updated to the 1998 model

Baron—representing the BE58 model

Beech 1900—representing the BE1900, D model

On Top Version 6.0 .... #ASA-ON-TOP-6.0 .... $99.95

ASA's AirClassics™ HS-1 Headset

The most advanced technology, high-quality components, sleek look, reasonable price, maximum comfort, and a lifetime warranty—your wish list is complete!

Features

- Earcups of high-density acoustic foam provide best passive noise attenuation
- High fidelity speakers for clear, natural sound
- Electret, noise-canceling microphone reduces background noise and allows clearest voice transmission
- Gold-plated microphone and headphone plugs ensure best connection and resist corrosion
- High-grade multi-strand tensile wire improves cable life

- Stereo/mono capability
- Large, dual controls for quick, easy volume adjustment for each ear
- Microphone muff
- Adjustable headband with easy thumb screws for maximum comfort, eliminates "hot spots," and accommodates eyeglasses and any head size
- Quality foam ear seals and light weight add to overall comfort
- Sleek, all-black design with ASA wings attractively silk-screened on each earcup
- Sturdy, high-quality and reliable—yet reasonably priced for maximum value

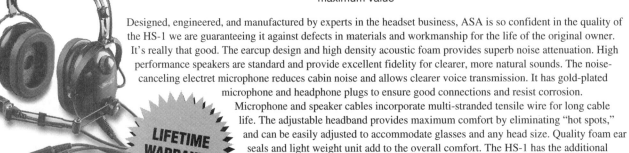

Designed, engineered, and manufactured by experts in the headset business, ASA is so confident in the quality of the HS-1 we are guaranteeing it against defects in materials and workmanship for the life of the original owner. It's really that good. The earcup design and high density acoustic foam provides superb noise attenuation. High performance speakers are standard and provide excellent fidelity for clearer, more natural sounds. The noise-canceling electret microphone reduces cabin noise and allows clearer voice transmission. It has gold-plated microphone and headphone plugs to ensure good connections and resist corrosion. Microphone and speaker cables incorporate multi-stranded tensile wire for long cable life. The adjustable headband provides maximum comfort by eliminating "hot spots," and can be easily adjusted to accommodate glasses and any head size. Quality foam ear seals and light weight unit add to the overall comfort. The HS-1 has the additional benefits of mono/stereo capability, dual volume controls, and a microphone muff.

#ASA-HS-1 .... $149.95

ASA's **Products**   Call **1-800-ASA-2-FLY** for the retailer nearest you.